Novel Food and Feed Safety

# Safety Assessment of Foods and Feeds Derived from Transgenic Crops, Volume 2

**Please cite this publication as:**
OECD (2015), *Safety Assessment of Foods and Feeds Derived from Transgenic Crops, Volume 2*, Novel Food and Feed Safety, OECD Publishing, Paris.
*http://dx.doi.org/10.1787/9789264180338-en*

ISBN 978-92-64-18032-1 (print)
ISBN 978-92-64-18033-8 (PDF)

Series: Novel Food and Feed Safety
ISSN 2304-9499 (print)
ISSN 2304-9502 (online)



**Photo credits:** Cover ©danyssphoto/Shutterstock.com; ©withGod/Shutterstock.com; © Aleksandar Mijatovic – Fotolia.com.

Corrigenda to OECD publications may be found on line at: *www.oecd.org/about/publishing/corrigenda.htm*.

# *Foreword*

From their first commercialisation in the mid-1990s, genetically engineered crops (also known as transgenic crops) have been increasingly approved for cultivation, and for entering in the composition of foods or feeds, by a number of countries. To date, genetically engineered varieties of over 25 different plant species (including agricultural crops, flowers and trees) have received regulatory approvals in OECD and non-OECD countries from all regions of the world. Up to now, the large majority of plantings remain for soybean, maize, cotton and rapeseed (canola), as outlined in the OECD's *The Bioeconomy to 2030: Designing a Policy Agenda.* Over the 19-year period from 1996 to 2014, the surface area grown with transgenic crops worldwide has constantly raised, resulting in a significant increase of their harvested commodities used in foods and feeds (often designated as "novel" foods and feeds). This is highlighted in analyses and statistics from several sources which, despite some differences in total estimates, all concur in underlining the general increasing trend in volumes produced, number of countries involved and growth potential.

For instance, James reports in the *Global Status of Commercialized Biotech/GM Crops: 2014* a record 181.5 million hectares of genetically engineered plants grown, representing an annual growth rate of more than 3.5% from 2013. According to this study, the five main producers in 2014 were the United States, followed by Brazil, Argentina, India and Canada, covering together almost 90% of the total area. Interestingly, developing countries grew more of global transgenic crops (53%) than industrial countries, at 47%. Among the 28 countries having planted transgenic crops in 2014, only 9 of them were OECD countries, listed by decreasing area as follows: the United States, Canada, Australia, Mexico, Spain, Chile, Portugal, the Czech Republic and the Slovak Republic. However, an additional group of countries does not produce transgenic crops but imports the produced commodities, for use in their feed industry in particular, as it is the case in several jurisdictions of Europe as well as some other economies worldwide.

Information of these transgenic crops which have been approved for commercial release in at least one country (for planting and/or for use in foods and feeds processing) can be found in the OECD *BioTrack Product Database* (www2.oecd.org/biotech). Each transgenic product and its Unique Identifier are described, as well as information on approvals in countries.

In parallel to the expansion of genetically engineered crops developed for their resistance to pests and diseases, varieties are being developed by breeders for new types of traits: adaption to climate change, improved composition (biofortification), enhanced meat productivity, easier processing and many other applications. The range of biotechnology applications to agricultural plant breeding is widening, and it seems that the trend will continue. Consequently, the volume of novel foods and feeds available on the market and exchanged internationally is expected to increase in the coming years.

Consumers from all over the world are requiring a high level of safety and full confidence in the products they eat. This is particularly important for the products of modern biotechnology, which are sometimes questioned and subject to diverse levels of acceptation among countries. The approvals of transgenic crops follow a science-based risk/safety assessment regarding their potential release in the environment (biosafety) and their use in foods or feeds (novel food and feed safety). The OECD has undertaken activities related to environmental safety aspects since the mid-1980s, while the development of scientific principles for food safety assessment was initiated in 1990. The OECD helps countries in their risk/safety assessment of transgenic organisms by offering national authorities a platform to exchange experience on these issues, identify emerging needs, collate solid information and data, and develop useful tools for risk assessors and evaluators.

To date, 26 consensus documents relating to the safety of novel foods and feeds have been published; 2 have been revised 10 years later. Most of these publications address compositional considerations of crops subject to plant breeding improvement with modern biotechnologies. These consensus documents are focused on key food and feed nutrients, anti-nutrients, toxicants and other constituents as relevant. They provide solid information commonly recognised by experts and collate the reliable range of data available in the scientific literature at the time of the publication. They can be used in the comparative approach to safety assessment. In addition, documents of a broader nature aiming to facilitate harmonisation have been developed: animal feedstuffs derived from transgenic commodities (2003), designation of an OECD "Unique Identifier" for transgenic plants (2002, revised in 2006) and molecular characterisation of transgenic plants (2010).

Volumes I and II of this series compile the consensus documents of the OECD Series on Safety of Novel Foods and Feeds issued since 2002 (Volume I covers 2002-08, Volume II covers 2009-14). The presentation of the OECD work, originally published in 2006, was used as a basis for the introduction section that explains the purpose of the consensus documents, their relevance to risk/safety assessment and their preparation by the relevant OECD task force. The present compendium offers ready access to those documents which have been published thus far. As such, it should be of value to applicants for uses of transgenic crop commodities in foods and feeds, regulators and risk/safety assessors in national authorities, as well as to the wider scientific community.

Each of the consensus documents may be updated in the future as new knowledge becomes available. Users of this book are therefore encouraged to provide information or an opinion regarding the contents of the consensus documents or any of the OECD's other harmonisation activities. Comments can be provided to: ehscont@oecd.org.

The published consensus documents are also available individually from the OECD's Biotrack website, at no cost: www.oecd.org/biotrack.

# *Acknowledgements*

This book results from the common effort of the participants in the OECD's Task Force for the Safety of Novel Foods and Feeds. Each chapter is composed of a "consensus document" which was prepared under the leadership of a participating country or several countries, as listed at the end of this volume. During their successive drafting, valuable inputs and suggestions for the documents were provided by a number of delegates and experts in the Task Force, whether from OECD member countries, non-member economies or observer organisations.

Each consensus document was issued individually, as soon as it was finalised and agreed for declassification, by the OECD Environment, Health and Safety Division in the Series on Safety of Novel Foods and Feeds. The manuscripts of Volumes I and II of this publication, containing the 2002-14 consensus documents, were prepared by Elisabeth Huggard, Arely Badillo, Carolina Tronco-Valencia and Jennifer Allain. They were edited by Bertrand Dagallier, under the supervision of Peter Kearns, at the Environment, Health and Safety Division of the OECD Environment Directorate.

# *Table of contents*

## Tables

## Figures

# Executive summary

This document constitutes the second volume of the OECD Series on Novel Food and Feed Safety. It is a compendium collating in a single issue the individual "consensus documents" published by the Task Force for the Safety of Novel Foods and Feeds from 2009 to 2014. The first volume of the series covered the documents issued from 2002 to 2008.

Modern biotechnologies are applied to plants, and also trees, animals and microorganisms. The safety of the resulting products represents a challenging issue, and in particular as genetically engineered crops are increasingly cultivated and foods or feeds derived from them are marketed worldwide. Modern biotechnology products should be rigorously assessed by governments to ensure high safety standards for environment, human food and animal feed. Such assessments are considered to be essential for a healthy and sustainable agriculture, industry and trade.

The OECD Task Force for the Safety of Novel Foods and Feeds ("the Task Force") was established in 1999. Its purpose is to assist countries in evaluating the potential risks of transgenic products, foster communication and mutual understanding of relevant regulations in countries, and facilitate harmonisation in risk/safety assessment of products from modern biotechnology. This is intended to encourage information sharing, promote harmonised practices and prevent duplication of efforts among countries. Therefore the Task Force's programme, while consolidating high food and feed safety standards, contributes to reducing costs and potential for non-tariff barriers to trade. Being focused on foods and feeds derived from genetically engineered organisms (also named "novel" foods and feeds), the Task Force's activities and outputs are directly complementary to those of the Working Group on Harmonisation of Regulatory Oversight in Biotechnology, which deals with environmental safety.

The Task Force is composed of delegates from OECD member countries, non-member economies, international bodies and observer organisations involved in these matters, from all regions of the world. National participants and experts are from those government ministries and agencies which have responsibility for the risk and safety assessment of novel foods and feeds in the respective countries. The Task Force provides a platform for delegates to exchange experience and information, identify new needs and develop practical tools for helping the food and feed safety assessment. The main outputs are the "consensus documents", which compile science-based information and data relevant to this task. The key composition elements (nutrients, anti-nutrients, toxicants and sometimes other constituents) that they contain can be used to compare novel foods and feeds with conventional ones. These documents are published after consensus is reached among countries.

Part I of this publication (Volume II) contains a document of broad application aimed to contribute to harmonised assessments of food and feed safety: molecular characterisation of plants derived from modern biotechnology, which was jointly developed by the OECD's Working Group on the Harmonisation of Regulatory Oversight in Biotechnology and the Task Force for the Safety of Novel Foods and Feeds.

Part II of the publication (Volume II) gathers the consensus documents prepared by the Task Force on compositional considerations for transgenic crops. Each chapter contains background information on the considered species: its production, process and uses of its products for foods and feeds, and a brief summary on appropriate comparators for testing new varieties and screening characteristics used by breeders. The core of the chapter is then constituted by detailed information on compositional elements: key nutrients and anti-nutrients, toxicants and allergens where applicable. The final sections suggest key products and constituents for analysis of new varieties for food use and for feed use. Volume II covers the following crops, presented in the order of their initial publication by the Task Force between 2009 and 2014: cotton, cassava, grain sorghum, sweet potato, papaya, sugarcane, low erucic acid rapeseed (canola), soybean and oyster mushroom.

This set of science-based information and data, agreed by consensus and published by the OECD, constitute a solid reference recognised internationally. It is already widely used in comparative approach as part of the risk/safety assessment of transgenic products. As such, this publication should be of value to applicants for commercial uses of genetically engineered crops, to regulators and risk assessors in national authorities in charge of granting approvals to transgenic plant products for their use as foods or feeds, as well as to the wider scientific community.

# Introduction

## OECD activities on novel food and feed safety

The OECD Task Force for the Safety of Novel Foods and Feeds (the "Task Force") was established in 1999, with primary goals to promote international regulatory harmonisation in the risk and safety assessment of biotechnology products among member countries.

The terms "novel foods and feeds" relate usually to foods and feeds derived from transgenic organisms, i.e. partly or fully composed of such ingredients. By extension, these terms could also be understood as foods and feeds containing products obtained from other modern biotechnology techniques. Regulatory harmonisation is the attempt to ensure that the information used in risk/safety assessments, as well as the methods used to collect such information, are as similar as possible. It could lead to countries recognising or even accepting information from one another's assessments. The benefits of harmonisation are clear: it increases mutual understanding among member countries, which avoids duplication, saves on scarce resources and increases the efficiency of the risk/safety assessment process. This, in turn, improves food and feed safety while reducing unnecessary barriers to trade (OECD, 2000).

The Task Force comprises delegates from the 34 member countries of the OECD and the European Commission. A number of observer delegations and invited experts also participate in its work, including Argentina and the Russian Federation, as well as the Food and Agriculture Organization of the United Nations (FAO), the World Health Organization (WHO), the Business and Industry Advisory Committee to the OECD (BIAC), and other organisms as relevant such as the United Nations Environment Programme, the World Bank, the Center for Environmental Risk Assessment of the ILSI Research Foundation (CERA) and the African Biosafety Network of Expertise. Since 2002, several other non-member countries (Bangladesh, Brazil, the People's Republic of China, Colombia, India, Indonesia, Kenya, Latvia, Moldova, Philippines, South Africa, Thailand and others) have participated in activities of the Task Force under the auspices of OECD Global Relations Secretariat and its Global Forum on Biotechnology.

Typically, delegates of the Task Force are from those government ministries and agencies which have responsibility for the food or feed safety assessment of products of modern biotechnology, including foods and feeds derived from transgenic organisms. In some OECD countries this is the Ministry of Health; in others it is the Ministry of Agriculture. Other countries have specialised agencies with this responsibility. Often, it is a shared responsibility among more than one ministry or agency. The expertise that these delegates have in common is related to their experience with food and/or feed safety assessment.

## The emergence of the concept of consensus documents

By 1997, several OECD countries had gained experience with safety assessment of foods derived through modern biotechnology. An OECD workshop in Aussois, France, examined the effectiveness of the application of substantial equivalence in safety assessment. It was concluded that the determination of substantial equivalence provides equal or increased assurance of the safety of foods derived from genetically modified plants, as compared with foods derived through conventional methods (OECD, 1997).

At this event, it was also recognised that a consistent approach to the establishment of substantial equivalence might be improved through consensus on the appropriate components (e.g. key nutrients, key toxicants and anti-nutritional compounds) on a crop-by-crop basis, which should be considered in the comparison. It is recognised that the components may differ from crop to crop.

Following the Aussois workshop, there was a detailed analysis of whether there was a need to undertake work on food/feed safety at the OECD, and if so, what that work would entail. This analysis was undertaken by an Ad Hoc Group on Food Safety (established by the Joint Meeting).[1] It took into account the results of national activities and those of previous OECD work, as well as the activities of the FAO and WHO.

As a result of the Ad Hoc Group on Food Safety's activities, the Joint Meeting established the Task Force, with a major part of its programme of work being the development of consensus documents on compositional data. These data are used to identify similarities and differences following the comparative approach as part of a food and feed safety assessment. They should be useful to the development of guidelines, both national and international, and to encourage information sharing among OECD countries as well as with non-members.

Participation from non OECD member economies is strongly encouraged by the Task Force. As transgenic crops are grown in several of these countries and economies, their commodities traded internationally and widely used for food and feeds. This exchange has increased over the years and now more actively involves their expertise. For example, the consensus documents on the composition of cassava, grain sorghum and papaya were developed in co-operation of non-member countries with leadership/co-leadership of South Africa for the two first and Thailand for the latter. Similarly, Brazil is co-ordinating the preparation of a future document on the common bean while the Philippines is actively involved in the revision of the rice composition document. This concrete enlargement to non-members' inputs and competence broadens the expertise available to the Task Force, while addressing a wider range of food and feed products that are of global interest.

## Background and principles surrounding the use of consensus documents

The OECD "consensus documents" are a compilation of current information that is important in food and feed safety assessment. Agreed by consensus among the Task Force participants, they provide a technical tool for regulatory officials, industry and other interested parties, as a general guide and reference source. They complement those of the OECD Working Group on Harmonisation of Regulatory Oversight in Biotechnology which deal with the environmental safety aspects (biosafety) (OECD, 2006a; 2006b; 2010a; 2010b). They are mutually acceptable to, but not legally binding on, member countries and are used as key references by other economies beyond the OECD for their assessment of novel foods and feeds. They are not intended to be a comprehensive description of all the issues considered to be necessary for a safety assessment, but a base set for an individual product that supports the comparative approach. In assessing an individual product, consideration of additional components may be required depending on the specific case in question.

The work of the Task Force builds on previous OECD experience in biotechnology safety-related activities, dating back to the mid-1980s. Initially, much of the work concentrated on the environmental and agricultural implications of the use of transgenic crops. By the end of 1990, however, work had been established to develop scientific principles for food safety assessment of products of modern biotechnology. This work was often undertaken in parallel to complementary activities of the FAO and WHO.

In 1990, a joint consultation of the FAO and WHO established that the comparison of a final product with one having an acceptable standard of safety provides an important element of safety assessment (WHO, 1991).

In 1993, the OECD further elaborated this concept and advocated the approach to safety assessment based on substantial equivalence as being the most practical approach to addressing the safety of foods and food components derived through modern biotechnology (as well as other methods of modifying a host genome, including tissue culture methods and chemical- or radiation-induced mutation).

A Joint FAO/WHO Expert Consultation on Biotechnology and Food Safety (1996) elaborated on compositional comparison as an important element in the determination of substantial equivalence. A comparison of critical components can be carried out at the level of the food source (i.e. species) or the specific food product. Critical components are determined by identifying key nutrients and key toxicants and anti-nutrients for the food source in question. The comparison of critical components should be between the modified variety and non-modified comparators with an appropriate history of safe use. The data for the non-modified comparator can be the natural ranges published in the literature for commercial varieties or those measured levels in parental or other edible varieties of the species (FAO/WHO, 1996). The comparator used to detect unintended effects for all critical components should ideally be the near isogenic parental line grown under identical conditions. While the comparative approach is useful as part of the safety assessment of foods derived from plants developed using recombinant DNA technology, the approach could, in general, be applied to foods derived from new plant varieties that have been bred by other techniques.

The Joint FAO/WHO Expert Consultation on Foods Derived from Biotechnology in 2000 (FAO/WHO, 2000) concluded that the safety assessment of genetically modified foods requires an integrated and stepwise, case-by-case approach, which can be aided by a structured series of questions. A comparative approach focusing on the determination of similarities and differences between the genetically modified food and its conventional counterpart aids in the identification of potential safety and nutritional issues and is considered the most appropriate strategy for the safety and nutritional assessment of genetically modified foods. The concept of substantial equivalence was developed as a practical approach to the safety assessment of genetically modified foods. It should be seen as a key step in the safety assessment process, although it is not a safety assessment in itself; it does not characterise hazard, rather it is used to structure the safety assessment of a genetically modified food relative to a conventional counterpart. The consultation concluded that the application of the concept of substantial equivalence contributes to a robust safety assessment framework.

Between 2000 and 2003, the *ad hoc* Intergovernmental Task Force on Foods Derived from Biotechnology to the Codex Alimentarius Commission ("Codex Task Force") undertook work to develop principles and guidelines for foods derived from genetically engineered plants. The full report of the Codex Task Force included:

- principles for the risk analysis of foods derived from modern biotechnology
- a guideline for the conduct of food safety assessment of foods derived from recombinant-DNA plants
- a guideline for the conduct of food safety assessment of foods produced using recombinant-DNA microorganisms (Codex Alimentarius Commission, 2003).

One notable feature of the principles is that they make reference to a safety assessment involving the comparative approach between the food derived from modern biotechnology and its conventional counterpart. Annex II (safety assessment of foods derived from recombinant-DNA plants modified for nutritional or health benefits) and Annex III (safety assessment in situation of low-level presence of recombinant-DNA plant material in food) were added to the guidelines in 2008.

The OECD Task Force is working closely with the Codex Task Force in order to strengthen their complementary activities.

## The process through which consensus documents are prepared

The consensus documents are prepared by the Task Force on official proposals by countries. Typically, the focus is a food crop or vegetable for which modern biotechnology can be used in the plant-breeding process. New improved varieties of these species are being developed by researchers for future release in at least one country, or even exist already at commercial level for some of them.

The Task Force establishes *ad hoc* drafting groups, composed of officials and scientific experts of the species in interested countries. These drafting groups work with all this diversity of inputs, under the co-ordination of "lead countries". The successive revised drafts are reviewed by the full Task Force, with careful examination of the proposed information, data, tables and figures. The several revisions and completions can require a few years, leading to a consensus from all delegations obtained on all elements. Following an OECD internal process for final approval, the document is published and becomes available online for worldwide users.

The OECD Biotrack website provides publications and news from the Task Force, the Series on Novel Food and Feed Safety, contact details of national safety systems and other information. It links to the biosafety (environmental safety) publications, the Series on Harmonisation of Regulatory Oversight in Biotechnology. It also gives free access to the OECD *BioTrack Product Database*. It is available at: www.oecd.org/biotrack.

## Current and future trends

With the growing development of products from modern biotechnology, the production of transgenic crops has increased drastically in the last 20 years. It might even be expanded in the future if new varieties adapted to new needs are adopted. Prospects encompass agriculture, industry and energy sectors.

Resistances to pests and diseases were introduced in plants from the early time of genetic engineering, and still constitute the essential feature of the varietal improvement for agriculture, horticulture and forestry. In parallel, breeders are also working on incorporating new traits in crops for gaining other types of beneficial effects. Some of these varieties are about to enter the market or start being grown. In recent years, drought-tolerant varieties (maize, and now sugarcane) are designed to contribute to climate change adaptation. "Innovation in plant breeding (including biotechnology) that aims to develop crop varieties that are more resilient to climate change impact (e.g. resistance to drought, soil salinity or temperature extremes) is part of a larger basket of possible adaptation options in agriculture" (Agrawala et al., 2012). Other innovative traits can have a direct beneficial impact on foods and feeds, and some are already promising: staple crops (rice, tubers, other species) offering nutritive improvements with increased content (biofortification) of elements such as pro-vitamins or micro-nutrients, feed plants (such as maize and alfalfa) modified for higher digestibility and meat productivity, and many other products under development. The range of biotechnology applications to plant breeding continues to widen, leading to an expected increase of derived foods and feeds used and exchanged internationally in the coming years.

A reliable risk/safety assessment of novel foods and feeds is therefore more than ever a necessity for many world economies, in the context of international trade of commodities. Release of such products should be based on solid information and appropriate tools for leading to national decision making. Harmonised regulations, common practices and easy access to solid science-based compiled information are sought. The tools developed by the OECD Task Force designed to promote international harmonisation in the field of food/feed safety assessment are recognised and appreciated, and they might play an increasing role for fulfilling these needs in the future.

The Task Force is continuing its work on a range of issues. New projects have begun recently on the composition of two new species, the common bean and apple. Further species might be subject to similar activity in the future. The main area of the 2013-16 programme of work remains the development of consensus documents on compositional considerations. Emerging topics are also considered for remaining reactive to key demand, e.g. other new biotechnology techniques, innovative feed ingredients, animal composition data, all of them to be considered regarding food and feed safety issues.

In parallel, the consensus documents are reviewed periodically and updated as necessary to ensure that scientific and technical developments are taken into account. Users of these documents have been invited to provide the OECD with new scientific and technical information, and to make proposals for additional areas to be considered. For example, the low erucic acid rapeseed (canola) and soybean documents, both published originally in 2001, were completed and revised by the Task Force, leading to updated issues in 2011-12. The rice document (2004) has initiated a revision process (a new version expected in 2015) and others might follow in the coming years.

# Note

1. The Joint Meeting was the supervisory body of the Ad Hoc Group and, as a result of its findings, established the Task Force as a subsidiary body. Today, its full title is the Joint Meeting of the Chemicals Committee and the Working Party on Chemicals, Pesticides and Biotechnology.

# *References and additional reading*


Agrawala, S. et al. (2012), "Adaptation and innovation: An analysis of crop biotechnology patent data", *OECD Environment Working Papers,* No. 40, OECD Publishing, Paris, http://dx/doi.org/10.1787/5k9csvvntt8p-en.

Codex Alimentarius Commission (2003; Annexes II and III adopted in 2008), *Guideline for the Conduct of Food Safety Assessment of Foods Derived from Recombinant DNA Plants,* CAC/GL 45/2003, at: www.codexalimentarius.net/download/standards/10021/CXG_045e.pdf (accessed 28 January 2015).

FAO/WHO (2000), *Safety Aspects of Genetically Modified Foods of Plant Origin,* Report of a Joint FAO/WHO Expert Consultation on Foods Derived from Biotechnology Food and Agriculture Organization, Rome, www.fao.org/fileadmin/templates/agns/pdf/topics/ec_june2000_en.pdf (accessed 28 January 2015).

FAO/WHO (1996), "Joint consultation report, Biotechnology and food safety", *FAO Food and Nutrition Paper No. 61,* Food and Agriculture Organization, Rome, at: www.fao.org/ag/agn/food/pdf/biotechnology.pdf (accessed 28 January 2015).

James, C. (2015), "Global status of commercialized biotech/GM crops: 2014", *ISAAA Brief No. 49,* International Service for the Acquisition of Agri-Biotech Applications: ISAAA, Ithaca, New York, www.isaaa.org/resources/publications/briefs/49 (executive summary accessed 28 January 2015).

OECD (2010a), *Safety Assessment of Transgenic Organisms: OECD Consensus Documents: Volume 4,* OECD Publishing, Paris, http://dx/doi.org/10.1787/9789264096158-en.

OECD (2010b), *Safety Assessment of Transgenic Organisms: OECD Consensus Documents: Volume 3,* OECD Publishing, Paris, http://dx/doi.org/10.1787/9789264095434-en.

OECD (2009), *The Bioeconomy to 2030: Designing a Policy Agenda,* OECD Publishing, Paris, http:/dx.doi.org/10.1787/9789264056886-en.

OECD (2006a), *Safety Assessment of Transgenic Organisms: OECD Consensus Documents: Volume 2,* OECD Publishing, Paris, http://dx.doi.org/10.1787/9789264095403-en.

OECD (2006b), *Safety Assessment of Transgenic Organisms: OECD Consensus Documents: Volume 1,* OECD Publishing, Paris, http://dx.doi.org/10.1787/9789264095380-en.

OECD (2000), "Report of the Task Force for the Safety of Novel Foods and Feeds", prepared for the G8 Summit held in Okinawa, Japan on 21-23 July 2000, C(2000)86/ADD1, OECD, Paris, www.oecd.org/chemicalsafety/biotrack/REPORT-OF-THE-TASK-FORCE-FOR-THE-SAFETY-OF-NOVEL.pdf (accessed 28 January 2015).

OECD (1997), "Report of the OECD Workshop on the Toxicological and Nutritional Testing of Novel Foods", held in Aussois, France, 5-8 March 1997, final text issued in February 2002 as SG/ICGB(98)1/FINAL, OECD, Paris, www.oecd.org/env/ehs/biotrack/SG-ICGB(98)1-FINAL-ENG%20-Novel-Foods-Aussois-report.pdf (accessed 28 January 2015).

WHO (1991), *Strategies for Assessing the Safety of Foods Produced by Biotechnology,* Report of a Joint FAO/WHO Consultation, World Health Organization of the United Nations, Geneva, out of print.

# Part I

# Towards harmonised assessments of food and feed safety

# *Chapter 1*

# Molecular characterisation of plants derived from modern biotechnology

*This chapter was jointly developed by the OECD Working Group on the Harmonisation of Regulatory Oversight in Biotechnology and the OECD Task Force for the Safety of Novel Foods and Feeds, with Canada serving as lead country of an Expert Steering Group. It addresses the issues linked to molecular characterisation in safety assessment of recombinant-DNA plants derived from modern biotechnology. Based on experience from the use of these procedures with advanced technology, it describes the background and purpose of molecular characterisation, the transformation methods, the inserted DNA, the insertion site and expressed material, the inheritance and genetic stability.*

## Introduction

The Working Group on the Harmonisation of Regulatory Oversight in Biotechnology (the "Working Group") and the Task Force for the Safety of Novel Foods and Feeds (the "Task Force") are implementing closely related programmes of work at the OECD. Both of them develop science-based consensus documents, which are mutually acceptable among member countries. These consensus documents contain information for use during the regulatory assessment of products derived from modern biotechnology.

In the area of plant biosafety (dealt with by the Working Group), consensus documents are being published on information on the biology of certain plant and animal species, selected traits that may be introduced into plant species, and environmental safety issues arising from certain general types of modifications made to crops, trees or microorganisms.

In the area of food and feed safety (dealt with by the Task Force), consensus documents are focused on the nutrients, anti-nutrients or toxicants, the use as a food/feed and other relevant information on particular products. Reference is made to the concept of substantial equivalence, as it is considered that a comparative approach focusing on the determination of similarities and differences between the genetically engineered food and its conventional counterpart aids in the identification of potential safety and nutritional assessment.

This chapter constitutes the first result from a joint collaborative project implemented from 2003 to 2010 by the Working Group and the Task Force. It addresses the issues linked to molecular characterisation in a risk/safety assessment. The first section describes the background and purpose of molecular characterisation, while the second section discusses transformation methods, inserted DNA, insertion site and expressed material, inheritance and genetic stability. The third section explains the scope of the text and a summary is provided in the final section of the chapter.

## Background

### *Molecular characterisation and risk/safety assessment*

Molecular characterisation is one component of the science-based multi-disciplinary approach used in food, feed and environmental risk/safety assessment of plants derived from modern biotechnology. The molecular characterisation of these plants is used to gain an understanding of the genetic material introduced and expressed in them. The purpose of this chapter is to explain the scientific basis underlying the application of molecular characterisation to the food, feed and environmental risk/safety assessment of these plants.

This chapter is meant to inform a risk/safety assessor on the use of molecular characterisation data and information, which is one component of an overall risk/safety assessment. The chapter does not discuss which data and information should be considered by the competent authority conducting the risk/safety assessment, because the use of the data and information considered may depend on the type of risk/safety assessment being performed as well as on the characteristics of the product. This chapter does not provide an exhaustive list of analytical techniques that may be used for molecular characterisation. Where examples of analytical techniques are given, these serve only to provide a better context for an aspect of molecular characterisation discussed and do not imply that specific techniques are recommended or necessary.

Modern biotechnology has been defined as "the application of a) [*i*]*n vitro* nucleic acid techniques, including recombinant deoxyribonucleic acid (DNA) and direct injection of nucleic acid into cells or organelles, or b) [f]usion of cells beyond the taxonomic family, that overcome natural physiological reproductive or recombination barriers and that are not techniques used in traditional breeding and selection," in the Cartagena Protocol on Biosafety (SCBD, 2000) and by the Codex Alimentarius Commission (Codex Alimentarius Commission, 2003a).

Notwithstanding the fact that plant varieties produced through all techniques, including conventional breeding methods, can pose risks, the scope of this chapter will be limited to plants produced using recombinant-DNA (rDNA) techniques and direct injection of nucleic acid into cells or organelles, referred to herein as recombinant-DNA plants.[1] More specifically, this chapter will examine the transformation process and vectors used during transformation; the genetic material delivered to the recipient plant; and the identification, inheritance and expression of the genetic material in the recombinant-DNA plant.

This chapter focuses on the subset of recombinant-DNA plants intended for commercialisation, unconfined or full release that is subject to risk/safety assessments.

For context, this subset of recombinant-DNA plants, subject to regulatory evaluation, has typically passed through a post-transformation screening and selection process. The development of new recombinant-DNA plants begins with the production of a large number of transformants (Padgette et al., 1995; Zhou et al., 2003; Heck et al., 2005). Plants derived from the initial transformants are cultivated over several propagation cycles in order to identify those plants that stably express and inherit the intended phenotype[2] while maintaining desirable agronomic characteristics such as growth characteristics, fertility and yield. This screening and selection process helps developers identify plants exhibiting pleiotropic effects resulting from the transformation process. With each successive propagation cycle, crop developers discontinue development of plants that have unexpected or undesired traits. This process results in the selection of recombinant-DNA plants intended for commercialisation, unconfined or full release; the risk/safety assessment is performed on these recombinant-DNA plants.

### *National and international experience*

Many national authorities with a history of regulating products of biotechnology have put in place standards and procedures for the pre-market assessment of recombinant-DNA plants and the products derived from them. The expertise and experience developed at the national level have been shared in a number of intergovernmental forums such as the OECD, the World Health Organization (WHO) and the Food and Agriculture Organization (FAO). The scientific principles and approach to risk/safety assessment, developed through consultation at the international level, are currently applied by regulatory agencies around the world. This chapter complements existing guidance developed by national authorities and international organisations in this area.

In the context of environmental risk/safety, several guidance documents have been developed that focus on an approach to evaluating environmental risk/safety, such as the *Safety Considerations for Biotechnology: Scale-up of Crop Plants* published by the OECD (1993). In addition, many other OECD documents, developed through the consensus of member countries, have provided the basis for environmental risk/safety assessment of recombinant-DNA plants.

In the context of food risk/safety, the Codex Alimentarius Commission, under the Joint FAO/WHO Food Standards Programme, has adopted several documents developed by the Codex Ad Hoc Intergovernmental Task Force on Foods Derived from Biotechnology, including the *Principles for the Risk Analysis of Foods Derived from Modern Biotechnology* (Codex Alimentarius Commission, 2003a) and the *Guideline for the Conduct of Food Safety Assessment of Foods Derived from Recombinant-DNA Plants* (Codex Alimentarius Commission, 2003b). In the context of feed risk/safety, the OECD has published "Considerations for the safety assessment of animal feedstuffs derived from genetically modified plants" (OECD, 2003). In addition, many other OECD documents, developed through consensus of member countries, have provided the basis for food and feed risk/safety assessment of recombinant-DNA plants.

## *The purpose of molecular characterisation*

The purpose of molecular characterisation is to inform the risk/safety assessment of plants derived from modern biotechnology. Such characterisation provides knowledge at the molecular level of the inserted DNA within the plant genome,[3] the insertion site and the expressed material (ribonucleic acid [RNA] and proteins), and may provide information on intended and possible unintended effects of the transformation. Molecular characterisation of the genotype[4] contributes to a rigorous assessment of the potential impacts of transformation on the food, feed and environmental risk/safety of a recombinant-DNA plant. It assists in the prediction of the phenotype and the phenotype will ultimately determine whether the recombinant-DNA plant poses any risk/safety concerns.

As it is generally considered by regulatory authorities, and in international consensus-building exercises, molecular characterisation encompasses a number of discrete considerations, including:

- The transformation method: A description of the transformation method, together with a detailed description of any DNA sequences that could be potentially inserted into the plant genome.
- The inserted DNA, the insertion site and expressed material: A description of the inserted DNA, including any genetic rearrangements, deletions or truncations that may have occurred as a consequence of the transformation, and the RNA and/or proteins expressed from the inserted DNA in various plant tissues and/or at different times during plant development.
- Inheritance and genetic stability: This addresses not only inheritance of the inserted DNA but also stability (e.g. translation or transcription) over multiple propagation cycles.

Molecular characterisation of the inserted DNA may be relevant in predicting possible unintended effects relevant to risk/safety, but it is not typically the primary means to detect such unintended effects. Other components of the risk/safety assessment, including allergenicity and toxicological assessment of new substances (e.g. proteins, metabolites), changes in the levels of nutrients and anti-nutrients and of endogenous toxicants and allergens, or changes in plant fitness, are integral for detecting unintended effects relevant to risk/safety.

Molecular characterisation for food, feed and environmental risk/safety assessment of recombinant-DNA plants is based on methods that target specific sequences and expressed products. New profiling technologies can provide information on many components at a particular level of biochemical/molecular organisation (e.g. transcriptomics – RNA; proteomics – proteins). While many of these new profiling technologies are under development, they are not as yet applied by national authorities in risk/safety assessment of recombinant-DNA plants. However, such technologies may serve as supplementary tools in risk/safety assessment in the future, provided they are sufficiently developed and validated. The potential applications of profiling technologies in the risk/safety assessment as well as the challenges associated with such applications have been discussed in several reviews (e.g. Kuiper et al., 2003; Chassy et al., 2004) and are not addressed further in this chapter.

For context, unintended effects could arise from any form of plant breeding. For recombinant-DNA plants, these unintended effects may be due to the disruption of genomic sequences by the insertions, the action of transformation-induced genomic deletions and rearrangements – including within the inserted DNA – or pleiotropic effects caused by the new trait. Unintended effects may result in off-types that would be eliminated during the post-transformation screening and selection process. While both recombinant-DNA plants and conventionally bred plants, including those generated using techniques of mutagenesis, may be evaluated and selected for agronomic and morphological traits, typically most conventionally bred plants do not undergo a risk/safety assessment comparable to that performed for recombinant-DNA plants.

In conclusion, molecular characterisation is considered an important part of risk/safety assessment; however, it is only one component in the overall approach to risk/safety assessment. Molecular characterisation complements other components of the risk/safety assessment, such as environmental, chemical, nutritional, allergenicity and toxicological data to compare the recombinant-DNA plant with its appropriate comparator. Of interest for the risk/safety assessment is whether plant transformation could inadvertently increase the potential toxicity or allergenicity of the recipient plant, alter its nutritional quality, have negative environmental impacts or confer other undesirable traits. The totality of the available information relevant to risk/safety enables regulatory authorities to determine if a recombinant-DNA plant meets appropriate risk/safety standards.

## Transformation methods

### *Introduction*

Transformation is the process of inserting DNA sequences of interest into a plant genome. Different transformation methods are available and each method has associated characteristics that could influence the inserted DNA sequences that are integrated into the plant genome. For instance, the integration process could lead to rearrangements, deletions or multi-copy insertions as well as the insertion of "other" sequences originating from either plasmid (vector) or chromosomal DNA. The presence of these "other" DNA sequences is relevant to risk/safety assessment insofar as such sequences may result in the presence of new substances in the recombinant-DNA plant and may also lead to altered levels of RNAs and proteins. In this section, focus is put on DNA integration that might occur as a result of the particular transformation method employed.

Various methods are available for introducing DNA into the plant genome (reviewed by Hansen and Wright, 1999). The most commonly used bacterial-mediated plant transformation methods employ disarmed *Agrobacterium* spp. Other plant-associated bacteria outside the *Agrobacterium* genus might become important in plant transformation (Broothaerts et al., 2005). Direct transformation methods include particle bombardment (also termed biolistics) and electroporation. Alternative methods (e.g. microinjection, electrophoresis) have been specifically designed for recalcitrant plant species or specific target tissues (Hansen and Chilton, 1996; reviewed by Rakoczy-Trojanowska, 2002). This section will focus on the most widely practiced transformation methods.

## Agrobacterium*-mediated transformation*

During *Agrobacterium*-mediated transformation, a DNA region, termed T-DNA, flanked by short specific DNA stretches (i.e. T-DNA borders), is transferred and integrated in the plant genome (for a review see Gelvin, 2003). Besides the T-DNA border sequences, virulence (*vir*) genes play a key role in the processing, export and integration of the T-DNA from the bacterium to the plant. In addition to their naturally *cis*-acting function, *vir* proteins have been shown to be able to act in *trans*. Based on the latter finding, the so-called binary vector system, comprising: *i)* a plasmid containing the DNA construct[5] flanked by T-DNA border sequences; and *ii)* a disarmed helper plasmid delivering the *vir* gene functions, has been developed. In order to disarm helper plasmids, T-DNA regions are removed. The binary vector system is nowadays most frequently applied in *Agrobacterium*-mediated transformation (Hellens et al., 2000).

The *Agrobacterium* strain and helper plasmid used can be identified, and if previously uncharacterised a description can be provided. Information can also be provided on how the helper plasmid used was disarmed. In addition, the plasmid containing the DNA construct can be described. This information will reveal DNA sequences potentially transferred.

*Agrobacterium*-mediated transformation of plant tissue usually results in a low copy number of the DNA construct at a single insertion site. In some recombinant-DNA plant varieties reaching commercialisation T-DNAs have been found to be inserted as tandem repeats (direct or inverted in structure) at a single locus (reviewed by Smith et al., 2001). Integration of incomplete T-DNA sequences is also occasionally seen. Integration may be accompanied by several types of rearrangements of the DNA construct (duplications, inversions and interspersion with plant DNA) and of plant genomic DNA at the insertion site (duplications, inversions and translocations). The insertion of plasmid backbone sequences from outside the T-DNA borders is also sometimes observed (reviewed by Smith et al., 2001), either with the right or the left T-DNA border sequences or as an independent unit unlinked from the T-DNA (Kononov et al., 1997). Further consideration of the risk/safety assessment of these phenomena is given in the following section.

## *Direct transformation*

Direct transformation of plant cells involves introducing the DNA sequences of interest directly to plant cells with the use of various techniques (e.g. particle bombardment, electroporation) that allow transport of the exogenous material across the cell wall and cell membrane. There is a possibility of introducing other DNA sequences not intended for transfer such as bacterial chromosomal DNA, depending on the purity of the DNA used for transformation. A description of the vector DNA, its preparation and its purity can be provided to reveal DNA sequences potentially transferred.

Direct transformation can be used with plant species not amenable to *Agrobacterium*-mediated transformation to successfully introduce new traits (see Taylor and Fauquet, 2002). Single integrants may be obtained if minimal expression cassettes (promoter, open reading frame and terminator) are used (Fu et al., 2000). Particle bombardment may lead to insertion of multiple copies of the DNA construct (in direct or inverted repeat structure) at a single or multiple loci (Jackson et al., 2001; reviewed by Smith et al., 2001). Multiple copies of the DNA construct at a single insertion site may have short stretches of plant genomic DNA interspersed between them. In some cases, the introduced DNA may have undergone deletions or rearrangements, such as concatamerisation (reviewed by Smith et al., 2001). Vector backbone DNA might also be present in recombinant-DNA plants produced using whole plasmids or in cases where purified expression cassettes were used for transformation and the expression cassettes were not sufficiently purified.

### *Conclusions*

A description of the transformation method employed provides information about the DNA sequences potentially transferred to the plant genome and can be valuable for identifying changes to the plant in order to focus subsequent aspects of the risk/safety assessment.

## Inserted DNA, the insertion site and expressed material

### *Inserted DNA and insertion site*

In a risk/safety assessment, the analysis of the inserted DNA can be used to characterise the genotype arising from the transformation. Data defining whether deletions and/or rearrangements have occurred in the DNA construct or at the insertion site can be used to identify whether there may be potential effects other than the intent of the original transformation. In this section, information on the inserted DNA and the changes at the insertion site resulting from the transformation are discussed.

It should be noted that in this section the analysis of the inserted DNA is considered to be part of an assessment where the inserted DNA is stably inherited in recombinant-DNA plants intended for commercialisation, unconfined or full release, as discussed in the sixth paragraph of the Background section.

#### *Integration and copy number*

Insertion of a DNA construct can either occur in the nuclear plant genome or in the genome of organelles, such as chloroplasts. Information on whether an insertion is located in the nucleus or an organelle can inform the environmental risk/safety assessment with regard to the potential dispersal of the gene of interest in relation to the reproductive biology of the recombinant-DNA plant. If the inserted DNA is located in the chloroplasts, it will most likely only be inherited maternally (most higher plants transmit their chloroplast DNA [predominantly] maternally rather than through pollen dispersal [Bock, 2007]). Inserted DNA will be inherited both maternally and paternally when located in the nucleus. Molecular analysis and inheritance studies can provide information on the location of the inserted DNA (see also the next section).

Depending on the transformation method used, the number of insertion sites might vary. In addition, there may be multiple copies of the DNA construct at each insertion site (see previous section). Although plants with a single copy of the DNA construct are typically selected, in some cases plants with multiple copies of the DNA construct may be more efficacious as they result in higher expression levels. Copy number may influence gene silencing; however, copy number may not be as relevant as the homology of the introduced DNA to endogenous genes (Flavell, 1994).

Using appropriate controls, experimental data (e.g. Southern blot analysis) may reveal information such as the number of insertion sites, the copy number at each site and the genetic elements (e.g. promoters, enhancers) that have been inserted.

### *Presence of plasmid backbone sequences*

Integration of DNA vector backbone sequence into the plant genome can occur with both *Agrobacterium*-mediated and direct transformation methods (see above). Incorporation of DNA vector backbone sequences may be important if it results in the expression of additional proteins (for discussion see first paragraph of section "Expressed Material" below) or alters endogenous gene expression. Therefore, Southern blots of genomic DNA may be probed with DNA sequences from vector backbone(s) to determine if these elements have been inserted.

### *Organisation of transforming DNA and sites of insertion*

The DNA used for transformation may be rearranged during the process of integration into the plant genome. Sequence analysis, polymerase chain reaction (PCR) analysis of the inserted DNA and Southern blotting are techniques that can be used to identify such rearrangements. If experimental results indicate a complex insert, such as one with rearrangements or deletions, further analysis may be useful to characterise the inserted DNA for the purposes of determining whether new substances may be present in the plant that could be relevant to the phenotype of the plant. These rearrangements may not necessarily be significant with regard to food, feed and/or environmental risk/safety.

T-DNA integration into an endogenous gene's coding or regulatory sequence and deletions or rearrangements of plant genomic DNA at the insertion site may cause loss of endogenous gene function or alteration of endogenous gene expression. This may result in changes in the plant which may or may not be significant with respect to risk/safety. Analysis of the regions flanking the inserted DNA may be used to determine if the DNA construct has been inserted in an endogenous gene's coding or regulatory sequence, and for the identification of any potential effects on plant gene function. The ability to analyse changes at the insertion site regarding the loss of plant gene function is, however, often compromised by lack of knowledge of most gene functions. Characterisation of insertion sites could inform the subsequent analyses for unintended effects that are part of the agronomic, phenotypic and compositional assessment of the plant (as discussed above).

New open reading frames (ORFs) might be formed as a result of transformation, potentially leading to the production of new proteins. DNA sequence analysis of the regions spanning the inserted DNA-genomic DNA junctions may reveal the presence of new ORFs as well as the presence of regulatory sequences upstream or downstream of the new ORF.

### *Expressed material*

Expression of the inserted DNA is taken into account in order to evaluate the risk/safety of the new gene products on food, feed and the environment. Expression of vector backbone sequences and new ORFs may also be considered. Data obtained through molecular analysis should reveal whether the inserted vector DNA can be transcribed and translated. If potential new ORFs are identified, bioinformatics tools can assist to determine the likelihood of RNA formation, the possibility for transcription and translation to occur, and the amino acid sequence of the putative new protein. If it is found that new proteins are likely produced, their potential impact on risk/safety should be fully characterised. The risk/safety assessment of any new protein is outside the scope of the present chapter.

In some cases, the intended goal of the insertion of the DNA construct is to suppress or down regulate the transcription of an endogenous target gene. In these cases, protein expression of the endogenous target gene will be reduced or inhibited. In some cases, gene silencing constructs may also influence, as an unintended effect, the transcription or translation of other endogenous genes sharing significant sequence similarity.

#### *Transcription and translation*

Successful transfer of a DNA construct into a new plant variety does not necessarily mean the construct will be expressed (Gelvin, 2003). Several factors can influence the level and stability of expression of the inserted DNA. The copy number of the insert, the structure of the inserted DNA (e.g. presence of inverted repeats) and the insertion site have been shown to affect transcription (Flavell, 1994; Gelvin, 1998; Matzke and Matzke, 1998). Moreover, where and when the inserted DNA is actively transcribed depends, in part, on the promoters used (e.g. tissue-specific promoters may limit expression to desired tissues), the developmental stages (e.g. flowering, seed setting) of the plant and the environment in which the recombinant-DNA plant is grown (Bregitzer and Tonks, 2003; Zhu et al., 2004).

Expression of the inserted DNA can be determined by use of either nucleic acid techniques such as northern blotting to detect recombinant RNA or by antibody-based methods such as western blotting to detect protein encoded by the inserted DNA. When performing analyses to characterise the expression of the inserted DNA, care should be taken to ensure that the conditions used for analysis (such as the tissues examined and the growth conditions used) are relevant to the risk/safety assessment. Once identified, the expression products from the inserted DNA can be characterised and assessed for risk/safety.

Expression of the inserted DNA in relevant tissues and under relevant environmental conditions is taken into consideration when assessing exposure and is considered as part of the subsequent risk/safety assessment. The stable integration in the plant genome does not imply that inserted DNA expression would, nor should, be expected to occur at steady state levels through the life cycle of the recombinant-DNA plant. Analysis of plant tissues at key developmental stages for proteins encoded by the inserted gene would reveal the amount of proteins produced at those developmental stages relevant to the risk/safety assessment, such as whether the protein is present in food and feed, or at which developmental phases environmental exposure will be most significant (e.g. expression of the protein in pollen).

### *Post-translational modification*

Following translation, a protein can undergo further modifications. Identifying and characterising the proteins encoded by the inserted gene(s) can provide information useful in confirming that the substances expressed are those that the developer intended to express. Characterising these proteins can create a link to the history of safe use, where relevant, by showing that the proteins expressed *in planta* are not meaningfully different from the proteins when expressed in their native hosts. This is necessary in order to ensure that the data and information about the proteins in their native hosts that may be referenced in the risk/safety assessment of the recombinant-DNA plant are relevant. Algorithms to identify potential post-translation modification such as N- and O-glycosylation sites, Ser/Thr/Tyr phosphorylation sites and (iso)prenylation have been developed (Blom et al., 2004; Maurer-Stroh and Eisenhaber, 2005). Protein analysis studies applying specific staining methods, radioactive labelling studies or matrix-assisted laser desorption/ionisation time-of-flight mass spectrometry (MALDI-TOF MS) may demonstrate the presence of the predicted post-translational modifications (Jensen, 2000) that are deemed relevant to the risk/safety assessment. While some of these post-translational modifications might impact on the risk/safety of the protein, these considerations fall beyond the scope of molecular characterisation but should be considered as part of the overall risk/safety assessment.

### ***Conclusions***

The analysis of the inserted DNA can be useful in the characterisation of the genotype arising from the transformation. Deletions and/or rearrangements that may have occurred in the DNA construct or at the insertion site may result in effects other than the intent of the original transformation. Analysis of expressed products is important for the assessment of the phenotype; however, it must be considered in the context of a complete risk/safety assessment.

## Inheritance and genetic stability

### ***Introduction***

Information regarding the inheritance and genetic stability of the inserted DNA is used to extend the conclusions of a risk/safety assessment conducted for a particular propagation cycle of the recombinant-DNA plant to subsequent genetic descendants. Therefore, information regarding the inheritance and genetic stability of the inserted DNA is important and necessary in the assessment of food, feed and environmental risk/safety.

Inheritance is defined as the pattern of transmission of genotype and phenotype into genetic descendants. The stability of a genetic modification is defined as maintenance of the integrity of the original structure and function of the modification over time and over propagation cycles. Genetic stability can be confirmed by conducting genotypic analysis at the insertion site and/or by phenotypic analysis for expression of the desired trait in the course of plant production and propagation.

### ***Inheritance and genetic stability in risk/safety assessment***

Genetic stability and inheritance of introduced traits within and across propagation cycles are considered as part of the risk/safety assessment. Analysis of inheritance includes consideration of whether the inserted DNA is located on a nuclear plant chromosome or in plant organelles and whether it is transferred into offspring maternally or paternally. Demonstrating that the inserted DNA has been stably integrated into the genome provides some assurance that a risk/safety assessment performed on an early propagation cycle of the plant is applicable to future propagation cycles of the plant. For context, when selecting plants for commercialisation, unconfined or full release, developers typically look for plants in which the inserted DNA has been stably integrated into the genome.

### *Patterns of inheritance*

In the case of insertion of the DNA construct into the nucleus, predictable patterns of inheritance are typically reflected in Mendelian segregation ratios for phenotype and genotype. Deviations from Mendelian inheritance are potential indicators of genetic instability, especially for chromosomal genetic modifications of the nuclear genome in diploid sexual plants that form the majority of new plants typically encountered by regulators. However, the patterns of inheritance applicable to a particular plant species depend on the mechanisms of inheritance that exist for the subject plant species such as the reproductive strategy, the ploidy and whether nuclear or organelle genomes are involved.

Mendelian inheritance would not be expected for all asexual, vegetatively propagated plants, some polyploids and all genetic modifications of plastid or mitochondrial genomes. Such expected instances of non-Mendelian inheritance should not be interpreted as genetic instability.

### *Factors of genetic stability*

As in all plants, genotypic change may occur over the course of mitotic or meiotic cell division and the transmission of genes into resulting progeny. Spontaneous mutations could occur due to errors in base pair incorporation during DNA replication and chromosome doubling prior to mitotic cell division. The pairing of homologous chromosomes during meiosis can lead to crossing over, a recombination that may result in a new grouping of genes. The stability of the inserted DNA may also depend on the sequence and structure of the introduced or modified genes and on characteristics of the insertion site.

### *Methods to determine the stability of a genetic modification*

The stability of a genetic modification may be analysed at the phenotypic and/or the genotypic level. The stability of phenotypic expression may be determined by trait characterisation or by analysis of sufficient samples, where appropriate, of RNA or protein expression. Some phenotypic traits (e.g. resistances) may be quantified under testing conditions with the intact plant. As with other plant genes, expression of inserted DNA will be influenced by the environment. This should be taken into account during a phenotypic consideration of stability. Changes in patterns of expression or expression levels can be quantified in a biochemical reaction mediated by an expressed enzyme or by detection of the expressed protein with specific antibodies (e.g. enzyme-linked immunosorbent assay [ELISA], western blot analysis).

The stability of a genetic modification at the genotypic level may be documented through comparative analyses of the structure of the genetic modification using techniques such as Southern blot, polymerase chain reaction (PCR) or other types of genetic analysis of multiple plants within and across propagation cycles. Genotypic changes across propagation cycles in the recombinant-DNA plant should be considered in the context of the normal variation that occurs with plant breeding.

### *Conclusions*

Inheritance and genetic stability can inform the food, feed and environmental risk/safety assessment. This information is important in extending the conclusions of a risk/safety assessment conducted for particular propagation cycles of the recombinant-DNA plant to subsequent genetic descendants.

## Summary

Molecular characterisation encompasses consideration of the transformation method employed, the inserted DNA and expressed material, and the inheritance and genetic stability of the inserted DNA. Molecular characterisation in and of itself is not a sufficient means of predicting the risk/safety of recombinant-DNA plants. However, molecular characterisation may be useful in focusing other components of the risk/safety assessment that assess the phenotype of the plant, such as characterisation of the levels of nutrients, anti-nutrients, endogenous toxicants or allergens, or changes in plant fitness. To date, the most appropriate available scientific procedures and technology have been used in the molecular characterisation of recombinant-DNA plants. Experience from the use of these procedures and technology form the basis of this chapter. Based on the current pace of technological advancement, it is expected that new methodologies may be applied to the molecular characterisation of recombinant-DNA plants should such technologies prove to have added value as a mechanism of hazard identification in food, feed and environmental risk/safety assessments.

# Notes

1. Other terms such as genetically modified plants, genetically engineered plants, transgenic plants and transformed plants are often used interchangeably with the term recombinant-DNA plant. For the purposes of this chapter, the term recombinant-DNA plant will be used specifically as defined at the beginning of this chapter.

2. Phenotype is defined as an observable characteristic or trait of an organism that is determined by interactions between its genotype and the environment, and may include, but is not limited to, physical, morphological, physiological and biochemical properties.

3. Genome includes genetic material from both the nucleus and organelles.

4. Genotype is defined as the genetic constitution of an organism.

5. For the purposes of this chapter the term "DNA construct" refers to the DNA intended for insertion into the plant genome.

# *References*


Blom, N. et al. (2004), "Prediction of post-translational glycosylation and phosphorylation of proteins from the amino acid sequence", *Proteomics*, Vol. 4, No. 6, pp. 1 633-1 649.

Bock, R. (2007), "Structure, function, and inheritance of plastid genomes", in: Bock, R. (ed.), *Cell and Molecular Biology of Plastids. Topics in Current Genetics*, Vol. 19, pp. 29-63, Springer-Verlag, Berlin, Germany.

Bregitzer, P. and D. Tonks (2003), "Inheritance and expression of transgenes in barley", *Crop Science*, Vol. 43, pp. 4-12, http://dx.doi.org/10.2135/cropsci2003.4000.

Broothaerts, W. et al. (2005), "Gene transfer to plants by diverse species of bacteria", *Nature*, Vol. 433, pp. 629-633, http://dx.doi.org/10.1038/nature03309.

Chassy, B. et al. (2004), "Nutritional and safety assessments of foods and feeds nutritionally improved through biotechnology", *Comprehensive Reviews in Food Science and Food Safety*, Vol. 3, pp. 35-104.

Codex Alimentarius Commission (Codex) (2003a, with amendment in 2008), *Principles for the Risk Analysis of Foods Derived from Modern Biotechnology*, CAC/GL 44-2003, available at: www.codexalimentarius.net/download/standards/10007/CXG_044e.pdf.

Codex (2003b; with Annexes II and III adopted in 2008), *Guideline for the Conduct of Food Safety Assessment of Foods Derived from Recombinant DNA Plants*, CAC/GL 45/2003, available at: www.codexalimentarius.net/download/standards/10021/CXG_045e.pdf (accessed 28 January 2015).

Flavell, R.B. (1994), "Inactivation of gene expression in plants as a consequence of specific sequence duplication", *Proceedings of the National Academy of Sciences of the United States of America*, Vol. 91, No. 9, pp. 3 490-3 496.

Fu, X. et al. (2000), "Linear transgene constructs lacking vector backbone sequences generate low-copy-number transgenic plants with simple integration patterns", *Transgenic Research*, Vol. 9, No. 1, pp. 11-19.

Gelvin, S.B. (2003), "*Agrobacterium*-mediated plant transformation: The biology behind the 'gene-jockeying' tool", *Microbiology and Molecular Biology Reviews*, Vol. 67, No. 1, pp. 16-37.

Gelvin, S.B. (1998), "The introduction and expression of transgenes in plants", *Current Opinion in Biotechnology*, Vol. 9, pp. 227-232.

Hansen, G. and M.D. Chilton (1996), "'Agrolistic' transformation of plant cells: Integration of t-strands generated *in planta*", *Proceedings of the National Academy of Sciences of the United States of America*, Vol. 93, No. 25, pp. 14 978-14 983.

Hansen, G. and M.S. Wright (1999), "Recent advances in the transformation of plants", *Trends in Plant Science*, Vol. 4, No. 6, pp. 226-231.

Heck, G.R. et al. (2005), "Development and characterization of a CP4 EPSPS-based glyphosate-tolerant corn event", *Crop Science*, Vol. 45, No. 1, pp. 329-339, http://dx.doi.org/10.2135/cropsci2005.0329.

Hellens, R. et al. (2000), "A guide to *agrobacterium* binary ti vectors", *Trends in Plant Science*, Vol. 5, No. 10, pp. 446-451, October, available at: http://plant-tc.cfans.umn.edu/listserv/2002/log0205/pdf00000.pdf.

Horvath, H. et al. (2001), "Stability of transgene expression, field performance and recombination breeding of transformed barley lines", *Theoretical and Applied Genetics*, Vol. 102, pp. 1-11, January.

Jackson, S.A. et al. (2001), "High-resolution structural analyis of biolistic transgene integration into the genome of wheat", *Theoretical and Applied Genetics*, Vol. 103, No. 1, pp. 56-62, http://dx.doi.org/10.1007/s001220100608.

Jensen, O.N. (2000), "Modification-specific proteomics: Systematic strategies for analysing post-translationally modified proteins", *Trends in Biotechnology*, Vol. 18, Supplement 1, pp. 36-42, http://dx.doi.org/10.1016/S0167-7799(00)00007-X.

Kononov, M.E. et al. (1997), "Integration of T-DNA binary vector 'backbone' sequences into the tobacco genome: Evidence for multiple complex patterns of integration", *The Plant Journal*, Vol. 11, No. 5, pp. 945-957.

Kuiper, H.A. et al. (2003), "Exploitation of molecular profiling techniques for GM food safety assessment", *Current Opinion in Biotechnology*, Vol. 14, No. 2, pp. 238-243.

Matzke, A.J.M. and M.A. Matzke (1998), "Position effects and epigenetic silencing of plant transgenes", *Current Opinion in Plant Biology*, Vol. 1, No. 2, pp. 142-148.

Maurer-Stroh, S. and F. Eisenhaber (2005), "Refinement and prediction of protein prenylation motifs", *Genome Biology*, Vol. 6, R55, http://dx.doi.org/10.1186/gb-2005-6-6-r55.

Muskens, M.W.M. et al. (2000), "Role of inverted DNA repeats in transcriptional and post-transcriptional gene silencing", *Plant Molecular Biology*, Vol. 43, Issue 2-3, pp. 243-260.

OECD (2003), "Considerations for the safety assessment of animal feedstuffs derived from genetically modified plants", ENV/JM/MONO(2003)10, *Series on the Safety of Novel Foods and Feeds No. 9*, OECD, Paris, available at: www.oecd.org/science/biotrack/46815216.pdf.

OECD (1993), *Safety Considerations for Biotechnology: Scale-Up of Crop Plants*, OECD Publishing, Paris.

Padgette, S.R. et al. (1995), "Development, identification, and characterization of a glyphosate-tolerant soybean line", *Crop Science*, Vol. 35, No. 5, pp. 1 451-1 461, http://dx.doi.org/10.2135/cropsci1995.0011183X003500050032x.

Rakoczy-Trojanowska, M. (2002), "Alternative methods of plant transformation – A short review", *Cellular & Molecular Biology Letters*, Vol. 7, No. 3, pp. 849-858.

SCBD (2000), *Cartagena Protocol on Biosafety to the Convention on Biological Diversity: Text and Annexes*, Secretariat of the Convention on Biological Diversity, Montreal, Canada.

Smith, N. et al. (2001), "Superfluous transgene integration in plants", *Critical Reviews in Plant Sciences*, Vol. 20, Issue 3, pp. 215-249, http://dx.doi.org/10.1080/20013591099218.

Tang, J. et al. (2004), "Stability of the expression of acyl-ACP thioesterase transgenes in oilseed rape doubled haploid lines", *Crop Science*, Vol. 44, No. 3, pp. 732-740, http://dx.doi.org/10.2135/cropsci2004.7320.

Taylor, N.J. and C.M. Fauquet (2002), "Microparticle bombardment as a tool in plant science and agricultural biotechnology", *DNA and Cell Biology*, Vol. 21, No. 12, pp. 963-977.

Yin, Z. et al. (2004), "Differences in the inheritance stability of kanamycin resistance between transgenic cucumbers (*Cucumis sativus* L.) containing two constructs", *Journal of Applied Genetics*, Vol. 45, No. 3, pp. 307-313.

Zhang, Y. et al. (2005), "Stability of inheritance of transgenes in maize (*Zea mays* L.) lines produced using different transformation methods", *Euphytica*, Vol. 144, pp. 11-22, http://dx.doi.org/10.1007/s10681-005-4560-1.

Zhou, H. et al. (2003), "Field efficacy assessment of transgenic Roundup ready wheat", *Crop Science*, Vol. 43, No. 3, pp. 1 072-1 075, http://dx.doi.org/10.2135/cropsci2003.1072.

Zhu, B. et al. (2004), "Stable *Bacillus thuringiensis* (Bt) toxin content in interspecific $F_1$ and backcross populations of wild *Brassica rapa* after Bt gene transfer", *Molecular Ecology*, Vol. 13, No. 1, pp. 237-241.

# Part II

# Compositional considerations for transgenic crops

# *Chapter 2*

# Cotton (*Gossypium hirsutum* and *G. barbadense*)

*This chapter, prepared by the OECD Task Force for the Safety of Novel Foods and Feeds with the United States as the lead country, deals with the composition of cotton (*Gossypium hirsutum *and* G. barbadense*). It contains elements that can be used in a comparative approach as part of a safety assessment of foods and feeds derived from new varieties. Background is given on the production and processing of cotton and derived products (fibres, oil, linters, hulls and meal), followed by appropriate varietal comparators and characteristics screened by breeders. Nutrients and anti-nutrients are then detailed for the whole cottonseed and its main products. The final sections suggest key products and constituents for analysis of new varieties for food use and for feed use.*

## Introduction

Compared to the original 2004 version, the present chapter was amended in 2009 for revising Table 2.8 "Levels of minerals in hulls and meal", as agreed at the 16th meeting of the OECD Task Force for the Safety of Novel Foods and Feeds held in November 2009. In addition, Table 2.10 was slightly amended regarding the level of malvalic acid in whole cottonseed.

## Background

### *Production of cotton for food and feed*

Cotton is of the *Gossypium* genus that is grown on every major continent and on West Indies and Pacific Basin islands. Cotton is cultivated in areas of intense heat. In the dryer climates, irrigation produces high-quality cotton. Cotton is considered the most prominent source of textile fibre in the world, making up over 40% of the total fibre used in the world (USDA ERS, 2002). It is one of the oldest cultivated crops, dating back some 5 000 years ago. Documentation of cotton cloth in ancient times has been achieved in Egypt, Pakistan and the south central United States. Explorations from Europe were stimulated during the 15th and 16th century by a desire to locate more sources of cotton (National Cottonseed Products Association, 1999). Natives wearing cotton garments were found in the West Indies and Mexico. There are over 40 species of cotton, but only 4 are important economically. In the United States, two primary types of cotton are grown, *Glossypium hirsutum*, which has a staple length of 2.5-3.2 cm being the dominant variety, and *Gossypium barbadense*, with a staple length of 2.5-3.8 cm, having limited production (USDA ERS, 2002).

Cotton plant contains a central stem with many branches. There are typically five separate petals per flower and stamens surround the style part of the plant. The ovary of the plant develops into a boll as a dry structure and when dried splits open along four or five lines. The fibres and seeds are contained within the boll. Each fibre grows as a single cell hair from the epidermis of the coat of the seed. Layers of cellulose form around the cell wall. Cell hairs develop into two lengths, long (lint) and short (fuzz) with the lint being the fibre of choice for textiles.

Only the cotton boll is useful for either textile fibres or for food or feed. The remainder of the plant is left in the field for decomposition as fertiliser. Historically cotton was hand picked, but today in industrialised countries most cotton is picked with a mechanical harvester. Following picking, the cotton boll is usually mechanically compressed into modules for transport to a processing plant called a cotton gin. The moduled cotton is usually quite high in moisture and must be processed in a timely manner to avoid spoilage. With spindle pickers and stripper harvestors, about 15% and 48% respectively, of the harvested product is a waste product called gin trash. Gin trash consists of stems, leaves, pieces of bolls and sand picked up in the field. Prior to ginning, gin trash is removed from the cotton by cleaning screens, shakers and air equipment. In the ginning process of the cotton boll, the fibre, for textile use, is separated from the seed and compressed into 217.7 kg bales (National Cottonseed Products Association, 1999). The separated seed at this point is called fuzzy cottonseed and makes up about 60% of the cotton boll. The resulting cottonseed can either be further processed or be used directly as cattle feed.

Table 2.1. **World cotton production, 2001/02**

| | Production ('000 tonnes) | % of total |
|---|---|---|
| China (People's Republic of) | 5 313 | 25 |
| United States | 4 421 | 21 |
| India | 2 569 | 12 |
| Pakistan | 1 785 | 8.4 |
| Uzbekistan | 1 067 | 5 |
| Turkey | 849 | 4 |
| Brazil | 784 | 3.7 |
| Australia | 675 | 3.1 |
| Greece | 457 | 2.1 |
| Syrian Arab Republic | 348 | 1.6 |
| Egypt | 310 | 1.2 |
| Mali | 250 | 1.1 |
| Other countries | 2 499 | 11.7 |
| Total world | 21 327 | |

*Source:* Adapted from USDA Foreign Agricultural Service (2003).

## *Processing of fuzzy cottonseed*

Fuzzy cottonseed is processed into four major products: oil, meal, hulls and linters. Cherry and Leffler (1984) list typical yields as 45% meal, 26% hulls, 16% oil, 9% linters and 4% lost in processing.

Upon arrival of fuzzy cottonseed at the processing plant, fuzzy cottonseed is delinted, by a machine which has a series of fine circular saws that cuts off the fibres, producing linters that are used for human food (National Cottonseed Products Association, 1999). Linters are highly processed (alkaline pH, high temperature) to remove non-cellulose components. Linters are a major source of cellulose for chemical and food use. The delinted cottonseed is then dehulled by machines equipped with knife blades cutting the hulls away from the seed. Separators sift out the seeds from the hull. Hulls are used in animal feed.

The resulting dehulled cottonseed (meats) are processed through a series of iron rollers to produce flakes. The flakes are cooked, reducing the moisture level. The flaked cooked seed moves to the presser to remove the oil. Modern high-pressure screw presses are employed but solvent extraction is also commonly included for maximum efficiency. Oil is pumped, filtered and stored in tanks. Oil goes for further processing for human consumption. The flake remnants are collected, cooled and ground into meal. The process is 96-97% efficient in removing oil, but can leave 3-4% of the oil in the meal. The meal is used for animal feed.

## *Processing of cottonseed oil*

Cottonseed oil requires further processing for food use. Sodium hydroxide is added after heating and forms soapstock or foot that is removed by centrifugation. Both soapstock and crude oil are used to produce fatty acids. To get clear oil, bleaching clay is added and combined with coloring material that can be separated from the oil by filtration. Stearine, a component of cottonseed oil, is further removed from the oil by reducing the temperature to 3.3-4.4°C, at which point the stearine crystalizes, lending itself to separation by filtration. All cottonseed oil is further treated with steam under a partial vacuum to remove off-flavors. This produces a very highly refined and quality product. Because of its superior flavour stability, most of the pure oil is used as cooking or salad oil.

The stearine that was separated by solidification is used in margarine and shortening products. For the pure oil to be used in shortening and margarine, it must be solidified by hydrogenation in the presence of a catalyst. Following hydrogenation, the product is again filtered to remove the catalyst. To make margarine, the solidified oil is mixed with cultured pasteurised skim milk, salt and minor ingredients. Shortening is prepared by chilling and aerating the solidified oil under pressure.

The processing of cotton is schematized in Figure 2.1.

Figure 2.1. **Processing of cotton**

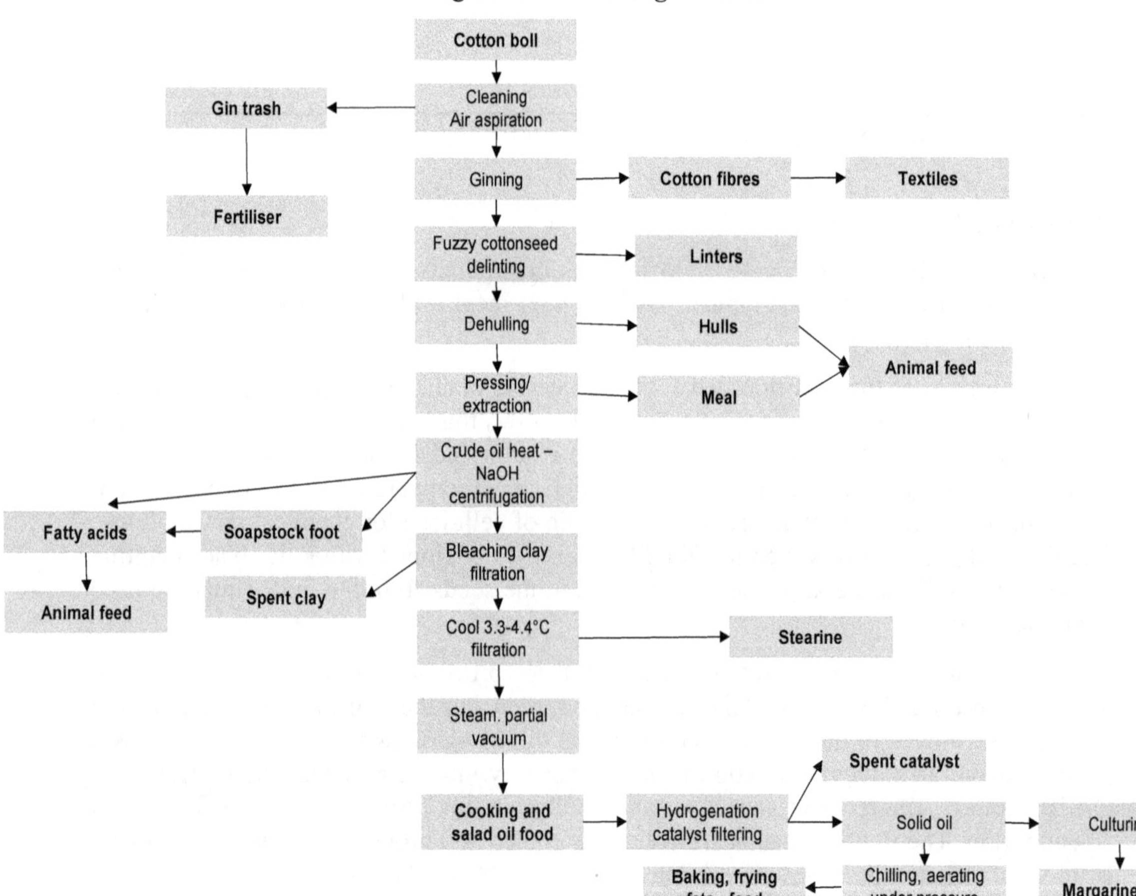


## *Appropriate comparators for testing new varieties*

This chapter suggests parameters that cotton developers should measure. Measurement data from the new variety should ideally be compared to those obtained from the near isogenic non-modified variety. A developer can also compare values obtained from new varieties with literature values present in this chapter.

Critical components include key nutrients, toxicants and anti-nutrients for the food or feed source in question. Key nutrients are those components in a particular product, which may have a substantial impact in the overall diet. These may be major constituents (fats, proteins, and structural and non-structural carbohydrates) or minor compounds (vitamins and minerals). Key toxicants are those toxicologically significant compounds known to be inherently present in the species, i.e. compounds whose toxic potency and levels may impact human and animal health. Similarly, the levels of known anti-nutrients and allergens should be considered. As part of the comparative approach, selected secondary plant metabolites, for which characteristic levels in the species are known, are analysed as further indicators of the absence of unintended effects of the genetic modification on the metabolism.

### *Traditional characteristics screened by cotton developers*

Phenotypic characteristics provide important information related to the suitability of new varieties for commercial distribution. Selecting new varieties is initially based on parent data. Plant breeders developing new varieties of cotton evaluate many parameters at different stages in the developmental process.

In the early stages of growth, breeders evaluate stand count and seedling vigour. As the plant matures, pesticide resistance and disease data are evaluated, e.g. root rot, leaf spots, blight, bollworm/tobacco budworm, cotton aphid, and *Verticillium* and *Fusarium* wilt (University of Georgia, 2002; Texas A&M University, 2002).

The harvested cottonseed is measured for yield, staple length and strength (Bourland, 2002).

In some cases, plants are modified for specific increases in certain components, and the plant breeder would be expected to analyse for such components.

## Nutrients in whole cottonseed and cottonseed products

### *Cottonseed*

Fuzzy or whole cottonseed is the linted cottonseed remaining after the ginning process to remove cotton fibres for textile production (National Cottonseed Products Association, 2002).

However, cottonseed is sometimes delinted and not further processed. Also, there are varieties, notably Pima, that have no linters. These products currently make up only a small percentage of cottonseed available for livestock feeding. Not a lot of data are available on the delinted cottonseed or the Pima varieties except that they contain more gossypol than other varieties (Kirk and Higginbotham, 1999). Arana et al. (2000) indicated that they found lower neutral detergent fibre and acid detergent fibre levels in the delinted products than for whole cottonseed.

The nutrient composition of whole cottonseed is shown in Tables 2.2-2.5.

Table 2.2. **Proximate analysis of cottonseed**

| Reference | | USDA ARS[2] | Ensminger et al. | NCPA | NRC[3] | Commercial range[4] | Range of all reported values |
|---|---|---|---|---|---|---|---|
| Moisture | % of fw | 4.7 | 9.0 | 8.4 | 8.0-9.9 | 4.0-8.7 | 4.0-9.9 |
| Protein | % of dw | 34.2 | 24.0 | 22.5 | 23.0-4.4 | 21.8-28.2 | 21.8-34.2 |
| Total fat | % of dw | 36.3 | | 29.5 | 17.2-23.1 | 15.4-23.8 | 15.4-36.3 |
| Ash | % of dw | 4.8[d] | | 3.8 | 4.2-5.0 | 3.8-4.9 | 3.8-5.0 |
| Neutral detergent fibre (total fibre) | % of dw | | | 47.2 | 40.0-50.3 | 42.1-54.8 | 40.0-54.8 |
| Acid detergent fibre (cellulose) | % of dw | | | 38.8 | 29.0-40.1 | 35.5-37.7 | 29.0-40.1 |
| Crude fibre | % of dw | | 21.4 | | 20.8-24.0 | 15.4-28.2 | 15.4-28.2 |
| Total dietary fibre | % of dw | 5.77 | | | | 5.77 | 5.77 |
| Nonfibrous carbohydrates[1] | % of dw | | | | 23.0 | 45.6-53.6 | 23.0-53.6 |

*Notes:* fw: fresh weight; dw: dry weight.

Proximate analysis of cotton usually includes acid detergent fibre (ADF) and neutral detergent fibre (NDF). The terms ADF and NDF are commonly used in the feed industry and values for comparison are available. Crude fibre, though not the preferred constituent, is still used by some. For food use, however, the concept of dietary fibre is preferred, although different definitions and methods of analysis are being used (see: USA Panel on the Definition of Dietary Fibre [NRC, 2001b]). The value for total dietary fibre from Souci et al. (1989) is obtained using a modification of the analytical method recommended by the Association of Official Analytical Chemists (AOAC). Total dietary fibre determined this way includes lignin and non-starch polysaccharides (including cellulose, hemicellulose and pectin).

1. Non-fibrous carbohydrate = 100 – (% NDF + % CP + % fat + % ash).

2. Cottonseed kernels roasted; dry weight data was converted from g/100g edible portion using the stated moisture content; possibly including modified varieties.

3. Possibly including modified varieties.

4. Commercial range on non-modified controls, compiled from data from acid delinted cottonseed (Monsanto, 2000 and Bayer CropScience, 2002).

*Sources*: USDA ARS (Agricultural Research Service) (2004); Ensminger et al. (1990); NCPA (National Cottonseed Products Association) (1999); NRC (1982, 1989, 1994, 2000, 2001a); Monsanto (2000); Bayer CropScience (2002).

Table 2.3. **Levels of minerals and vitamins in cottonseed**

| Reference | | USDA[1] | NRC[2] | NCPA[3] | Commercial range[4] | Range of all reported values |
|---|---|---|---|---|---|---|
| Sodium (Na) | mg/100g | 26.2 | 10-290 | 8.0 | 5.4-300 | 5.4-300 |
| Potassium (K) | mg/100g | 1 417 | 1 210-1 240 | 1 140 | 1 080-1 250 | 1 080-1 417 |
| Calcium (Ca) | mg/100g | 105 | 160-170 | 140 | 120-330 | 105-330 |
| Phosphorus (P) | mg/100g | 839 | 600-750 | 560 | 610-860 | 560-860 |
| Magnesium (Mg) | mg/100g | 461 | 320-380 | 350 | 370-490 | 320-490 |
| Iron (Fe) | mg/100g | 5.7 | 9.4-16.0 | 5.0 | 4.2-7.2 | 4.2-16.0 |
| Copper (Cu) | mg/100g | 1.3 | 0.7-5.4 | 0.7 | 0.4-1.0 | 0.4-5.4 |
| Selenium (Se) | mg/100g | | 0.00-0.01 | | | 0.00-0.01 |
| Zinc (Zn) | mg/100g | 6.3 | 3.7-3.8 | 3.3 | 2.7-5.1 | 2.7-6.3 |
| Manganese (Mn) | mg/100g | 2.3 | 1.0-1.3 | | 1.1-1.8 | 1.0-2.3 |
| Vitamin A | mg/kg RE[5] | 442 | | | | 442 |
| Vitamin B1 (Thiamin) | mg/kg | 7.5 | | | | 7.5 |
| Vitamin B2 (Riboflavin) | mg/kg | 2.6 | | | | 2.6 |
| Vitamin B6 (Pyridoxine) | mg/kg | 7.8 | | | | 7.8 |
| Vitamin C (Ascorbic acid) | mg/kg | 90 | | | | 90 |
| Vitamin E | mg ATE[6] | 30 | | | | 30 |
| Folate, total | mcg/100g | 2.0 | | | | 2.0 |
| Niacin (Nicotinic acid) | mg/100g | 3.0 | | | | 3.0 |

*Notes:* Values are expressed on a dry weight basis.

1. Cottonseed kernels roasted; values calculated from given values on total weight-basis, using reported moisture content of 4.65%; possibly including modified varieties.

2. Possibly including modified varieties.

3. Possibly including modified varieties.

4. Monsanto (2000).

5. RE (retinol equivalent).

6. 1 mg ATE (alpha tocopherol equivalent) equals 1.1 international units of vitamin E.

*Sources*: USDA ARS (Agricultural Research Service) (2004); NRC (1982, 2000, 2001a); NCPA (National Cottonseed Products Association) (1999); Monsanto (2000).

Table 2.4. **Amino acid composition of cottonseed in percentage of dry weight**

| Reference | USDA ARS[1] | NRC[2,3] | Commercial range[4] | Range of all reported values |
|---|---|---|---|---|
| Methionine | 0.53 | 0.40 | 0.35-0.54 | 0.35-0.54 |
| Cystine | 0.86 | 0.41 | 0.38-0.48 | 0.38-0.86 |
| Lysine | 1.65 | 1.02 | 1.01-1.33 | 1.01-1.65 |
| Tryptophan | 0.49 | 0.30 | 0.23-0.36 | 0.23-0.49 |
| Threonine | 1.21 | 0.81 | 0.74-0.96 | 0.74-1.21 |
| Isoleucine | 1.17 | 0.75 | 0.71-0.88 | 0.71-1.17 |
| Histidine | 1.03 | 0.73 | 0.62-0.82 | 0.62-1.03 |
| Valine | 1.67 | 1.10 | 1.01-1.28 | 1.01-1.67 |
| Leucine | 2.23 | 1.38 | 1.27-1.65 | 1.27-2.23 |
| Arginine | 4.40 | 2.71 | 2.38-3.23 | 2.38-4.40 |
| Phenylalanine | 2.03 | 1.25 | 1.13-1.45 | 1.13-2.03 |
| Glycine | 1.58 | | 0.93-1.19 | 0.93-1.58 |
| Alanine | 1.51 | | 0.85-1.13 | 0.85-1.51 |
| Aspartic acid | 3.55 | | 2.09-2.66 | 2.09-3.55 |
| Glutamic acid | 8.16 | | 4.33-5.28 | 4.33-8.16 |
| Proline | 1.39 | | 0.82-1.14 | 0.82-1.39 |
| Serine | 1.63 | | 0.94-1.32 | 0.94-1.63 |
| Tyrosine | 1.17 | | 0.48-0.79 | 0.48-1.17 |

*Notes:* 1. Cottonseed kernels roasted; possibly including modified varieties. 2. Possibly including modified varieties. 3. Values from NRC (1994 and 1998) were calculated from given values on total weight basis; values from NRC (2001a) were calculated from reported percentage of crude protein, using given crude protein content on a dry basis. 4. Bayer CropScience (2002) and Monsanto (2000).

*Sources*: USDA ARS (Agricultural Research Service) (2004) NRC (1994, 1998, 2001a); Bayer (2002) and Monsanto (2000).

Table 2.5. **Fatty acid composition of cottonseed in percentage of dry weight**

| Reference | USDA ARS[1] | Monsanto 1994[2] | Monsanto 1995[3] | Range |
|---|---|---|---|---|
| 14:0 Myristic | 0.36 | 0.35 | 0.32 | 0.32-0.36 |
| 16:0 Palmitic | 8.84 | 9.41 | 8.88 | 8.84-9.41 |
| 16:1 Palmitoleic | 0.27 | 0.24 | 0.21 | 0.21-0.27 |
| 18:0 Stearic | 0.89 | 0.88 | 0.88 | 0.88-0.89 |
| 18:1 incl. Oleic | 6.93 | 6.09 | 5.13 | 5.13-6.93 |
| 18:2 incl. Linoleic | 18.74 | 20.12 | 16.01 | 16.01-20.12 |
| 18:3 incl. Linolenic | 0.07 | 0.07 | 0.07 | 0.07 |

*Notes:* 1. Cottonseed kernels roasted; possibly including modified varieties; data converted from g/100 g edible portion to percentage of dry weight using stated moisture content of 4.65%. 2. Non-transgenic parent variety; values converted from percentage of total lipid to percentage of dry weight using mean lipid level in cottonseed of 39.2%. 3. Non-transgenic parent variety; values converted from percent total lipid to percent dry weight using mean lipid level in cottonseed of 33.5%.

*Sources*: USDA ARS (Agricultural Research Service) (2004); Monsanto (1994, 1995).

## *Oil*

Cottonseed oil was the first oilseed oil produced in the United States (White, 2000).

The crude oil contains about 2% nonglyceride materials, which are mostly removed during processing. Included in these materials are terpenoid phytoalexin, cyclopropenoid fatty acids (CPFA), phospholipids, sterols, resins, carbohydrates and related pigments. The most notable terpenoid phytoalexin is gossypol (Hanson, 2000). The toxic effects of gossypol and the CPFAs will be discussed later in this chapter. Processing of the oil as described above removes most of the gossypol. Also the deodorisation step removes most of the CPFAs. Cottonseed oil is a pure source of fatty acids. The fatty acid composition of refined cottonseed oil is shown in Table 2.6.

Table 2.6. **Relative fatty acid composition of refined cottonseed oil**

% of total fatty acids

| Reference | USDA ARS[1] | NCPA | White | Monsanto[2] | Bayer[3] | Range |
|---|---|---|---|---|---|---|
| 14:0 Myristic | 0.8 | 0.8 | 0.9 | 0.8-2.4 | 0.6 | 0.6-2.4 |
| 16:0 Palmitic | 23.8 | 24.4 | 24.7 | 24.3-28.1 | 21.1 | 21.1-28.1 |
| 16:1 Palmitoleic | 0.8 | 0.4 | 0.7 | 0.4-1.0 | 0.6 | 0.4-1.0 |
| 18:0 Stearic | 2.4 | 2.2 | 2.3 | 2.1-3.1 | 2.9 | 2.1-3.1 |
| 18:1 Oleic | 17.8 | 17.2 | 17.6 | 12.9-20.1 | 14.9 | 12.9-20.1 |
| 18:2 Linoleic | 54.0 | 55.0 | 53.3 | 46.0-57.1 | 58.2 | 46.0-58.2 |
| 18:3 Linolenic | 0.2 | 0.3 | 0.3 | 0.1-0.3 | 0.2 | 0.1-0.3 |

*Notes:* 1. Cottonseed kernels roasted; possibly including modified varieties; values converted from g/100 g oil to percentage of total fatty acids. 2. Non-transgenic commercial varieties. 3. Non-transgenic parent variety.

*Sources*: USDA ARS (Agricultural Research Service) (2004); NCPA (National Cottonseed Products Association) (1999); White (2000); Monsanto (2000); Bayer CropScience (2002).

## *Meal, linters and hulls*

Cottonseed meal, hulls and linters are by-products of the cottonseed oil industry.

Of these, cottonseed meal is the most abundant and is produced by pressing and solvent extraction. It is produced with and without hulls. The most common is a 41% crude protein product, but some official feed definitions require a minimum of 36% crude protein for all cake and meal cottonseed products. In order to be sold as a low gossypol product, the gossypol content is limited to 0.04% (400 ppm) (AAFCO, 2003).

Linters are composed of almost pure cellulose. The highest quality linters are purified in a chemical treatment of digesting, bleaching, washing and drying (National Cottonseed Products Association, 1999).

Hulls are very high in indigestible fibre.

The proximate analysis, mineral content and amino acid content of meal and hulls are shown in Tables 2.7 through 2.9, respectively.

Table 2.7. **Proximate analysis of meal and hulls in percentage of dry weight**

| Component | Meal[1] | | Hulls[2] |
|---|---|---|---|
| | Mechanical | Solvent | |
| Moisture | 7.7-9.2 | 8.0-10.9 | 10.0-11.0 |
| Protein | 41.7-46.1 | 41.7-48.9 | 4.2-6.2 |
| Fat | 3.9-11.4 | 0.8-3.5 | 2.5 |
| Crude fibre | 11.4-12.6 | 11.2-12.7 | 47.8-48.6 |
| Neutral detergent fibre | 28-32.3 | 20.8-30.8 | 89.0 |
| Acid detergent fibre | 18.1 | 17.3-19.9 | 64.9 |
| Ash | 6.0-7.2 | 6.2-7.5 | 2.8 |

*Notes:* 1. Values from NRC (1998) were converted from an "as fed" basis to a dry matter basis; meal was prepared by mechanical extraction or by solvent extraction.

*Sources*:

1. NRC (1998, 2000, 2001a); National Cottonseed Products Association (1999); Tanksley (1990).

2. NRC (2001a); National Cottonseed Products Association (1999).

Table 2.8. **Levels of minerals in hulls and meal**

| Mineral | | Meal[1] | | Hulls[2] |
|---|---|---|---|---|
| | | Mechanical | Solvent | |
| Sodium (Na) | mg/100g | 0.7-40 | 30-140 | 150-180 |
| Potassium (K) | mg/100g | 1 240-1 680 | 1 200-1 720 | 1 130-1 160 |
| Calcium (Ca) | mg/100g | 160-230 | 160-222 | 150-180 |
| Phosphorus (P) | mg/100g | 760-1 140 | 760-1 200 | 120-150 |
| Magnesium (Mg) | mg/100g | 350-650 | 350-660 | 80-170 |
| Iron (Fe) | mg/100g | 10.7-16.0 | 12.6-16.2 | 3.01-6.8 |
| Copper (Cu) | mg/100g | 1.09-5.39 | 2.6-4.4 | 0.5-3.6 |
| Zinc (Zn) | mg/100g | 3.77-6.28 | 6.1-7.4 | 0.99-1.7 |

*Notes:* Data are presented on a dry weight basis. Data for iron, copper and zinc were corrected in December 2009.

1. Data possibly contain modified varieties. Meal was prepared by mechanical extraction or by solvent extraction. 2. Data possibly contain modified varieties.

*Sources*:

1. USDA ARS (Agricultural Research Service) (2004); NRC (2000, 2001a); Tanksley (1990).

2. NRC (2001a); National Cottonseed Products Association (1999).

Table 2.9. **Amino acid composition of cottonseed meal in percentage of meal dry weight**

| Amino acid | Meal – mechanically extracted | Meal – solvent extracted |
|---|---|---|
| Methionine | 0.62-0.73 | 0.62-0.74 |
| Cystine | 0.64-0.78 | 0.69-0.90 |
| Lysine | 1.57-1.79 | 1.85-2.01 |
| Tryptophan | 0.51-0.57 | 0.53-0.56 |
| Threonine | 1.44-1.52 | 1.45-1.58 |
| Isoleucine | 1.27-1.56 | 1.29-1.59 |
| Histidine | 1.15-1.45 | 1.27-1.50 |
| Valine | 1.80-2.05 | 1.83-2.20 |
| Leucine | 2.50-2.74 | 2.62-2.67 |
| Arginine | 4.40-4.63 | 4.71-4.96 |
| Phenylalanine | 2.14-2.35 | 2.21-2.38 |
| Glycine | 1.83 | 1.87 |
| Tyrosine | 1.01 | 1.27 |
| Serine | 1.84 | 2.01 |

*Notes:* Data possibly include modified varieties. Values from NRC (1998) were converted from an "as fed" basis to a dry matter basis.

*Sources*: National Cottonseed Products Association (1999); NRC (1982, 1998, 2001a).

## Anti-nutrients in cotton

### *Gossypol*

Cotton contains a number of terpenoid phytoalexins. Phytoalexins are antibiotics that, in cotton, accumulate in the pigment glands. They play a critical role in their resistance to potential pathogens that attack cotton. Terpenoid phytoalexins common to cotton include gossypol, hemigossypol, desoxyhemigossypol, 2,7-dihydroxy cadalene, hemigossypolone and heliocides H1 and H2 (Stipanovic, 1994). Gossypol is the most notable of the terpenoid phytoalexins and was first isolated from the pigment glands in cottonseed. It is particularly toxic to non-ruminants and has male anti-fertility properties. Gossypol is either free or bound. Free gossypol is the toxic compound. Sudweeks (2002) reported a gossypol toxicity incident where large amounts of cottonseed meal were fed, estimated to be 24 mg gossypol per head per day. Based on a review of the data, Sudweeks (2002) has suggested that 18 mg of free gossypol (equivalent to 0.1% free gossypol) is the maximum that should be fed to dairy cows. Bailey et al. (2000) and Ziehr et al. (2000) have shown that gossypol exists as two isomers, (+) and (-). The (-) isomer is the more toxic one. However, researchers are also investigating gossypol as an anti-viral and anti-carcinogenic drug (NIH, 2002; Reidenberg, 2003). Typical total and free gossypol levels reported for cottonseed are shown in Table 2.10.

### *Cyclopropenoid fatty acids*

Cotton contains several cyclopropenoid fatty acids (CPFA) that are associated with the oil. Those identified that can be measured are malvalic, sterculic and dihydrosterculic acids (Wood et al., 1994). These CPFAs elevate the melting point of fats in animals fed whole cottonseed and cottonseed meal. The mechanism of action appears to be inhibition of desaturation of saturated fatty acids. In chickens, egg yolk discoloration and reduced hatchability are two detrimental effects, and consequently, the industry limits the use of cottonseed meal and cottonseed oil in poultry diets (Phelps et al., 1965). CPFAs have also been implicated in a high incidence of liver cancer in trout fed whole cottonseed (Hendricks et al., 1980), although it is known that aflatoxin, a common mycotoxin contaminant of cotton, also causes liver cancer in rainbow trout (Park and Price, 2001). Typical levels for these CPFAs in cottonseed are shown in Table 2.10.

Table 2.10. **Levels of gossypol and cyclopropenoid fatty acids in whole cottonseed, cottonseed meal and cottonseed oil**

| Fatty acid | Unit | Whole cottonseed[1,2,4,5,6] | Cottonseed oil[1,6] (refined) | Cottonseed meal[1,2,3] |
|---|---|---|---|---|
| Gossypol (total) | % of dry weight | 0.51-1.43 | 0.00-0.09 | 0.93-1.43 |
| Gossypol (free) | % of dry weight | 0.47-0.70 | ND | 0.02-1.77 |
| Malvalic acid | % of fatty acids | 0.17-0.66* | 0.22-1.44 | |
| Sterculic acid | % of fatty acids | 0.13-0.70 | 0.08-0.58 | |
| Dihydrosterculic acid | , % of fatty acids | 0.11-0.50 | 0.00-0.22 | |

*Notes:* ND: non-detectable.

* Data corrected in December 2009; *ILSI Crop Composition Database* (www.cropcomposition.org; accessed 2009).

*Sources*: 1. Monsanto (2000). 2. Martin (1990). 3. Tanksley (1990), converted values to a dry matter basis. 4. Arana et al. (2000), converted values assuming a 91% dry matter. 5. Bayer (2002). 6. Berberich et al. (1996).

### *Other compounds*

The leaves of cotton contain flavonoids, tannins and anthocyanin. Some of the leaves are harvested with the cotton bolls and these are removed during the ginning process. Under exceptional circumstances, e.g. drought conditions, cotton plants in the form of gin trash or cotton stubble are sometimes used for cattle feed. However, because of this limited exposure, flavonoids, tannins and anthocyanin are not considered key anti-nutrients/natural toxicants.

## Food use

### *Identification of key cotton products consumed by humans*

Cottonseed oil is the primary cotton product used for human consumption. Cottonseed oil ranks a distant third behind soybean and corn oil for human consumption, making up only 5-6% of the total US domestic fat and oil supply (National Cottonseed Products Association, 1999). Crude cottonseed oil contains about 2% of non-glyceride materials such as gossypol and CPFAs, most of which are removed in processing as previously discussed (White, 2000). About 56% of the oil is used for salad or cooking oil, 36% is used for baking and frying fats, and the remaining 8% goes into margarine and other uses. Cottonseed oil is one of the most unsaturated oils, ranking with canola, corn, soybean, safflower and sunflower seed oils. Its mild, nut-like taste makes it highly desirable for use as a salad oil.

The processed linter pulp product is used in food mainly in the production of casings for bologna, sausages and frankfurters. However, the total amount of linters used is very small. Cotton fibre is also used in ice cream and salad dressings to increase viscosity (National Cottonseed Products Association, 1999).

A food grade cottonseed flour product is mixed with corn flour, torula yeast and fortified with niacin, riboflavin, vitamin A and iron and is given to children throughout Central America in their first year of age to combat protein deficiency. Similar products have been marketed in other Latin American countries and India (Franck, 1989; Ensminger et al., 1994). However, the product may be prone to contamination with aflatoxin, making it unsuitable for human consumption (FDA, 1998). Another cottonseed flour product is used as a color additive for foods with restrictions as to its arsenic, lead and gossypol content (FDA, 2002).

### *Identification of key products and suggested analysis for new varieties*

For human nutrition, it is important to assess the fatty acid composition of the oil. Cottonseed oil should also be assessed for its tocopherol content. Tocopherol (vitamin E) is an important micronutrient and antioxidant that prolongs the shelf life of the oil and food products containing the oil. It is also important to measure the levels of gossypol and CPFAs (sterculic, malvalic and dihydrosterculic acids) either in cottonseed or the cottonseed oil. Because other cottonseed products are used to some extent in human food, the proximate analysis of cottonseed is recommended. Table 2.11 lists the key products and suggested analysis for new varieties.

Table 2.11. **Suggested nutritional and compositional parameters to be analysed in cottonseed matrices for human food**

| Parameter | Oil | Cottonseed |
|---|---|---|
| Proximates[1] | | X |
| Tocopherol (vitamin E)[2] | X | X |
| Fatty acids | X | X |
| Gossypol (total and free) | X | X |
| Malvalic acid | X | X |
| Sterculic acid | X | X |
| Dihydrosterculic acid | X | X |

*Notes:* 1. Proximates include protein, fat, ash, total dietary fibre, carbohydrate (calculated) and moisture. 2. One IU of vitamin E is the activity of 1 mg of DL-alpha-tocopherol.

## Feed use

### *Identification of key cottonseed products consumed by animals*

Cottonseed meal is an excellent source of protein for ruminant animals. It is the most valuable animal product of cottonseed, making up over a third of the value. The presence of free gossypol, its lower content and digestibility of the limiting amino acid lysine and its low energy digestibility limits its use primarily for ruminant feed. However, recent research indicates it can also be used in non-ruminant feed, but the level has to be less than 50% of the total protein (Tanksley, 1990). High-quality proteins, such as soybean meal or fish meal, are necessary to include in the diet with the cottonseed meal in order to obtain the best performance for swine (Dove, 1998). It has also been suggested that ferrous sulphate be added in a 1:1 ratio of the free gossypol content. Solvent extracted meals tend to contain the least amount of gossypol (< 0.05%) (Tanksley, 1990). Improvements in the efficacy of removing the oil from cottonseed have produced a less valuable meal product because of lowered oil content of the meal, which means it is a poorer source of energy. For ruminant animals, proximate analysis is important to delineate its nutrient value. For non-ruminant animals, amino acid content is important in addition to the proximate analysis. It is limiting in the amino acid lysine.

Whole cottonseed is a very important dairy feed, and a lesser important beef and sheep feed. It is added to dairy feed as a concentrated source of protein, fat and energy at levels of up to 15% of the total diet or at a total dietary amount of 1.8-3.2 kg per head per day. Higher levels usually decrease feed intake. The important nutritional parameters are proximates, amino acids and fatty acids. The minerals, calcium and phosphorus are also important. The level of gossypol and, to some extent, CPFAs, limits the level of cottonseed that can be added to dairy cow feeds.

Cottonseed hulls are very palatable for ruminant animals and are commonly used in combination with limited amounts of corn silage or hay. Fuzzy cottonseed hulls are preferred over delinted cottonseed hulls. They are also preferred in starter rations for newly weaned calves. Ration texture and palatability appear to be improved by the inclusion of hulls in the diet.

### *Identification of key products and suggested analysis for new varieties*

Proximate analyses are commonly conducted on animal feedstuffs, including the amounts of nitrogen, ether extract, ash and crude fibre. Carbohydrates are measured as starch or nitrogen-free extract. Nitrogen-free extract includes starch, sugars, some cellulose, hemicellulose and lignin, and is calculated using the equation: 100 – CP% – EE% –ash% – CF%. Crude protein is calculated by multiplying the nitrogen content by 6.25, a conversion factor based on the average amount of nitrogen in protein. Fat is considered to be acid-ether-extractable material (Ensminger et al., 1990). In the case of ruminants and swine, the traditional analysis for crude fibre is considered obsolete and has been replaced by analyses for acid detergent fibre and neutral detergent fibre. For amino acids, the ten essential amino acids plus glysine, cystine, tyrosine, serine and proline are the key nutrients. Linoleic is the fatty acid of key importance for the meal, while the relative fatty acid spectrum is more important for the oil.

Other minerals such as selenium are also important, but the amount in plants has been shown to reflect the amount of the mineral in the soil. Nutritionists incorporate supplemental sources of calcium, phosphorus, sodium chloride, magnesium, iron, zinc, copper, manganese, iodine and selenium as needed to balance diets. Again, nutritionists supplement swine diets with vitamins A, D, E, K, B12, riboflavin, niacin and pantothenic acid (NRC, 1998); and ruminant diets with vitamins A, D, E and K (NRC, 2000, 2001a).

In considering the anti-nutrients and natural toxins in cottonseed and cottonseed products, gossypol, malvalic, sterculic and dihydrosterculic acids are significant to the animal feed.

When one considers the cottonseed products that might be fed to animals, their nutrient content would not be expected to change if the content of whole cottonseed is not changed. Hence, only the whole cottonseed and cottonseed meal are suggested to be analysed (Table 2.12). However, for amino acids and fatty acids, either whole cottonseed or cottonseed meal would yield equivalent results.

Table 2.12. **Suggested nutritional and compositional parameters to be analysed in cotton matrices for animal feed**

| Parameter | Cottonseed | Meal |
|---|---|---|
| Proximates[1] | X | X |
| Amino acids[2] | X | |
| Fatty acids[3] | X | |
| Calcium (Ca) | X | X |
| Phosphorus (P) | X | X |
| Gossypol (total and free) | X | X |
| Sterculic acid | X | X |
| Dihydrosterculic acid | X | X |
| Malvalic acid | X | X |

*Notes:* 1. Proximates include protein, fat, ash, neutral detergent fibre, acid detergent fibre and moisture. 2. See first paragraph of the above section for the key amino acids to be measured. 3. See first paragraph of the above section for the key fatty acids to be measured.

# *References*


AAFCO (2003) Official Publication, American Feed Control Officials, Oxford, Indiana, p.261.

Arana, M. et al. (2000), *Comparing Cotton By-Products for Dairy Feed Ration,* Dairy Business Communications.

Bailey, C.A. et al. (2000), "Cottonseed with a high (+) to (-) gossypol enantiomer ratio favorable to broiler production", *Journal of Agricultural Food Chemistry,* Vol. 48, No. 11, pp. 5 692-5 695.

Bayer CropScience (2002), "Premarket biotechnology notice for the safety, compositional and nutritional aspects of glufosinate-tolerant cotton transformation event LLCotton25", US FDA/CFSAN. BNF 86.

Berberich, S.A. et al. (1996), "The composition of insect-protected cottonseed is equivalent to that of conventional cottonseed", *Journal of Agricultural and Food Chemistry,* Vol. 44, No. 1, pp. 365-371, http://dx.doi.org/10.1021/jf950304i.

Bourland, F.M. (2002), "University of Arkansas Cotton Breeding Program – 2001 progress report, 8 summaries of Arkansas cotton research 2001", *Arkansas Agricultural Experiment Station Report,* Research Series 497, p.19.

Cherry, J.P. and H.R. Leffler (1984), "Seed. Cotton", in: Koher, R.J. and C.F. Lewis (eds.), *Agronomy,* Vol. 24, p.512.

Dove, C.R. (1998), "Cottonseed meal for nursery pigs. 1997 Annual Report", pp. 224-231, Department of Animal and Dairy Science, University of Georgia, Athens, Georgia.

Ensminger, A.K.H. et al. (1994), *Foods Nutrition Encyclopaedia, 2nd Edition,* Ann Arbor, Michigan.

Ensminger, M.E. et al. (1990), *Feeds and Nutrition, 2nd Edition,* The Ensminger Publ. Co., Clovis, California.

FAO (2000), *Safety Aspects of Genetically Modified Foods of Plant Origin,* Report of a Joint FAO/WHO Expert Consultation on Foods Derived from Biotechnology Food and Agriculture Organization, Rome, available at: www.fao.org/fileadmin/templates/agns/pdf/topics/ec_june2000_en.pdf (access 28 Jan. 2015).

FAO (1996), "Joint consultation report, Biotechnology and food safety", *FAO Food and Nutrition Paper No. 61*, Food and Agriculture Organization, Rome, available at: www.fao.org/ag/agn/food/pdf/biotechnology.pdf (accessed 28 Jan. 2015).

FDA (2002), "Listing of color additives exempt from certification: Foods: Toasted partially defatted cooked cottonseed flour", *Code of Federal Regulations (U.S.) Title 21: Chapter 1: Part 73*, pp. 73-140, United States Food and Drug Administration, Washington, DC.

FDA (1998), *Import Alert*, United States Food and Drug Administration, Washington, DC.

Franck, A.W. (1989), *Food Uses of Cottonseed Protein. Development in Food ProteinsVol. 5*, pp. 31-80, New York.

Hanson, L.E. (2000), "Reduction of verticillium wilt symptoms in cotton following seed treatment with *Trichoderma virens*", *Journal of Cotton Science*, Vol. 4, pp. 224-231, available at: www.cotton.org/journal/2000-04/4/upload/jcs04-224.pdf.

Hendricks, J.D. et al. (1980), "Hepatocarecinogenicity of glandless cottonseeds and cottonseed oil to rainbow trout (*Salmo gairdnerii*)", *Science*, Vol. 208, No. 4441, pp. 309-311, http://dx.doi.org/10.1126/science.6892734.

ILSI (2009), *Crop Composition Database*, International Life Sciences Institute,: www.cropcomposition.org.

Kirk, J.H. and G.E. Higginbotham (1999), U. Cal., Davis Cooperative Extension Service.

Martin (1990), "Drought strategies for dairy producers: Guidelines for use of aflatoxin containing feeds in dairy rations", *Feedstuffs*, August.

Monsanto (2000), "Safety, compositional, and nutritional aspects of bolegard II cotton event 15985", US FDA/CFSAN. BNF 74.

Monsanto (1995), "Safety, compositional and nutritional aspects of bollgard cotton lines 757 & 1076: Conclusion based on studies and information evaluated according to FDA's policy on foods from new plant varieties", US FDA/CFSAN. BNF 13.

Monsanto (1994), "Safety, compositional and nutritional aspects of bollgard cotton line 531: Conclusion based on studies and information evaluated according to FDA's policy on foods from new plant varieties", US FDA/CFSAN. BNF 13.

National Cottonseed Products Association (2002), *Cottonseed Feed Products Guide*, National Cottonseed Products Association, Cordova, Tennessee, at: www.cottonseed.com/publications/feedproductsguide.asp.

National Cottonseed Products Association (1999), *Cottonseed and Its Products CSIP 10th ed.*, National Cottonseed Products Association, Cordova, Tennessee, available at: www.cottonseed.com/publications/cottonseedanditsproducts.asp.

NIH (2002), "Prospective grant of exclusive license: Gossypol, gossypol acetic acid and derivatives thereof and the use thereof for treating cancer", *U.S. Federal Register (67FR9762d-9763)*, National Institutes of Health, Washington, DC.

NRC (2001a), *Nutrient Requirements of Dairy Cattle*, Seventh Revised Edition, National Research Council, National Academies Press, Washington, DC, available at: www.nap.edu/catalog/9825/nutrient-requirements-of-dairy-cattle-seventh-revised-edition-2001.

NRC (2001b), *Dietary Reference Intakes: Proposed Definition of Dietary Fiber*, Institute of Medicine, National Academies Press, Washington, DC, available at: www.nap.edu/catalog/10161/dietary-reference-intakes-proposed-definition-of-dietary-fiber.

NRC (2000), *Nutrient Requirements of Beef Cattle*, Seventh Revised Edition: Update 2000, National Research Council, National Academies Press, Washington, DC, available at: www.nap.edu/catalog/9791/nutrient-requirements-of-beef-cattle-seventh-revised-edition-update-2000.

NRC (1998), *Nutrient Requirements of Swine*, Tenth Revised Edition, National Research Council, National Academies Press, Washington, DC, available at: www.nap.edu/catalog/6016/nutrient-requirements-of-swine-10th-revised-edition.

NRC (1994), *Nutrient Requirements of Poultry*, Ninth Revised Edition, National Research Council, National Academies Press, Washington, DC, available at: http://books.nap.edu/catalog/2114.html.

NRC (1989), *Nutrient Requirements of Dairy Cattle (Sixth Revised Edition, Update 1989)*, National Academies Press, Washington, DC.

NRC (1982), *United States – Canadian Tables of Feed Composition: Nutritional Data for United States and Canadian Feeds*, Third Revision, National Academies Press, Washington, DC, available at: www.nap.edu/catalog/1713/united-states-canadian-tables-of-feed-composition-nutritional-data-for.

OECD (2000), "Report of the Task Force for the Safety of Novel Foods and Feeds", prepared for the G8 Summit held in Okinawa, Japan on 21-23 July 2000, C(2000)86/ADD1, OECD, Paris, available at: www.oecd.org/chemicalsafety/biotrack/REPORT-OF-THE-TASK-FORCE-FOR-THE-SAFETY-OF-NOVEL.pdf.

OECD (1997), "Report of the OECD Workshop on the Toxicological and Nutritional Testing of Novel Foods", held in Aussois, France, 5-8 March 1997, final text issued in February 2002 as SG/ICGB(98)1/FINAL, OECD, Paris, available at: www.oecd.org/env/ehs/biotrack/SG-ICGB(98)1-FINAL-ENG%20-Novel-Foods-Aussois-report.pdf.

OECD (1993), *Safety Evaluation of Foods Derived by Modern Biotechnology: Concepts and Principles*, OECD, Paris, available at: www.oecd.org/science/biotrack/41036698.pdf.

Park, D.L. and W.D. Price (2001), "Reduction of aflatoxin hazards using ammonization", *Reviews of Environmental Contamination and Toxicology*, Vol. 171, pp. 139-175.

Phelps, R.A. et al. (1965), "A review of cyclopropenoid compounds; Biological effects of some derivatives", *Poultry Science*, Vol. 44, pp. 358-394.

Reidenberg, M. (2003), *Clinical Pharmacology and Pharmacology of Gossypol*, N.Y. Weith Cornel Medical Center.

Souci, S.W. et al. (1989), *Food Composition and Nutrition Tables*, Wiss. Verlagsges, Stuttgart, Germany.

Stipanovic, R.D. (1994), "Gossypol in cotton", *Biochemistry of Cotton Workshop Proceedings*, Cotton Incorporated, Raleigh, North Carolina.

Sudweeks, E.M. (2002), *Feeding Whole Cottonseed to Dairy Cows and Replacements*, Texas A&M University, College Station, Texas, Publication 13277101 – cottonsd.wp6.

Tanksley, T.D. Jr. (1990), "Cottonseed meal", Chapter 15 in: Thacker, P.A. and R.N. Kirkwood (eds.), *Nontraditional Feed Sources for Use in Swine Production*, Butterworths, Stoneham, Massachussetts.

Texas A&M University (2002), *Entomology. Aphid and Stick Cotton Research*, Texas A&M AggriLife Research and Extension Center, Vernon, Texas.

University of Georgia (2002), *Cotton Diseases*, College of Agricultural and Environmental Sciences and Warner School of Forest Resources, available at: www.bugwood.org.

USDA Agricultural Research Service (2004), *USDA Nutrient Database for Standard Reference, Release 16.1*, Nutrient Data Laboratory, United States Department of Agriculture, available at: www.ars.usda.gov/main/site_main.htm?modecode=80-40-05-25.

USDA ERS (2002), *Cotton Background*, United States Department of Agriculture Economic Research Service, Washington, DC, at: http://www.ers.usda.gov/topics/crops/cotton-wool/background.aspx

USDA Foreign Agricultural Service (2003), "Section commodities and products: Cotton", United States Department of Agriculture, available at: www.fas.usda.gov/commodities/cotton.

White, P.J. (2000), "Fatty acids in oilseeds (vegetable oils)", Chapter 10 in: Chow, Ching Kuang (ed.), *Fatty Acids in Foods and their Health Implications (Second Edition)*, Marcel Dekker, Inc. New York.

WHO (1991), *Strategies for Assessing the Safety of Foods Produced by Biotechnology*, Report of a Joint FAO/WHO Consultation, World Health Organization, Geneva, out of print.

Wood, R. et al. (1994), "Studies designed to eliminate cyclopropenoid fatty acids in cottonseed. The development of a method for the rapid analysis of cyclopropenoid fatty acids", *Biochemistry of Cotton Sept. 28-30, 1994*, Galveston, Texas, Cotton Inc., Raleigh, North Carolina.

Ziehr, A.U. et al. (2000), *Cottonseed with a High (+)- to –(-) Gossypol Enantimer Ratio Favorable to Broiler Production*, USDA Agricultural Research Service.

# *Chapter 3*

# Cassava (*Manihot esculenta*)

*This chapter, prepared by the OECD Task Force for the Safety of Novel Foods and Feeds with South Africa as the lead country, deals with the composition of cassava (*Manihot esculenta*). It contains elements that can be used in a comparative approach as part of a safety assessment of foods and feeds derived from new varieties. Background is given on cassava production and processing for human and animal consumption, industrial uses and ethanol production, followed by appropriate varietal comparators and characteristics screened by breeders. Nutrients in fresh cassava roots and leaves and in processed products, anti-nutrients, toxicants and allergens are then detailed. The final sections suggest key products and constituents for analysis of new varieties for food use and for feed use.*

# Background

## *General description of cassava*

Wild cassava (*Manihot flabellifolia* Phol and *Manihot peruviana*) is native to tropical America (Olsen and Schaal, 2001; Chacón et al., 2008).

Cultivated cassava is known scientifically as *Manihot esculenta* Crantz. Cultivated cassava (referred to in this chapter as cassava) is also known amongst rural populations in various countries as yuca, manioc and mandioca. It was later introduced to Africa and Asia, where it forms the subsistence base of the poorer populations in the marginal areas of these continents. Recently, Chacón et al. (2008) came up with evidence suggesting that the different subspecies of *M. esculenta* are not monophyletic, most probably due to hybridisations between the cultivated crop and wild species.

Cassava is a perennial woody shrub that produces storage roots that can be harvested six months to three years after planting. It is propagated by mature woody stem cuttings, while seeds are used mainly in breeding programmes.

Under optimal environmental conditions cassava compares favourably in the production of energy with most other major staple crops due to its high yield potential (El-Sharkawy, 2004). The cultivars are traditionally characterised as high or low cyanide content. They can also be grouped into high and low starch varieties for commercial application, edible lines for human consumption and lines suitable for animal feed.

## *Production*

Cassava is the fourth most important crop grown in the developing world, with global production in 2006/07 estimated at 218 million tonnes (FAOSTAT, 2009).

Cassava is used for human consumption (60% of the worldwide production); animal feed industry (33%); other industrial purposes such as textile, food and beverages (Soccol, 1996); as well as ethanol production for a short while. Cassava is a major source of energy in the tropics (Cock, 1982); based on kcal consumption per capita per day, it ranks eighth among the major food crops (FAOSTAT, 2009).

Cassava is the staple food of nearly 1 billion people in 105 countries, providing as much as a third of daily calories. Globally, production of cassava is expected to increase by over 50% during the period from 1993 to 2020, at an annual growth rate of around 2.5% in Africa and 1.2% in Latin America (Scott et al., 2000). World cassava production increased from 188.4 million tonnes (Mt) in 2002/03 to 217.9 Mt in 2006/07 (Table 3.1). The five countries with the highest production of cassava in 2006/07 were Nigeria (40.1 Mt), Brazil (26.6 Mt), Thailand (24.7 Mt), Indonesia (20 Mt) and the Democratic Republic of the Congo (15 Mt) (FAOSTAT, 2009).

Table 3.1. **Estimated global cassava production**

Million tonnes, Mt[1]

| | 1983-85 | 1993-95 | 2000-01 | 2002-03 | 2004-05 | 2006-07 |
|---|---|---|---|---|---|---|
| Africa | 55.3 | 83.2 | 96.7 | 101.9 | 110.9 | 111.2 |
| Asia | 47.9 | 49.1 | 51.1 | 53.6 | 57.7 | 70.1 |
| Americas | 28.5 | 31.0 | 31.6 | 32.6 | 35.9 | 36.4 |
| Oceania | 0.2 | 0.2 | 0.2 | 0.2 | 0.2 | 0.2 |
| World | 132.0 | 163.5 | 179.6 | 188.4 | 204.6 | 217.9 |

*Note:* 1. In comparing cassava and grain crops production figures, it should be noted that cassava figures are reported at 70% moisture content, while those of most grain crops are reported at approximately 15% moisture content.

*Source:* FAOSTAT (2009).

Cassava is produced, mainly by small stakeholders, in the humid, sub-humid and semi-arid conditions of tropical and subtropical areas of Asia, Latin America and the Caribbean, and Africa.

In 2007, the world average yield of fresh cassava roots was 11.6 tonnes per hectare (t/ha), with an average of 19.1 t/ha in Asia, 13 t/ha in the Americas and 8.8 t/ha in Africa. The yield varies with the cultivar, season of planting, soil type and fertility. The average cassava yields in 2000 were estimated to be barely 20% of those obtained under optimum conditions, which can result in yields ranging from 25-40 t/ha.

The global estimated harvest area of cassava increased from 13.85 million hectares in 1983-85 to 18.44 million hectares in 2006/07 (Table 3.2).

Table 3.2. **Estimated global cassava harvest area**

Million hectares

| | 1983-85 | 1993-95 | 2000-01 | 2002-03 | 2004-05 | 2006-07 |
|---|---|---|---|---|---|---|
| Africa | 7.53 | 10.18 | 11.02 | 11.43 | 11.80 | 11.86 |
| Asia | 3.74 | 3.80 | 3.45 | 3.41 | 3.47 | 3.76 |
| Americas | 2.57 | 2.62 | 2.53 | 2.55 | 2.95 | 2.80 |
| Oceania | 0.02 | 0.02 | 0.02 | 0.02 | 0.02 | 0.02 |
| World | 13.85 | 16.62 | 17.01 | 17.41 | 18.24 | 18.44 |

*Source:* FAOSTAT (2009).

## Processing and use

For the purpose of this chapter, cassava products are defined in Table 3.3.

Table 3.3. **Terms commonly found in literature to describe parts, types and uses of cassava**

| Term | Definition in this chapter |
|---|---|
| Cassava roots | The enlarged starch-filled root portion of cassava plant, sometimes wrongly called starchy tuber. |
| Cassava peels | Outer cover of the starchy root that is usually removed manually with a sharp knife with little or no pulp. |
| Cassava leaves | The vegetative part of the plant used as vegetable and leaf meal. |
| Sweet cassava | Edible cassava variety (low cyanogenic potential). |
| Bitter cassava | Poisonous cassava variety (high cyanogenic potential). |
| Cassava flour/meal | Dried milled cassava roots used mainly for human consumption. Includes flour, meal and flakes. |
| Cassava chips | Dried un-milled cassava. |
| Cassava starch | Complex carbohydrate from peeled root used in paper, textile and food industries. |
| Dried cassava | Includes peeled, sliced and sun-dried (chips) and ground and compressed cassava (pellets) used mainly as livestock feed. |
| Tapioca | Cassava starch used in the preparation of puddings and infant feed. |

*Source:* Adapted from Purdue University, Center for New Crops & Plant Products (1995).

### General human and animal consumption

Sweet cassava cultivars, which contain low cyanogenic glycoside levels (< 180 ppm dry weight basis), are used for human consumption, while bitter cultivars are mainly used for industrial purposes (FAO, 2009) but can also be used for human consumption after special processing (e.g. "*gari*" in West Africa, or "*farinha*" in Brazil). Based on kcal per capita per day consumption, cassava ranks eighth among the major food crops, after rice, wheat, sugar cane, maize, soybean, potatoes and palm oil (FAOSTAT, 2009). Staple food of nearly 1 billion people, cassava brings as much as a third of their daily calories (Eggum, 1970; Awoyinka et al., 1995; Tonukari, 2004; Izuagie et al., 2007).

Cassava tubers are valued as an energy source in human and animal diets (Babu and Chatterjee, 1999) having a carbohydrate content of about 92% (dry weight), mainly in the form of starch (Oke, 1968). Cassava roots are low in protein (Babu and Chatterjee, 1999). The leaves are also consumed and are a source of vitamin A, vitamin C, minerals (iron and calcium) and proteins (Nweke et al., 2002). Cassava shoots (young stem, leaves and petioles) are also an edible source of proteins and minerals; widely used as food in Africa, they constitute a major component of the diet in the cassava growing regions (Hahn, 1992; Achidi et al., 2001; FAO, 2009).

A factor limiting the human and animal consumption of cassava is its content of cyanogenic glycosides (Kakes, 1990). In the plant cells they are found in the cytoplasm and are accompanied by relatively specific hydrolytic β-glucosidases and hydroxynitrile lyase, able to degrade the cyanogenic glycosides and form bioactive toxic compounds, most notably hydrocyanic acid (HCN). However, the enzymes are sequestered from the cyanogenic glycosides and remain inactive by cellular compartmentalization to prohibit the formation of HCN at normal conditions (Conn, 1973, 1979). On cell damage, the cyanogenic glycosides and the enzymes are brought in contact and HCN is formed. Cyanide ($CN^-$) is largely removed by traditional processing methods such as grating, fermentation, boiling or drying (Hahn, 1989). Cooking the roots inactivates the enzymes and slowly destroys the cyanogens (Nweke et al., 2002). The cyanogenic potential, and dry matter (DM) content of cassava roots, as well as the pasting properties, influence the safety and quality of processed foods and industrial products.

Cassava products are rarely eaten on their own, but commonly in combination with relatively protein-rich food. However, certain processing techniques may reduce or enhance the protein, vitamin or mineral contents of the cassava product to be consumed. Nutrients such as vitamin C are reduced during processing and cooking (Berry, 1993). Cyanogenic potential ranged from 14 ppm to 3 275 ppm in a large study including more than 4 000 clones with an average of 327 ppm (Sánchez et al., 2009).

Post-harvest physiological deterioration often begins within 24-48 hours after harvest and quickly spoils the roots (Beeching et.al., 1998). It is not a microbial process but a self-inflicted reaction by genes active in the root. Therefore roots need to be consumed or processed shortly after harvesting. Physiochemical and functional properties of the storage root primarily determine the quality of cassava-based products. Chávez et al. (2005) studied the association between carotene content and post-harvest physiological deterioration and obtained a negative correlation, although further study is still required.

Processing of cassava leaves has a marginal effect on the majority of the compositional nutrients. In a study by Achidi (2003), leaves of two varieties of *Manihot esculenta* Crantz were subjected to processing (heat treated, pounded and cooked and crushed, ground and cooked) and compared for proximate composition, minerals, vitamins and anti-nutritional factors. The processing methods had no significant effect on ash, lipids, protein, fibre, total carbohydrate, carotene, calcium, magnesium, potassium, sodium, phosphorus, copper, zinc and manganese, but produced a significant reduction in the levels of free sugars, ascorbic acid, thiamine, cyanogenic potential and tannin levels. Ravindran et al. (1987) determined the crude protein content of cassava leaf meal (including petiole) after different periods of wilting, methods of drying, and chopping or not chopping. These processing methods had little influence on the crude protein content of leaf meal, except chopping of leaves, which resulted in consistently reduced crude protein content. The mean crude protein level was 23.1 g/100 g DM.

Fasuyi (2005) studied the nutrient profile of leaves of three genetically improved varieties of cassava plants that were harvested and subjected to different processing methods (sun drying, oven drying, steaming, shredding, steeping and a combination of these methods). The level of protein and several minerals (calcium, zinc, nickel and potassium) were found to be high.

### *Human food processing*

Cassava is consumed by humans as fresh processed roots, fermented roots, cassava flour-based products or cooked leaves. Traditionally, roots and leaves are processed by diverse methods on the different continents, offering a range of food products which include dried cassava chips, flour used for a variety of baked products and snacks, etc. Fresh cassava roots can be frozen, fried or boiled (Agrocadenas, n.d.; Cock, 1985; Cereda, 2003; Howeler, 2004; Embrapa, 2005).

Cassava processing involves a combination of step-wise activities, including: *i)* peeling; *ii)* chipping, crushing, milling, slicing or grating; *iii)* dehydration by pressing, decanting or drying in the sun or over a hearth; *iv)* fermenting by soaking in water, heaping or stacking; *v)* sedimentation; *vi)* sieving; *vii)* cooking, boiling, toasting or steaming. The number of steps required and the sequence varies with the product being made. This sequence of activities also generates a wide range of intermediate products, which can be either sold or stored until the need arises for conversion into the final product. Some of the processed products can be eaten without further cooking, while others require some extra preparation (Nweke, 1990). The most commonly used processing methods are presented in Figure 3.1.

The most important processed product of cassava is fermented ("bitter") starch, which on a dry weight basis (12% moisture) consists of 96% carbohydrates and 3% proteins. The starch is good for making bread, because it expands during baking. Fermented starch is very important in the snack industry to produce local products such as *pandebono* and *pandeyuca* (cheese breads), *rosquillas* (small, baked and crunchy doughnuts) and *besitos* (small, baked and crunchy puffs) in Brazil (Agrocadenas, n.d.). Another product, cassava flour, can be used as a substitute (up to 30%) for wheat flour in baking (Grace, 1977, on FAO website)

### *Animal feed processing*

All cassava varieties can be used in animal feed, but it is necessary to process them because of the presence of cyanogenic glycosides, otherwise hydrolysis of the cyanogenic glycoside linamarin would form HCN. Less than 100 g HCN/kg cassava product is considered as acceptable for animal feed.

Cassava leaves and roots are a useful alternative energy source for animal production. Fresh cassava foliage for a balanced animal feed has potential and could be as high as 100 t/ha per year depending on fertility of soil and rainfall (Ospina et al., 2002). Because of poor post-harvest life of the tubers, rapid processing is important (Padmaja, 2000).

Silage can be produced from forage and from cassava roots (Chauynarong et al., 2009). The moisture content has to be reduced when forage is used for silage production. Based on experiences and data collated in different countries, CIAT-Colombia established formula for silage shown in Table 3.4. Addition of nitrogen (usually in the form of urea) is recommended when roots constitute an important share of the silage formula. Fermentable carbohydrates brought with added molasses (or with other added sources such as corn meal; see Ubalua, 2007) can facilitate rapid fermentation, especially when forage is the main silage component.

Table 3.4. **Formula for silage from different sources**

| Silage component | Content (%) | | | |
|---|---|---|---|---|
| Cassava forage | 80.0 | 65.5 | 92.0 | |
| Cassava roots | 20.0 | 33.0 | | 98.2 |
| Urea | | 1.5 | | 1.8 |
| Molasses (or corn meal) | | | 8.0 | |
| Total | 100.0 | 100.0 | 100.0 | 100.0 |

*Source:* Personal communication from CIAT (unpublished data).

Figure 3.1. **Schematic representation of cassava processing into different food and feed products**

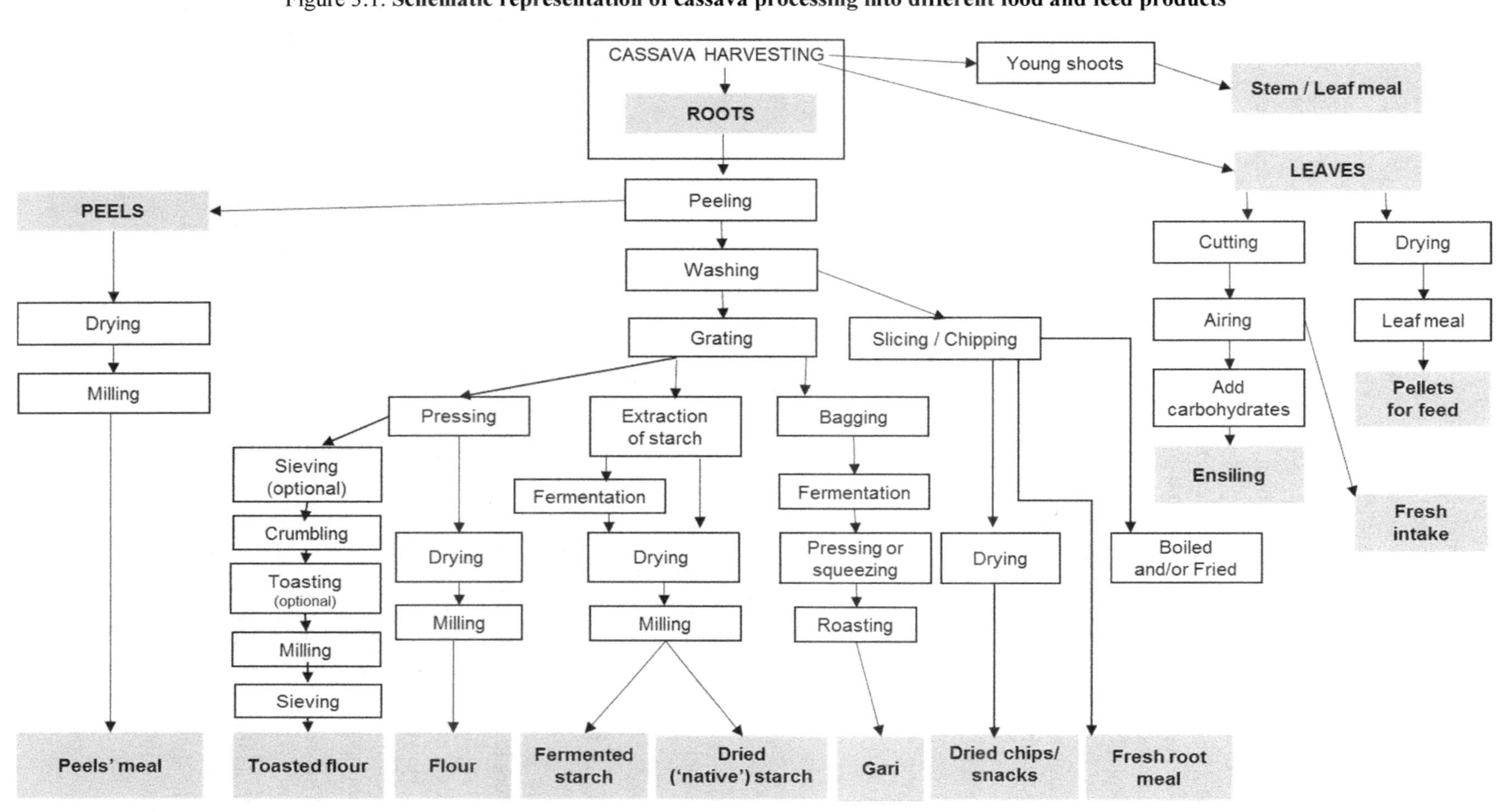

In Colombia and some other tropical countries, the aerial parts of cassava are used for animal feed, especially in ruminants. The leaf is characterised by a high level of crude protein (22% on average). Cassava foliage provides pigmentation because it contains a considerable concentration of total xanthophylls (605 mg/kg DM) and xanthophylls (508 mg/kg DM) (Ceballos and Ospina, 2002).

Solid wastes produced from cassava processing – comprising peelings from initial processing, fibrous by-products from crushing and sieving, and starch residue after starch settling – require specific management practices for their use as feed (Sackey and Bani, 2007; Chauynarong et al., 2009). Peels are used for animal feed after adequate fermentation in many South American, African and Asian countries; cassava peels are also reported as a medium for mushroom cultivation and to produce compost. Protein enrichment of cassava wastes by development of microorganisms can provide high-quality feedstuffs (Ubalua, 2007). The fibrous residual material, constituting around 30% of the original tubers, forms the cassava pulp left after starch extraction; dewatered in a screen press and dried in a flash dryer, it is sold to the feed industry (TIME IS, 2005). Solid residues can also be ensiled. The ensiling process contributes to lower the cyanide level to a non-toxic level thus reducing the pH to about 4.0 and allowing lactic acid to build up, and the product can be used as animal feed (Sackey and Bani, 2007).

### *Range of food products and other industrial outputs*

Modified starch derived from cassava can be used widely in the food industry. Cassava starch has unique properties, such as high viscosity and resistance to freezing. Industrial markets include those for unmodified starch for glucose products used in food binders and thickeners, and for animal feed. There is also great potential for cassava starch utilisation in the sweetener and alcoholic beverages industries. Large volumes of "native" or modified cassava starches are used for many different non-food industrial uses, such as in the paper and textile industry.

### *Ethanol production and animal feed by-products*

Apart from its traditional role as a food crop, cassava can be used as a carbohydrate source to produce ethanol which is used by the pharmaceuticals and beverages industry (TIME IS, 2005). It has become an important crop for bio-fermentation in Brazil, the People's Republic of China, Thailand and countries in sub-Saharan Africa, especially Nigeria. In a series of steps, starch is converted to glucose that is then fermented to produce ethanol. A flowchart showing the process is depicted in Figure 3.2 (IITA -ICP, n.d.). However, recent technological developments are simplifying the process to produce ethanol from starchy crops by merging some of the stages detailed in Figure 3.2 (Chamsart et al., 2007).

The cassava pulp (solid waste resulting from starch extraction) is also considered by some studies as a potential source for low-cost ethanol production (JIRCAS, 2006).

In addition to ethanol production, the manufacturing process provides for marketable feed by-products, e.g. cassava cake and bagass (Suthsmma and Sorapipatana, 2007). Selling these by-products to the feed industry is an important economic outlet for ethanol manufacturers.

Figure 3.2. **Schematic presentation of the ethanol production process from cassava**

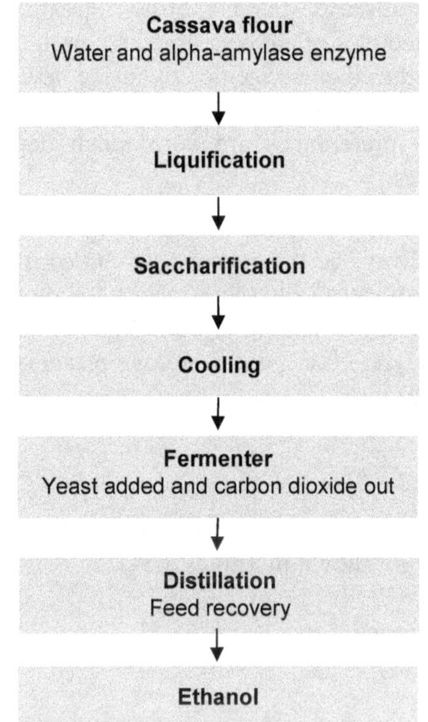


*Source*: Adapted from IITA-ICP (n.d.).

## *Appropriate comparators for testing new varieties*

This chapter suggests parameters that cassava breeders should measure when developing new modified varieties. The data obtained in the analysis of a new cassava variety should ideally be compared to those obtained from an appropriate near isogenic non-modified variety, grown and harvested under the same conditions.[1] The comparison can also be made between values obtained from new varieties and data available in the literature, or chemical analytical data generated from other commercial cassava varieties.

Components to be analysed include key nutrients, toxicants and allergens. Key nutrients are those which have a substantial impact in the overall diet of humans (food) and animals (feed). These may be major constituents (fats, proteins, and structural and non-structural carbohydrates) or minor compounds (vitamins and minerals). Similarly, the levels of known anti-nutrients and allergens should be considered. Key toxicants are those toxicologically significant compounds known to be inherently present in the species, whose toxic potency and levels may impact human and animal health. Standardised analytical methods and appropriate types of material should be used, adequately adapted to the use of each product and by-product. The key components analysed are used as indicators of whether unintended effects of the genetic modification influencing plant metabolism has occurred or not.

## *Breeding characteristics screened by developers*

About 98 species of genus *Manihot* are recognised. Cultivars have been developed through domestication of natural hybrids of wild species and maintained through vegetative reproduction (Allem, 1994). The most important commercial quality trait for cassava breeders in Asia is starch yield. In Africa, breeders focus on disease and pest resistance (cassava mosaic disease, cassava brown streak, green mites and bacterial blight). Other important objectives of cassava breeding are increased root protein content, β-carotene content, and reduced cyanogenic glycoside levels. Cassava varieties exist that are either "bitter" (high in cyanogenic glycosides) or "sweet" (low in cyanogenic glycosides) in taste. Taste preferences are very much dependent on the local rural communities within countries.

Since cassava is vegetatively propagated, and irregular flowering and low seed set occurs, breeding may not always be the appropriate choice for developing new varieties. Genetic engineering can provide an alternate means for developing improved or novel varieties (Taylor et al., 2004). For example, transgenic cyanogen-free cassava has been developed (Siritunga and Sayre, 2003), and a storage protein (ASP1) has been expressed in cassava (Zhang et al., 2003).

## Nutrients

The range of mean values for the nutrient composition of cassava roots, leaves and processed cassava products are shown in Tables 3.5-3.21.

### *Unprocessed roots and leaves*

#### *Proximate composition*

Representative data on nutrient composition of fresh cultivated cassava roots and leaves are presented in Tables 3.5 and 3.6, respectively. In some studies only average values are presented. When available, the variation in each parameter, as indicated by the minimum and maximum values, is given. The variation in values can be attributed to genetic, agricultural and environmental factors. Thus, the composition of fresh cassava roots and leaves varies with cultivar/variety, age of the plant tissue, geographical location, agricultural conditions and climate (Fasuyi and Aletor, 2005). For example, in six cassava varieties from South Viet Nam, India and Japan, cassava leaves were shown to vary substantially in composition, *viz.* (g/100 g DM): 23.9-34.7 g crude protein, 13.3-15.6 g fat, 9.7-14.6 g crude fibre, 31.7-45.5 g nitrogen-free extract and 5.0-7.9 g ash (Nhu Phuc et al., 2000; see Table 3.6). In many reports where it was not stated whether the roots were peeled or not, it was assumed that they were not peeled. This would have a significant effect on composition, because of elevated fibre values in unpeeled roots. In unpeeled roots, root size will also affect the proportion of fibre to non-fibre carbohydrate because small unpeeled roots should contain proportionally more fibre than large roots. Also, crude fibre and nitrogen-free extract analysis (Maynard et al., 1979) have been replaced by neutral detergent fibre and acid detergent fibre analysis. However, only one author reported acid detergent fibre and neutral detergent fibre values.

Observed differences may also reflect to some extent the analytical method used, but to a large extent investigators used standard methods, such as those published by the Association of Official Analytical Chemists (AOAC, 1990).

A wide variation in moisture content of roots has been reported. Bradbury and Holloway (1988) reported an average moisture content of 62.8 g/100g (fresh weight basis) for roots, while a range of 56.4-76 g/100 g sample was reported by Yeoh and Truong (1996). Using the specific gravity method, Chávez et al. (2005) measured the moisture content of 2 022 root samples collected from all over the world.

The DM values ranged from 10.7-57.2 g/100 g fresh sample weight, with a mean of 34.3 g/100 g sample. Sánchez et al. (2009) recently reported a range of DM content (from 4 000 genotypes) of cassava roots from 14.3-48.1%. The DM content is not presented in Tables 3.5 and 3.6 because, in most studies, no clear indication was given at what stage of the processing or harvesting the moisture determination was done.

Bradbury and Holloway (1988) reported an average moisture content of 74.8 g/100 g sample (fresh weight basis) for leaves. A study on young cassava leaves obtained from 19 different cassava varieties showed that the DM content of this plant tissue ranged from 23-27.8% (Achidi, 2003), while Gomez and Valdivieso (1984) recorded dry matter values between 30.8 g/100 g and 35.7 g/100 g in leaves (plus petioles) harvested 9-12 months after planting.

Table 3.5. **Proximate composition of fresh cassava roots**[1]

g/100 g of dry matter

| References | Akinfala et al. | USDA[2] | FAO | Oguntimein | Tien Dung et al. | Smith | Range of mean values |
|---|---|---|---|---|---|---|---|
| Crude protein | 4.7 | 3.4 | 1.9 | 2.0 | 2.4 | 1.5-3.5 | 1.5-4.7 |
| Crude fibre | 2.1 | | 0.8 | 4.0 | | 1.3-7.7 | 0.8-7.7 |
| Total dietary fibre | | 4.5 | | | | | 4.5 |
| Crude fat[3] | 2.5 | 0.7 | 0.3 | 0.7 | 2.2 | 0.8-3.2 | 0.3-3.2 |
| Ash | 8.4 | 1.5 | 0.5 | 5.0 | 1.5 | 1.6-4.1 | 0.5-8.4 |
| Nitrogen-free extract | 75.3 | (94.4)[4] | 56.0 | 75.5 | | 88.0-94.1 | 56.0-94.1 |
| Neutral detergent fibre | | | | | 5.5 | | 5.5 |
| Acid detergent fibre | | | | | 2.5 | | 2.5 |

*Notes:* 1. It is assumed the roots were not peeled (not always reported). 2. The data was converted to a dry matter basis, using the level of water content of 59.68 g/100 g given in the USDA table. 3. Ether extractable fat. 4. USDA lists this value as carbohydrate, by difference.

*Sources:* Akinfala et al. (2002); USDA Agricultural Research Service (2008); FAO (2001), table "Proximate composition of food"; Oguntimein (1988); Tien Dung et al. (2005); Smith (1988).

Table 3.6. **Proximate composition of fresh cassava leaves**

g/100 g of dry matter

| References | Akinfala et al.[1] | Nhu Phuc et al.[2] | Oguntimein | Smith | Hang and Preston | Range of mean values |
|---|---|---|---|---|---|---|
| Crude protein | 18.0 | 23.9-34.7 | 24.1 | 14.7-36.4 | 20.0-30.01 | 14.7-36.4 |
| Crude fibre | 14.1 | 9.7-14.6<br>11.5 | 26.0 | 4.8-15.4 | | 4.8-26.0 |
| Crude fat | 9.4 | 13.3-15.6<br>14.3 | 5.0 | 4.0-15.2 | 5.9 | 4.0-15.6 |
| Ash | 7.9 | 5.0-7.9<br>6.5 | 8.0 | 5.5-16.1 | 10.0 | 5.0-16.1 |
| Nitrogen-free extract | 43.3 | 31.7-45.5<br>38.8 | 39.9 | 31.7-45.5 | 44.2 | 31.7-45.5 |
| Neutral detergent fibre | | | | | 29.6 | 29.6 |
| Acid detergent fibre | | | | | 24.1 | 24.1 |

*Notes:* 1. Composite sample prepared for trial. 2. Average of six cultivars, sun-dried.

*Sources:* Akinfala et al. (2002); Nhu Phuc et al. (2000); Oguntimein (1988); Smith (1988); Hang and Preston (2005).

Crude protein is widely determined using the Kjeldahl technique, in which the nitrogen content is measured and multiplied by 6.25 to estimate crude protein. In cassava, and possibly other crops, not all the nitrogen is incorporated in proteins. Differences in genetics (germplasm) and growth conditions create huge variations in free amino acids and non-protein nitrogen (Yeoh and Truong, 1996). Chávez et al. (2005) analysed the roots (assumed unpeeled) of 600 cassava genotypes collected worldwide and reported a mean crude protein content of 3.06 g/100 g DM, ranging from 0.77-8.31 g/100 g DM. Ceballos et al. (2006) searched for varieties containing high protein levels and reported (using a conversion factor of 6.25 to go from total nitrogen to crude protein) a crude protein content ranging between 0.95 g and 6.42 g/100 g DM. These investigators also measured the hydrocyanic acid (HCN) produced in the cassava and found no correlation between HCN content and crude protein content, perhaps because most of the HCN is removed from the plant in sample preparation. Fifteen cassava varieties from Asia showed a lower root protein content, ranging from 0.5-1.9 g/100 g DM (Hock-Hin and Van-Den, 1996). The value of 6.42/100 g and 8.3/100 g in roots as shown in some landraces from South America is high, but most cassava cultivars worldwide (Table 3.5) have a lower level of crude protein in the roots (1.5-4.7/100 g DM) (Babu and Chaterjee, 1999; Ceballos et al., 2006). *Note: Preliminary data from CIAT would suggest that the N-to-protein conversion factor is considerably lower than the standard value of 6.25. A reliable and relatively simple method for a direct quantification of total soluble proteins based on Bradford's approach would be much more precise than indirect quantification based on N.*

The crude protein content of leaves of cassava ranges between 14.7 g and 36.4 g/100 g dry matter (Table 3.6).

The crude fat was measured as ether extract. Cassava roots contain low concentrations of fat, ranging from 0.3-3.2 g/100 g DM (Table 3.5). However, the leaves contain relatively high levels of fat, ranging from 4.0-15.6 g/100 g DM (Table 3.6).

Ash is what remains after the organic part of the plant material has been oxidized through combustion, and is a measure of the total amount of inorganic matter in the samples. For cassava roots, ash varies between 0.5 g and 8.4 g/100 g DM, and is higher in leaves, ranging between 5 g and 16.1 g/100 g DM (Tables 3.5 and 3.6). The extent to which variation is due to soil contamination is not clear, because in some references it was stated explicitly that the roots were washed while in others no mention was made about the preparation of the material. Fresh leaves have an ash content of 10 g/100 g DM according to studies conducted by Eggum (1970) and Luyken et al. (1961).

Nitrogen-free extract (NFE), representing the non-fibre carbohydrates, is usually determined by difference (moisture, fat, ash, crude fibre and proteins are measured and the remainder is attributed to NFE) and constitutes a heterogeneous complex of compounds, including the starch. NFE levels in cassava roots vary considerably depending on the cultivar, ranging between 56 g to 94 g/100 g DM (Table 3.5). In addition, peeling of roots may have an effect on the proportion of proximates, since non-fibre carbohydrates are present in the roots. Leaves contain lower levels of NFE than roots, ranging between 31.7 g and 45.5 g/100 g DM (Table 3.6).

Cassava normally contains 0.8-7.7 g/100 g dry weight crude fibre, a component that reduces its digestibility. Digestibility is important in both human and animal cassava-based diets. Excess fibres interfere with the utilisation of phosphorous and zinc (Oke, 1978). The crude fibre content, like the ash content, is highly dependent on growth conditions and germplasm of cassava. Cassava bagasse (solid waste from industrial processing) is a fibrous residue which contains 14.88-50.55 g of crude fibre/100 g dry weight and can be used in bioconversion processes using microbial cultures (Pandey et al., 2000). Fresh leaves have a high fibre content (an average of 17 g/100 g dry weight) and digestibility is low (70-80% in young leaves, decreasing to 67% in old leaves) (Eggum, 1970; Luyken et al., 1961).

Acid detergent fibre (ADF) and neutral detergent fibre (NDF) provide much more accurate fibre values for feeds containing high levels of lignin as part of the fibre than the proximate analysis of crude fibre and NFE; however, only one author reported values for these parameters. ADF and NDF are strictly not grouped as proximate (Tables 3.6 and 3.7).

### *Carbohydrates*

Cassava roots are a good source of energy, with carbohydrate contents reported as high as 91% on a dry weight basis (Oke, 1968; Sánchez et al., 2009). Szylit et al. (1978) determined that cassava roots contained 74.7 g starch/100 g DM and 0.6 g ethanol-soluble carbohydrates/100 g DM and a mean starch granule diameter of 12 µg. On average, 73-85% of dry root weight of cassava is starch (Rickard et al., 1991). Starch content varies in different cassava germplasm, such as improved clones and landraces (Sánchez et al., 2009). The high starch content, ranging from 18-24% amylose and 70% amylopectin, makes for ideal digestion (Johnson and Raymond, 1965). Cassava starch is classified as easily degradable, since 20% is degraded in six hours when exposed to bacterial α-amylase *in vitro* (Szylit et al., 1978). Average amylose content in starch from a large sample of cultivars was 20.7% (Sánchez et al., 2009) and can be used as a standard reference point. Amylose-free natural mutation and induced mutation for high-amylose (36%) cassava starch have been reported (Ceballos et al., 2007, 2008). Amylose-free starch is easily digestible and better for ethanol production. High-amylose can lead to the production of resistant starches, which have distinctive advantage in health, particularly in diabetes management and stimulation of butyrate production in the large intestine that has been found to be beneficial to colon health (Jobling, 2004; Lehman and Robin, 2007).

The metabolisable energy of cassava roots varies with the genotype (variety), age of the root, harvesting time and climatic conditions, and is also dependent on the method of processing. Differences might also be due to the state of processing of cassava, the raw sample having a digestibility of about 48.3% and the cooked sample of 77.9%. The higher amylopectin content of cassava relative to maize makes it a more suitable source of energy for ruminants than for monogastric animals (Oke, 1978). Analysing 1 755 samples, Chávez et al. (2005) recorded an average total root sugar content of 8.4 g/100 g DM and an average content of reducing sugars of 2.2 g/100 g DM, while Sánchez et al. (2009) reported total and reducing sugars in cassava roots ranging from 0.2-18.8% and 0.0-15.7%, respectively, on a dry weight basis. A group of interesting "sugary" mutations in cassava that result in storage roots with high free sugars (mostly glucose) and a glycogen-like molecule was reported by Carvalho et al. in 2004. The roots from these genotypes have reduced levels of amylose.

A study by Nhu Phuc et al. (2000) on leaves of six cassava varieties from India, Japan and South Viet Nam showed that the free sugars range varied from 2.2-4.4 g/100 g DM, starch from 4.7-6.1 g/100 g DM, total non-fibre carbohydrates from 7.1-10.4 g/ 100 g and food energy from 307.0-376.2 x $10^3$ joules/kg DM.

### *True protein (amino acids)*

Cassava roots contain so little protein (0.7-2%) that the amino acid composition is of little significance in nutrition. Of the small amount of nitrogen in cassava roots, only about 60% is protein nitrogen, while 30-40% is non-protein nitrogen, comprising of free amino acids, nitrate, nitrite and cyanogeic glycosides. The traditional formula for calculation of crude protein (nitrogen measured by Kjeldahl method and multiplied by 6.25) therefore overestimates the true protein content. Based on analysis of crude protein content of 15 varieties of cassava roots, Yeoh and Truong (1996) estimated that 51-75% of the nitrogen in cassava roots consists of true protein, i.e. nitrogen incorporated as protein-associated amino acids.

The amino acid composition of the true protein in cassava roots, as well as the concentration of individual amino acids per 100 g DM, are presented in Table 3.7. However, the latter will depend on the concentration of true protein per 100 g DM. In Table 3.7, the mean values reported in the two studies on cassava roots were 0.404 g and 0.827 g/100 g DM, while that of the individual cultivars analysed by Nassar and Vale de Sousa (2007) varied between 0.25 g5 and 1.654 g/100 g DM.

Table 3.7. **Amino acid composition in the protein of cassava roots**

g/100 g of dry sample powder

| References | Nassar and Vale de Sousa | Oke | USDA[1] | Range of mean values |
|---|---|---|---|---|
| Arginine | 0.145 | 0.178 | 0.340 | 0.145-0.340 |
| Histidine | 0.020 | 0.034 | 0.050 | 0.020-0.050 |
| Isoleucine | 0.031 | 0.046 | 0.067 | 0.031-0.067 |
| Leucine | 0.055 | 0.064 | 0.097 | 0.055-0.097 |
| Lysine | 0.043 | 0.067 | 0.109 | 0.043-0.109 |
| Methionine | 0.019 | 0.022 | 0.027 | 0.019-0.027 |
| Phenylalanine | 0.065 | 0.041 | 0.064 | 0.041-0.065 |
| Threonine | 0.030 | 0.043 | 0.069 | 0.030-0.069 |
| Tryptophan | 0 | 0.019 | 0.047 | 0.019-0.047 |
| Valine | 0.056 | 0.054 | 0.087 | 0.054-0.087 |
| Alanine | 0.048 | | 0.094 | 0.048-0.094 |
| Aspartic acid | 0.068 | | 0.196 | 0.068-0.196 |
| Cystine | 0.026 | 0.023 | 0.069 | 0.023-0.069 |
| Glutamic acid | 0.124 | | 0.511 | 0.124-0.511 |
| Glycine | 0.038 | | 0.069 | 0.038-0.069 |
| Proline | 0.020 | | 0.082 | 0.020-0.082 |
| Serine | 0.040 | | 0.082 | 0.040-0.082 |
| Tyrosine | 0 | | 0.042 | 0.000–0.042 |
| AA/100 g DM | 0.827 | 0.404 | 2.103 | 0.404-2.103 |

*Note:* 1. The data were converted to a dry matter basis, using the level of water content of 59.68 g/100 g given in the USDA table.

*Sources:* Nassar and Vale de Sousa (2007); Oke (1978); USDA Agricultural Research Service (2008).

Cassava leaves are rich in proteins and essential amino acids. Studies have shown a range of leaf protein content of 29.3-38.6 g/100 g DM (Yeoh and Chew, 1976). Nhu Phuc et al. (2000) analysed leaves from six cassava varieties from India, Japan and South Viet Nam for their amino acid profile to determine protein quality. The results indicate that, on average, the amino acids glutamic acid and leucine were highest, with values above 4 g/100 g true protein, followed by aspartic acid, arginine and alanine with values above 3 g/100 g true protein. The amino acid composition of leaf protein (g/100 g protein) (Eggum, 1970; Devendra, 1977; Cereda, 2001) is presented in Table 3.8.

Table 3.8. **Amino acid composition in the protein of cassava leaves and meal**

g/100 g of protein

| References | Eggum[1] | Devendra[2] | Cereda[3] | Range of mean values |
|---|---|---|---|---|
| Arginine | 4.0-5.7 | 5.1 | | 4.0-5.7 |
| Histidine | 1.1-2.5 | 2.7 | 2.2 | 1.1-2.7 |
| Isoleucine | 3.9-5.0 | 4.3 | 5.0 | 3.9-5.0 |
| Leucine | 7.2-8.9 | 4.7 | 9.1 | 4.7-9.1 |
| Lysine | 3.8-7.5 | 7.1 | 6.3 | 3.8-7.5 |
| Methionine | 1.3-2.0 | 1.1 | (4.8)[1] | 1.1-2.0 |
| Phenylalanine | 5.3-5.4 | 3.6 | (8.8)[2] | 3.6-5.4 |
| Threonine | 3.2-5.0 | 4.7 | 4.8 | 3.2-5.0 |
| Tryptophan | 2.0 | 1.0 | | 1.0-2.0 |
| Valine | 5.1-5.7 | 6.4 | 6.4 | 5.1-6.4 |
| Cystine | 0.7-1.4 | 1.0 | | 0.7-1.4 |
| Glycine | | 4.6 | | 4.6 |

*Notes:* 1. Leaves: methionine + cysteine. 2. Meal: phenylalanine + tyrosine. 3. Dried leaves.

*Sources:* Eggum (1970); Devendra (1977); Cereda (2001).

## *Lipids*

Lipid composition of cassava roots has not been studied extensively, as it occurs in such low concentrations. Total lipids in fresh cassava roots average at 0.25% (Lalaguna and Agudo, 1988). Figures for phospholipids, glycolipids and neutral lipids are presented in Table 3.9.

Polar lipids plus sterols and steryl esters constitute the major portion (77.9%) of the extracted lipids. Of the seven phospholipids identified, phosphatidylcholine occurred in the highest concentration (265.4 nmol/g fresh weight), while of the six glycolipids identified, digalactosyldiacylglycerol was the most abundant glycolipid (333.2 nmol/g fresh weight). Free sterols averaged 304.3 nmol/g fresh weight and triacylglycerol was measured at 444.4 nmol/g fresh weight. Young cassava leaves have a low content of lipids (3.02%), of which 22.4%, 25.1% and 48.2% were non-polar lipids, glycolipids and phospholipids, respectively.

Non-polar lipids of the leaves contained 2.1% fatty acids, and with the exception of steryl esters, all leaf lipids have a high content of polyunsaturated fatty acids (Khor and Tan, 2006).

Table 3.9. **Lipid composition of cassava roots**

| Lipid | nmol/g fresh weight[1] |
|---|---|
| Total phospholipids | 706.0 |
| Total glycolipids | 818.6 |
| Total neutral lipids | 892.6 |

*Note:* 1. The concentrations are not converted to dry weight because the lipids are presented as lipid combinations.

*Source:* Adapted from Lalaguna and Agudo (1998).

The major fatty acids of cassava root meal lipid are oleic and palmitic acids.

The other fatty acids found in raw cassava roots are linoleic, linolenic, palmitoleic, stearic, myristic, pentadecanoic, heptadecanoic and nonadecanoic acids (Ezeala, 1985).

The fatty acid composition of raw cassava roots is summarised in Table 3.10.

Table 3.10. **Fatty acid composition of raw cassava roots**

| Fatty acid | g per 100 g dry matter[1] |
|---|---|
| Palmitic acid (16:0) | 0.17 |
| Stearic acid (18:0) | 0.01 |
| Oleic acid (18:1) | 0.19 |

*Notes:* 1. The data were converted to a dry matter basis, using the level of water content of 59.68 g/100 g given in the USDA table.

*Source:* Adapted from USDA Agricultural Research Service (2008).

Fermentation of processed roots of cassava does not alter the profile of the composition of fatty acids but causes an increase in the concentration of saturated fatty acids. Stearic acid increased by about 92.6%, while linoleic acid was reduced by 72% (Ezeala, 1985).

Table 3.11 illustrates fatty acid composition of fermented and unfermented cassava tuber meal.

Table 3.11. **Fatty acid composition and content of unfermented and fermented cassava tuber meal**

| Fatty acid | Content (g/kg dry tuber meal)[1] | | Content % total fatty acid)[1] | |
|---|---|---|---|---|
| | Unfermented | Fermented | Unfermented | Fermented |
| Myristic acid (14:0) | 0.06 | 0.08 | 1.2 | 1.2 |
| Pentadecanoic acid (15:0) | 0.03 | 0.06 | 0.6 | 0.9 |
| Palmitic acid (16:0) | 1.50 | 2.10 | 31.0 | 31.4 |
| Palmitoleic acid (16:1) | 0.20 | 0.22 | 4.1 | 3.3 |
| Margaric acid (17:0) | 0.02 | 0.03 | 0.5 | 0.5 |
| Stearic acid (18:0) | 0.13 | 0.34 | 2.7 | 5.2 |
| Oleic acid (18:1) | 1.80 | 2.46 | 37.5 | 37.2 |
| Linoleic acid (18:2) | 0.70 | 1.02 | 14.5 | 15.4 |
| A-linolenic acid (18:3) | 0.38 | 0.30 | 7.9 | 4.6 |
| Nonadecanoic acid (19:0) | Trace | 0.02 | Trace | 0.3 |

*Note:* 1. Means of three different determinations.

*Source:* Ezeala (1985).

### *Minerals*

The mineral content of cassava roots and leaves is shown in Tables 3.12 and 3.13, respectively.

In addition, the Food and Agriculture Organization (FAO) gives the calcium and iron content of processed root flour as 0.74 g/kg DM and 4.0 mg/g DM, respectively.

Evaluations by the CIAT of iron and zinc contents in cassava roots found some genetic variation, with average values of 15.7 mg and 6.35 mg/kg (dry weight basis) respectively. Quantification for these two elements can result from iron and zinc contaminations coming from the soil attached to the roots. The pH of the soil where the cassava was planted was found to have a high impact on iron and zinc contents in roots (CIAT, 2005).

Hung Nguyen et al. (2002) tested the influence of different levels of NPK fertiliser on the mineral composition of cassava leaves four months after planting. Increased rates of NPK in a ratio of 2:1:2 ($N:P_2O_5:K_2O$) significantly increased the concentrations of nitrogen, phosphorus, potassium, sulfur, manganese and copper in cassava leaves, while the concentrations of magnesium and calcium were reduced.

Table 3.12. **Mineral composition of cassava roots**

| References / Mineral | FAO | USDA[1] | Chávez et al. | Range of mean values |
|---|---|---|---|---|
| | | | g/kg dry matter | |
| Calcium (Ca) | 1.1 | 0.40 | 0.31-2.5 | 0.31-2.5 |
| Phosphorus (P) | | 0.67 | 0.71-3.2 | 0.67-3.2 |
| Magnesium (Mg) | | 0.52 | 0.52-2.4 | 0.52-2.4 |
| Potassium (K) | | 6.72 | | 6.72 |
| Sodium (Na) | | 0.35 | 0.02-1.23 | 0.02-1.23 |
| | | | mg/kg dry matter | |
| Iron (Fe) | 3.0 | 6.70 | 6.0-230.0 | 3.0-230.0 |
| Manganese (Mn) | | 9.52 | 0.45-5.0 | 0.45-9.52 |
| Copper (Cu) | | 2.48 | 0.79-40.3 | 0.79-40.3 |
| Zinc (Zn) | | 8.43 | 2.63-37.5 | 2.63-37.5 |
| Aluminium (Al) | | | 4.4-330 | 4.4-330 |
| Selenium (Se) | | 0.02 | | 0.02 |

*Note:* 1. The data were converted to a dry matter basis, using the level of water content of 59.68 g/100 g given in the USDA table.

*Sources:* FAO (2009); USDA Agricultural Research Service (2008); Chávez et al. (2005).

Table 3.13. **Mineral composition of dried cassava leaves and processed leaf meal**

| References / Mineral | Dried leaves | | | Dried leaf meal | | |
|---|---|---|---|---|---|---|
| | Cereda | Hung Nguyen et al.[1] | Range of mean values | Yousuf et al. | Vongsamphanh and Wanapat | Range of mean values |
| | | g/kg dry matter | | | g/kg dry matter | |
| Calcium (Ca) | 16 | 3.6-6.2 | 3.6-16 | 17.4 | 9.2 | 9.2-17.4 |
| Phosphorus (P) | 2.9 | 1.6-2.8 | 1.6-2.9 | 3.6 | 3.0 | 3.0-3.6 |
| Magnesium (Mg) | 3.8 | 2.0-4.1 | 2.0-4.1 | | | |
| Potassium (K) | 10 | 9.5-22.3 | 9.5-22.3 | | | |
| Sodium (Na) | | 0 | | | | |
| Sulphur (S) | 2.4 | 3.0-3.8 | 2.4-3.8 | | | |
| | | mg/kg dry matter | | | mg/kg dry matter | |
| Iron (Fe) | 442 | 800-2 000 | 442-2 000 | | 16.1-2 000 | 16.1-2 000 |
| Manganese (Mn) | 351 | 140-200 | 140-351 | | 12.9-200 | 12.9-200 |
| Copper (Cu) | 6 | 5.5-7.4 | 5.5-7.4 | | 3.6-17.7 | 3.6-17.7 |
| Zinc (Zn) | 40 | 61-81 | 40-81 | | 8.4-81.0 | 8.4-81.0 |

*Note:* 1. Fully expanded leaves, third and fourth from the top, four months after planting.

*Sources:* Cereda (2001); Hung Nguyen et al. (2002); Yousuf et al. (2007); Vongsamphanh and Wanapat (2004).

## *Vitamins*

Vitamin levels in mature cassava roots and leaves are low, with provitamin carotenes being the most important constituent. Total carotene levels in roots vary widely amongst cassava cultivars/varieties (Iglesias et al., 1997). A study by Chávez et al. (2005) of 1 789 accessions from the CIAT germplasm bank showed a range of total carotene in roots from 1.02-10.4 μg/g fresh weight, with an average of 2.457 μg/g fresh weight (7.17 μg/g dry weight). Maximum levels of total carotenoids in breeding populations range between 15-18 μg/g (fresh weight basis) and maximum levels of total β-carotene range between 12-13 μg/g also on a fresh weight basis (CIAT, 2009). FAO figures showed low levels of total carotene in roots of bitter cassava of 0.24 mg/kg DM. Among vitamins, ascorbic acid (vitamin C), thiamine, riboflavin and niacin are the most important. Again these vary according to the cultivar and age of the cassava plants (Table 3.14). Cassava leaves, however, are rich in vitamins, especially the young leaves that are usually eaten by humans (Awoyinka et al., 1995); in a study by Nhu Phuc et al. (2000), the leaves of six cassava varieties from India, Japan and South Viet Nam were found to be rich in ascorbic acid, thiamine and β-carotene.

Table 3.14. **ß-carotene and vitamin content of cassava roots and flour**

per kg dry weight

| References | Unit | Root | | | Flour |
|---|---|---|---|---|---|
| | | FAO | USDA[1] | Range of mean values | Grace |
| Provitamin A (β-carotene) | mcg | 24 | 198 | 24-198 | 0 |
| Vitamin B1 | mg | 48 | 2.16 | 2.16-48 | |
| Vitamin B2 | mg | 0.06 | 1.19 | 0.06-1.19 | 0.07 |
| Vitamin B6 | mg | 0.08 | 2.18 | 0.08-2.18 | 0.06 |
| Niacin | mg | 0.9 | 21.18 | 0.9-21.18 | 0 |
| Folic acid | µg | 38 | | 38 | |
| Folate | mcg | | 669.64 | 669.64 | |
| Vitamin C | mg | 50 | 510.91 | 50-510.91 | 4.5 |

*Notes:* 1. The data were converted to a dry matter basis, using the level of water content of 59.68 g/100 g given in the USDA table.

*Sources:* FAO (2001); USDA Agricultural Research Service (2008); Grace (1977).

### *Processed cassava products*

Table 3.15 illustrates the proximate composition of cassava peel meal, cassava meal and flour. Cassava tapioca and starch contain an average of 0.50 g protein and 0.33 g fat/100 g DM.

The proximate composition of processed cassava leaves and cassava hay is pesented in Table 3.16. Processed leaves are used in the feeding of monogastric animals such as pigs and poultry (Nhu Phuc et al., 2000; Du and Preston, 2005) as well as for ruminants, while the foliage (leaves and stems), is fed almost exclusively to ruminants (Wanapat et al., 1997; Tien Dung et al., 2005). The composition of the foliage, including the hay, would depend on the proportion of leaves to stems, the latter having a lower protein content (Tien Dung et al., 2005).

Table 3.17 shows the moisture and protein contents, and amount of metabolizable energy in fresh cassava roots and various cassava products and by-products used in animal feed.

### Table 3.15. **Nutrient composition of processed cassava roots**

g/100 g dry matter

| | Peel meals | | | Cassava meal | | | | Root meal | Flour |
|---|---|---|---|---|---|---|---|---|---|
| | Salami | Osei et al. Unfermented | Osei et al. Fermented | FAO | Sable et al. | Kozloski et al. | Range of mean values | An et al. | Aregheore et al. |
| Crude protein | 5.9 | 5.1 | 5.3 | 1.8 | 2.6 | 1.6 | 1.6-3.2 | 2.9 | 3.2 |
| Crude fibre | 13.4 | 11.3 | 11.5 | 0.5 | | | 0.5-6.3 | 3.2 | 6.3 |
| Fat[1] | 1.2 | 0.8 | 0.7 | 0.6 | | 0.3 | 0.3-1.9 | 1.9 | 0.8 |
| Ash | 10.8 | 4.9 | 9.3 | 0.3 | 0.8 | | 0.3-1.8 | 1.7 | 1.8 |
| Nitrogen-free extract[2] | 68.9 | 67.4 | 63.9 | 92 | | 87.7[1] | 83.8-92.0 | | 83.8[2] |
| Neutral detergent fibre | | | | | 3.1 | 9.8 | 3.1-9.8 | 9.3 | |
| Acid detergent fibre | | | | | 1.8 | 3.0 | 1.8-3.6 | 3.6 | |

*Notes:* 1. Ether extractable fat; non-fibre carbohydrates. 2. Starch; peeled tubers.

*Sources:* Salami (2000); Osei et al. (1990); FAO (2001); Sable et al. (1992); Kozloski et al. (2006); An et al. (2004); Aregheore et al. (1988).

### Table 3.16. **Proximate composition of processed cassava leaves and foliage**

g/100 g dry matter

| | Leaves | | | Foliage[1] | | | | |
|---|---|---|---|---|---|---|---|---|
| | Nhu Phuc et al. Silage | Nhu Phuc et al. Meal | Yousuf et al. Meal | Kiyothong and Wanapat Hay[2] | Wanapat et al. Hay | Vongsamphanh & Wan. Hay[3] | Tien Dung et al. Hay | Range of mean values |
| Crude protein | 27.6 | 26.0 | 26.8 | 20.6 | 24.9 | 27.3 | 18.9 | 18.9-27.3 |
| Crude fibre | 17.1 | 16.1 | | | | | | |
| Fat | 13.9 | 9.9 | 5.8 | | | | 9.8 | 9.8 |
| Ash | 10.3 | 10.9 | | 7.5 | 6.6 | 8.0 | 10.7 | 6.6-10.7 |
| Nitrogen-free extract | 31.1 | 37.1 | | | | | 39.5 | 39.5 |
| Neutral detergent fibre | 33.5 | 33.5 | 26.1 | 55.0 | 34.4 | 67.7 | 29.7 | 29.7-67.7 |
| Acid detergent fibre | | | 13.4 | 38.9 | 27.0 | 41.7 | | 27.0-41.7 |
| Acid detergent lignin | | | | 16.8 | 3.8 | 13.2 | | 3.8-13.2 |

*Notes:* 1. Leaves, stems and petiole. 2. Whole plant at three months after planting. 3. Harvested at three, five and seven months.

*Sources:* Nhu Phuc et al. (2000); Yousuf et al. (2007); Kiyothong and Wanapat (2003); Wanapat et al. (1997); Vongsamphanh and Wanapat (2004); Tien Dung et al. (2005).

Table 3.17. **Moisture, protein and energy content of cassava products/by-products used in animal feed**

| Products and by-products | Moisture | Metabolizable energy, Kcal/g, dry weight | Crude protein, %, dry weight |
|---|---|---|---|
| Fresh roots | 65% | 3.7 | 1.5-4.7[1] |
| Root silage | 60% | 3.5 | 3.5 |
| Dried roots | 13% | 3.6 | 3.5 |
| Fresh foliage | 72% | 1.1 | 21.8 |
| Foliage silage | 68% | 1.3 | 20.0 |
| Foliage dry | 13% | 1.3 | 25.3 |
| Fresh bran | 90% | 5.0 | 9.0 |
| Dry bran | 13% | 2.6 | 3.3 |
| Mancha fresca | 90% | 5.0 | 8.0 |
| Mancha seca | 13% | 3.1 | 3.2 |

*Note:* 1. Range of mean values from Table 3.5.

*Source:* Ceballos and Ospina (2002).

## Other constituents

### *Anti-nutrients*

#### *Tannins*

Tannins are considered anti-nutrients because they can interfere with the absorption of iron and other minerals as well as precipitate dietary proteins potentially rendering them indigestible (Brune et al., 1989).

Tannin concentrations are negligible in roots and also are low in fresh or dry leaves from most cassava varieties (Achidi, 2003; Rickard, 2006). In leaves, the highest tannin level (29.7 g/kg dry weight) has been found in fresh red cassava leaves (Awoyinka et al., 1995). After drying, tannin levels decline rapidly to a range of 2-3 g/kg dry matter (Table 3.18). Cassava leaves also contain complexes between tannins and proteins (Wanapat, 1995). Reed et al. (1982) showed that the processing of the cassava leaf affects the tannin content.

Vongsamphanh and Wanapat (2004) found that cassava foliage harvested at three, five and seven months after planting did not change much in condensed tannin content, with an average value of 3.48 (± 0.19) g/100 g DM. According to Kiyothong and Wanapat (2003) and Tien Dung et al. (2005), cassava hay contained 3.3 and 2.3 (± 0.65) g condensed tannin/100 g dry hay, respectively. Vongsamphanh and Wanapat (2004) and Tien Dung et al. (2005) used the Vanillin-HCI method for condensed tannin.

Table 3.18. **Tannin content (Vanillin-HCI assay) of cassava leaf meal as influenced by processing methods**

| | Oven-drying (g/kg dry matter) | | Sun-drying (g/kg dry matter) | |
|---|---|---|---|---|
| Wilting (days) | Full | Chopped | Full | Chopped |
| 0 | 2.8 | 2.6 | 2.9 | 2.7 |
| 1 | 2.5 | 2.5 | 2.4 | 2.4 |
| 2 | 2.5 | 2.4 | 2.4 | 2.4 |
| 3 | 2.4 | 2.4 | 2.2 | 2.3 |

*Source:* Ravindran et al. (1987).

## *Phytic acid*

The anti-nutritional effect of phytic acid (phytin or inositol hexakisphosphate), a phosphate-rich cassava constituent, arises from its ability to chelate divalent cations such as calcium, magnesium, iron and zinc (Forbes and Erdman, 1983). This renders the metals metabolically unavailable.

Non-ruminants (including humans) lack phytase to break down phytic acid so that phosphorus can be released for metabolism. When a high proportion of the phosphorus present in the feed occurs as the poorly digestible phytic acid, a considerable amount of dietary phosphorus may be voided in faeces.

Reed et al. (1982) have reported a phytic acid content of between 107 mg and 249 mg/100 g sample in fresh unprocessed leaves of different varieties. The authors also showed that processing of cassava leaves affects the phytic acid content considerably. Charles et al. (2005) found 95-136 mg/100 g of phytic acid in five varieties of peeled cassava roots. Favaro et al. (2008) found 258-365 mg phytic acid per 100 g dry weight in two varieties of peeled cassava roots.

## *Oxalate, nitrate, polyphenol, saponin, trypsin inhibitor*

Wobeto et al. (2007) studied the levels of several anti-nutrients in leaf meal of five different cultivars of cassava appropriate for human consumption at three different maturity stages of growth – 12, 15 and 17 month-old plants. The oxalate levels were lowest in the 12-month-old plants, except for the cultivars Ouro do Vale and Maracanã. Nitrate levels decreased with maturity of the plant.

Table 3.19 shows the polyphenol (tannin) content, and trypsin inhibitor and saponin activity of the five analysed cultivars. In general, the polyphenol content increased with the maturity of the plant. The polyphenol contents found in cassava leaf meal have been reported to vary from 2.1-120 mg/100 g dry matter (Wobeto et al., 2007).

Table 3.19. **Average polyphenol and trypsin inhibitor content, and saponin in activity in cassava leaf meal at three ages of the plant**

mg/100 g dry matter

| | Polyphenol | | | Trypsin inhibitor | | |
|---|---|---|---|---|---|---|
| Cultivars | 12 months | 15 months | 17 months | 12 months | 15 months | 17 months |
| Ouro do Vale | 61.49 | 52.29 | 92.31 | 2.75 | 1.88 | 2.80 |
| Maracanã | 43.37 | 75.31 | 106.43 | 1.09 | 2.54 | 2.46 |
| MANT.IAC | 48.58 | 60.51 | 95.78 | 1.48 | 1.98 | 2.61 |
| IAC 289-70 | 47.33 | 59.69 | 71.15 | 0.86 | 2.43 | 2.95 |
| Mocotó | 44.13 | 78.86 | 79.88 | 0.57 | 3.13 | 3.28 |
| | Saponin | | | | | |
| Ouro do Vale | 1.74 | 2.48 | 3.62 | | | |
| Maracanã | 2.28 | 3.20 | 4.43 | | | |
| MANT.IAC | 2.95 | 3.35 | 3.61 | | | |
| IAC 289-70 | 3.13 | 4.33 | 4.07 | | | |
| Mocotó | 4.41 | 4.73 | 4.38 | | | |

*Source:* Wobeto et al. (2007).

## *Toxicants*

The cassava plant produces two cyanogenic glycosides, linamarin and lotaustralin, in the edible portion of its roots and leaves. Linamarin is stored in the vacuoles of leaf and root cells. On stress, linamarin is released from the vacuole and interacts with the cell wall-localised enzyme linamarase which deglycosylates linamarin yielding acetone cyanohydrin, the precursor of cyanide (HCN) (Mkpong et al., 1990). Linamarin is synthesised in leaves and transported to roots where it serves as a source of nitrogen for protein synthesis. Levels of linamarin in the roots vary between 15-500 mg CN equivalents/kg fresh weight while levels in leaves vary less and are higher at 200-500 mg CN equivalents/kg fresh weight (Mkpong et al., 1990; Haque and Bradbury, 2004). In addition to linamarase, cassava leaves have hydroxynitrile lyase (located in the cell wall) that converts 79ndispe cyanohydrin into cyanide. At small doses, cyanide is detoxified to thiocyanate by means of the enzyme rhodanase, which use methionine that becomes the first limiting amino acid in cassava feed.

The amounts of cyanogenic glycosides vary considerably, according to cultivar and growing conditions, and the cyanogenic potential, therefore, varies greatly between studied varieties (Achidi, 2003), Roots frequently contain 10-500 mg CN equivalents/kg dry weight, and leaves 200-1 300 mg CN equivalents/kg dry weight. Chávez et al. (2005) reported an average of 263.7 (range 13.9-2 561.7) mg/kg dry weight HCN in cassava roots from cultivars in the CIAT breeding programme (Sánchez et al., 2009). This implies that for many cassava varieties, the cyanogenic potential results in levels exceeding the maximum recommended cyanide level in foods (10 mg CN equivalents/kg dry weight) established by the FAO. Thus, some varieties contain such high levels of cyanogenic glycosides that the cassava requires domestic processing in order to remove the toxins. Most of the cyanide can be eliminated by crushing or fermentation followed by heating. However, the detoxification product thiocyanate is a potent goitrogen. Moreover, the sugars in cassava may react with the ε-amino group of lysine in a Maillard reaction, making lysine unavailable (lysine is the second limiting amino acid in cassava protein).

The cyanogen content of cassava foods can be reduced to safe levels by maceration, soaking, rinsing and baking. However, short-cut processing techniques can yield toxic food products.

The hydrocyanic acid potential ($HCN_p$) of fresh cassava leaves is influenced by the stage of maturity (Table 3.20), and also by processing methods such as oven or sun drying. The $HCN_p$ can vary from 1 436 mg $HCN_p$/kg DM in freshly harvested cassava leaves before chopping to an average of 1 045 mg $HCN_p$/kg DM three hours after chopping. Fasuyi (2005) subjected cassava leaves to different processing to deliberately reduce the high level of cyanogenic glycosides present in the leaves. A combination of shredding and sun drying appear to be most effective to reduce the cyanide content. Calculating dry cassava as having 12% moisture, the estimated hydrocyanic acid potential of bitter cassava roots is approximately 110-1 300 mg/kg dry weight, while levels are much lower in sweet varieties (50-100 mg/kg dry weight) (Ogunsua 1989; Chiwona-Karltun et al., 2004; Mkumbira et al., 2003).

The HCN of cassava leaf meal, the $HCN_p$, is also influenced by storage time during which levels can decline by 14.2-58.2% of initial levels (Table 3.21). Many cassava products contain very low amounts of cyanogens, which can be efficiently eliminated by the body if the protein intake is adequate.

Table 3.20. **Hydrocyanic acid potential of fresh cassava leaves (at different maturity stages) and roots**

| Leaf | | Hydrocyanic acid potential/$HCN_p$ Vongsamphanh/kg dry weight) | | |
|---|---|---|---|---|
| Stage of maturity | Number from apex | **Petioles** | **Leaf blades** | **Whole leaves** |
| Expanding | 1-4 | 5 198 | 3 161 | 4 073 |
| Just fully expanded | 5-7 | 1 731 | 1 962 | 1 766 |
| Mature | 8-11 | 609 | 774 | 745 |

*Source:* Ravindran et al. (1987).

Table 3.21. **Hydrocyanic acid potential and crude protein contents of cassava leaf meal as influenced by storage time**

| Storage time (months) | **$HCN_p$** (mg/kg dry matter) | **HCN loss as a % of initial level** | **Crude protein** (g/kg dry matter) |
|---|---|---|---|
| 0 | 91 | | 227 |
| 1 | 78 | 14.2 | 0 |
| 2 | 68 | 25.3 | 226 |
| 3 | 59 | 35.2 | 0 |
| 4 | 49 | 46.2 | 217 |
| 5 | 43 | 52.7 | 0 |
| 6 | 40 | 56.0 | 209 |
| 7 | 38 | 58.2 | 0 |
| 8 | 38 | 58.2 | 203 |

*Source:* Ravindran et al. (1987).

### *Allergens*

Cassava is not a commonly allergenic food. However, in recent years there have been several reports that described seven individuals who had suffered adverse allergy-like symptoms after oral ingestion or topical exposure to cassava (Caraballo et al., 2001; Galvao et al., 2004; Gaspar et al., 2003, 2004; Ibero et al., 2004, 2007). One highly atopic allergy sufferer who was also allergic to milk, soy, wheat, corn, egg, nuts, peanut, multiple fruits and vegetables reacted to tapioca in a double-blind placebo controlled challenge.

The remainder of the subjects tested positive for latex-fruit allergy. Latex is a relatively recently characterised allergenic substance that shares cross-reactivity with proteins in many unrelated food plants. The latex-cassava sensitive subjects displayed positive skin prick tests with cassava extracts and their sera cross-reacted with latex allergens. Additionally, latex allergens inhibited IgE binding to cassava allergens. The latex allergens have been identified and characterised at the molecular level (Kurup et al., 2005); however, the number and sequences of the epitopes present in each of these allergens has not been reported. Cassava can thus be added to the list of fruits and vegetable to which latex allergy positive subjects could potentially cross-react.

## Suggested constituents to be analysed related to food use

### *Food uses and products*

Cassava is grown for its enlarged starch-filled roots which contain nearly the maximum theoretical concentration of starch on a dry weight basis among crops. Cassava varieties can be classified as either the "sweet" (edible) variety or a "bitter" (poisonous) variety. Nutritionally, the cassava is comparable to potatoes, except that it has twice the fibre content and a higher level of potassium.

Around the world, cassava is used in a variety of food products: as vegetables in dishes, grated to make pancakes, dried and ground into tapioca flour or sliced and made into snack chips, etc. Roots are prepared much like potato. They should be cooked before eating, and to reduce cyanogenic potential of potentially toxic concentrations of cyanogenic glycosides to an innocuous level. Thus they are usually peeled and boiled, baked or fried. After peeling, the roots are sometimes grated and the sap extracted through squeezing or pressing. The cassava mixture is then dried over a fire to make a meal or it is fermented and cooked. The dried meal can then be rehydrated with water or added to soups or stews. Roots for human consumption are eaten after cooking or in processed forms (see Figure 3.1). Bitter varieties are peeled, and the root grated to make a pulp that is left to ferment slightly before being pressed, dried and roasted. Some of the processed food products are known as *farinha*, *gari*, *foufou or gablek*. For example, *gari* accounts for 70% of Nigeria's total cassava consumption. In addition, alcoholic beverages can be made from the roots.

Leaves of the cassava plant can be cooked in a manner similar to spinach. The young leaves, up to leaf position nine or ten, and the tender petioles and stem, are harvested for human consumption as a green vegetable or as a constituent in a sauce eaten with main staple meals (Lancaster and Brooks, 1983). Cassava leaves are consumed to varying degrees in several countries in Africa, constituting a major component of the diet in some countries. Their role in the diet is very different from that of the roots. Despite its substantial importance, the level of cassava leaf production or consumption is not reported in current agricultural statistics. There are country to country variations in the preference for particular varieties based on petiole colour, taste (bitter or sweet) and lower pest and pathogen susceptibility. Prior to cooking, cassava leaves are usually pounded or ground, with pounding being the more popular method.

## *Suggested analysis for food use*

The key nutrients and anti-nutrients suggested to be analysed in roots and leaves of new varieties of cassava intended for human consumption are shown in Table 3.22. If a cassava breeding objective was to produce higher levels of a particular mineral or vitamin not normally analysed (possible aim of bio-fortification programmes) (Sautter et al., 2007), then in this case these constituents should be included in root and leaf analysis.

Since appropriate key comparators may vary with age and maturity of cassava, it is recommended that data to be compared are obtained from plants of about 12 months of age, since much of the nutritional data available is on 12-month-old harvested cassava (harvested between 9-18 months; average 12 months).

Although cassava roots are considered to be a poor protein source in regions where food is abundant, it serves as an important source of protein in other countries, e.g. in Africa. Protein is evaluated in relationship to its biological value, which is markedly influenced by the relative amounts of indispensable (essential) and dispensable (non-essential) amino acids and the form of nitrogen in the diet (WHO, 2007). WHO (2007) and National Academies of Sciences (2005) list the following nine amino acids as 82ndispensable, i.e. those that have carbon skeletons that cannot be synthesised to meet body needs from simpler molecules: histidine, isoleucine, leucine, lysine, methionine, phenylanine, threonine, tryptophan and valine. Additionally, National Academies of Sciences (2005) identifies six amino acids as "conditionally indispensable", i.e. those requiring a dietary source when endogenous synthesis cannot meet metabolic needs: arginine, cysteine, glutamine, glysine, proline and tyrosine. However, WHO (2007) indicated that the requirement for indispensable amino acids is not an absolute value, and one must consider the total nitrogen content of the diet, including the dispensable amino acids particularly at lower levels of nitrogen consumption. Also potassium and calcium are important minerals to consider for both cassava tubers and leaves. Leaves are a fair source of iron. The vitamins beta-carotene and C as well as thiamine and riboflavin, are also important. Raw storage roots and leaves also contain phytic acid that binds phosphorus, making that portion of the dietary phosphorus unavailable to the consumer.

Since all cassava food products used by consumers and industry are derived from fresh or processed material, it would be considered sufficient, in most circumstances, to analyse key constituents only in fresh roots and leaves. It would not be necessary to perform separate analysis in commodities such as dried cassava roots, cassava flour, starch or cassava pellets. Some constituents, such as fatty acids, do alter in fermented cassava products such as *gari*, but since there are: *i)* a variety of ways in which cassava carbohydrates are fermented; *ii)* a wide diversity of microorganisms used in these processes in a range of geographical areas; *iii)* a number of products produced during fermentation (Brauman et al., 1996; CIAT website), it would not be practical to attempt to measure key constituents in these fresh cassava-derived products. It should also be noted that most cyanogenic compounds are usually removed during cassava processing.

Table 3.22. **Suggested constituents to be analysed in fresh roots and leaves of cassava**

| Constituent | Fresh leaves | Fresh roots |
|---|---|---|
| Proximate | X | X |
| Starch | | X |
| Fatty acids | X | X |
| Amino acids | X | X |
| Minerals[1] | X | X |
| Vitamins[2] | X | X |
| Cyanogenic glycosides (linamarin and lotaustralin) | X | X |
| Hydrogen cyanide (HCN) | X | X |
| Tannins | X | |
| Phytic acid | X | |

*Notes:* 1. Calcium, phosphorus, magnesium and iron. 2. ß-carotene, ascorbic acid (vitamin C), thiamine, riboflavin, niacin.

## Suggested constituents to be analysed related to feed use

### *Livestock feed uses*

Cassava roots, leaves and by-products have long been recognised as appropriate feed for livestock. Cassava is used in most tropical areas for feeding of pigs, cattle, sheep and poultry. It is estimated that approximately 4 million tonnes of cassava peels are annually produced during processing of cassava roots in Nigeria alone (Hahn, 1989). In some countries, cassava is now used as a partial substitute for maize. By-products from cassava processing are widely used to feed chickens and goats in the traditional sector. In Brazil and many parts of Asia, cassava roots, stems and leaves are chopped and mixed into silage for feeding of cattle and pigs. In Asia, cassava production is focused on animal feed in the form of chips and pellets for export; while in Latin America, 30% of cassava produced is used for domestic animal feed, compared to less than 2% in Africa (FAO, 2008).

Cassava roots contain a very small amount of true protein (1.5-4.7/100 g DM), which is of poor quality; therefore, a supplementary source of protein is needed for animal feed (Oke, 1978). Leaf protein concentrate appears more effective, but fishmeal is still the protein source of choice. Supplementation with lysine and methionine is also suggested for maximum efficiency. Oils are also important in feed, and supplementing with palm oil is suggested as it is easily digestible, improves palatability and is readily metabolised. A combination of oil and molasses (or sugar) seems even more effective. Cassava may also affect the mineral balance resulting, for example, in parakeratosis in chicks, but this can be eliminated by the addition of zinc carbonate (Oke, 1978). As powdered starch can produce ulcerogenic effects upon the gastric mucosa of some animals, cassava-based feeds are best served as pellets. The high fibre and ash content of cassava are not only deleterious, but also limit the choice of other ingredients, high in these components.

Cassava leaf preparations have a relative high protein content, ranging from 18.9-27.3 g/100 g dry weight. Cassava leaf yields can be as much as 4.6 tonnes DM per hectare. In earlier times, most of the cassava forage material was returned to the soil as a "green manure" product. However, there is an increased interest in using leaf products for animal feed (Ravindran, 1991). Ruminant animals can be fed fresh cassava forage, including tender stems, with good results. However, monogastric animals should not be fed cassava leaf products unless they have been processed by heating or curing to lower the cyanogenic glycoside content to a negligible level. Cassava leaf meal is high in lysine, but deficient in methionine. There are also reports on less than optimal levels of tryptophan, isoleucine and threonine (Oguntimein, 1988). The comparatively high tannin content appears to cause lower amino acid utilisation, probably because of tannins forming indigestible complexes with proteins.

In ruminant nutrition, the extent of protein degradation in the rumen is an important criterion of protein quality of a feed. Using the *in sacco* technique, Wanapat et al. (1997) found the effective degradation of proteins in cassava leaves to be 47%, in branches to be 28%, in stems to be 56.9% and in the whole crop to be 48.8%. Promkot and Wanapat (2003) reported 54.6% effective crude protein degradability for cassava hay. This is a relatively low degradation compared to other plant protein sources, suggested to be due to the relatively high content of condensed tannin in cassava foliage.

Ravindran (1991) reported that there is a good potential for using low levels of cassava leaf meal in diets for poultry and swine. Considering that the diet of animals should contain calcium and phosphorus in a ratio of 1.5-2:1, it is clear that cassava roots and root meal are grossly deficient in calcium. Leaves, on the other hand, have a better calcium:phosphorus ratio, though from an animal nutritional point of view, it could even be considered deficient in phosphorus. In the case of monogastric animals, a proportion of the phosphorus would probably be bound in phytate and not be available to the animal, typical of most phosphorus in plants.

### *Suggested analysis for feed use*

The key nutrients suggested to be analysed in roots and leaves with appropriate methodology in new varieties of cassava, intended for animal consumption is shown in Table 3.23.

Since appropriate key comparators may vary with age and maturity of cassava, it is recommended that data to be compared are obtained from plants of about 12 months of age, since much of the nutritional data available is on 12-month-old harvested cassava (harvested between 9-18 months; average 12 months).

Since all feed products of cassava consumed by animals are derived from fresh or processed leaves and roots, it would be considered sufficient, in most circumstances, to analyse key constituents only in fresh roots and leaves. It would not be necessary to perform separate analysis of key constituents in commodities such as dried cassava roots, cassava flour, starch or cassava pellets.

The constituents of key importance are the proximates (crude protein, crude fat, crude fibre, ash), acid detergent fibre, neutral detergent fibre, starch, calcium, phosphorus, cyanogenic glycosides (linamarin and lotaustralin), phytic acid and tannins. Some constituents, such as fatty acids, either are found in low concentration in the root products, or in the case of the leaf products, are fed in such a low amount as to make only a neglible contribution to the total fatty acid intake of animals. Cassava is not grown for its minerals and vitamins, which occur in low amounts, and therefore it would not be necessary to analyse for these constituents, with the exception of calcium and phosphorus, unless the breeding objective is to produce higher levels of carotene and trace elements (possible aim of biofortification programmes) (Sautter et al., 2007).

While roots do not serve as a significant protein/amino acid source for animals, leaves – or products derived from leaves – would. Although there are 20 primary amino acids that occur in proteins, there are only 10 or 11 that are recognised as essential, i.e. a need has been shown to be supplied by the diet (National Academies of Sciences, 2005). According to the National Academies of Sciences (2005), the essential amino acids for swine include argenine, histidine, isoleucine, leucine, lysine, methionine, phenylalanine, tryptophan, valine and threonine. There is also a requirement for cystine and tyrosine, but these amino acids can be synthesised from methionine and phenylalanine, respectively. Content is also important, especially in swine and poultry diets. The National Academies of Sciences (2005) lists the same amino acids as essential for poultry, plus glycine that is also included.

In cattle and sheep, where microbial protein from the rumen has been considered the primary protein source for the animal, there is increased interest in proteins that escape rumen fermentation, particularly in high producing dairy cattle. Thus, nutritionists are taking a closer look at the potential for cattle to also have certain limiting amino acids. Methionine, lysine, phenylalanine and threonine have been suggested as being limiting amino acids for cattle.

Table 3.23. **Suggested constituents to be analysed in cassava matrices for animal feed**

| Constituent | Fresh leaves | Fresh roots |
|---|---|---|
| Proximate | X | X |
| Acid detergent fibre | X | X |
| Neutral detergent fibre | X | X |
| Starch | | X |
| Calcium | X | |
| Phosphorus | X | |
| Cyanogenic glycosides | X | X |
| Tannins | X | |
| Phytic acid | X | |

## Note

1. For additional discussion of appropriate comparators, see Codex Alimentarius Commission (2003: paragraphs 44 and 45).

# *References*


Achidi, A.U. (2003), *Nutritional Evaluation of Cassava (*Manihot esculenta *Crantz) Leaves and Effect of Processing on the Nutrient Composition and Utilization*, Ph. D. Thesis, University of Ibadan, Nigeria.

Achidi, A.U. et al. (2001), "Cassava leaf utilization as vegetable source for humans in Africa", *Root Crops, The Small Processor and Development of Local Industries for a Market Economy*, Proc. 8th Triennial Root Crops Symp. ISTRC AB, Ibadan, Nigeria, November, pp. 12-16.

Agrocadenas (Observatorio de Competitividad Agrocadenas Colombia) (n.d.), Ministero de Agricultura y Desarrolo Rural/IICA, Colombia, www.agronet.gov.co/agronetweb1/Agrocadenas.aspx.

Akinfala, E.O. et al. (2002), "Evaluation of the nutritive value of whole cassava plant as replacement for maize in the starter diets for broiler chicken", *Livestock Research for Rural Development*, Vol. 14, No. 6, available at: www.lrrd.org/lrrd14/6/akin146.htm.

Allem, A.C. (1994), "The origin of *Manihot esculenta* Crantz (Euphorbiaceae)", *Genetic Resources and Crop Evolution,* Vol. 41, Issue 3, pp. 133-150.

An, L.V. et al. (2004), "Ileal and total tract digestibility in growing pigs fed cassava root meal diets with inclusion of fresh, dry and ensiled sweet potato (*Ipomoea batatas* L. (Lam)) leaves", *Animal Feed Science and Technology*, Vol. 114, Issues 1-4, pp. 127-139, http://dx.doi.org/10.1016/j.anifeedsci.2003.12.007.

AOAC (1990), *Official Methods of Analysis of AOAC International*, 13th Edition, Association of Official Analytical Chemists, Washington, DC.

Aregheore, E.M. et al. (1988), "The maize replacement value of cassava flour in rations for growing ewe lambs", *Animal Feed Science and Technology*, Vol. 20, Issue 3, pp. 233-240, http://dx.doi.org/10.1016/0377-8401(88)90046-6.

Awoyinka, A.F. et al. (1995), "Nutrient content of young cassava leaves and assessment of their acceptance as a green vegetable in Nigeria", *Plant Foods for Human Nutrition*, Vol. 47, No. 1, pp. 21-28.

Babu, L. and S.R. Chaterjee (1999), "Protein content and amino acid composition of cassava tubers and leaves", *Journal of Root Crops*, Vol. 25, pp. 163-168.

Beeching, J.R. et al. (1998), "Wound and defence responses in cassava as related to post-harvest physiological deterioration", Chapter 12 in: Romeo, J.T. et al. (eds.), *Phytochemical Signals in Plant-Microbe Interactions: Recent Advances in Phytochemistry*, Vol. 32, pp. 231-248, Plenum Press, London, http://dx.doi.org/10.1007/978-1-4615-5329-8_12.

Berry, S. (1993), "Socio-economic aspects of cassava cultivation and use in Africa: Implications for the development of appropriate technology", *Collaborative Study of Cassava in Africa – Working Paper No. 8*, International Institute of Tropical Agriculture, Ibadan, Nigeria.

Bradbury, J.H. and W.D. Holloway (1988), *Chemistry of Tropical Root Crops: Significance for Nutrition and Agriculture in the Pacific*, ACIAR (Australian Centre for International Agricultural Research), Canberra, Australia, available at: http://aciar.gov.au/publication/mn006.

Brauman, A. et al. (1996), "Microbiological and biochemical characterization of cassava retting, a traditional lactic acid fermentation of foo-foo (cassava flour) production", *Applied and Environmental Microbiology*, Vol. 62, No. 8, pp. 2 854-2 858.

Brune, M. et al. (1989), "Iron absorption and phenolic compounds: Importance of different phenolic structures", *European Journal of Clinical Nutrition*, Vol. 43, No. 8, pp. 547-557.

Caraballo, L. et al. (2001), "The allergenic properties of cassava (*Manihot esculenta*): A case of cassava allergy", *Journal of Allergy and Clinical Immunology*, Vol. 107, S183.

Carvalho, L.J.C.B. et al. (2004), "Identification and characterization of a novel cassava (*Manihot esculenta* Crantz) clone with high free sugar content and novel starch", *Plant Molecular Biology*, Vol. 56, Issue 4, pp. 643-659.

Ceballos, H. and B. Ospina (2002), *La Yuca en el Tercer Milenio: Sistemas Modernos de Producción, Procesamiento, Utilización y Comercialización*, CLAYUCA-CIAT, Palmira, Colombia.

Ceballos, H. et al. (2008), "Induction and identification of a small-granule, high-amylose mutant in cassava (*Manihot esculenta* Crantz)", *Journal of Agricultural and Food Chemistry*, Vol. 56, No. 16, pp. 7 215-7 222, http://dx.doi.org/10.1021/jf800603p.

Ceballos, H. et al. (2007), "Discovery of an amylose-free starch mutant in cassava (*Manihot esculenta* Crantz)", *Journal of Agricultural and Food Chemistry*, Vol. 55, No. 18, pp. 7 469-7 476, http://dx.doi.org/10.1021/jf070633y.

Ceballos, H. et al. (2006), "Variation in crude protein content in cassava (*Manihot esculenta* Crantz) roots", *Journal of Food Composition Analysis*, Vol. 19, Issues 6-7, pp. 589-593, http://dx.doi.org/10.1016/j.jfca.2005.11.001.

Cereda, M.P. (2003), "Tecnologia, usos e potencialidades de tuberosas amiláceas Latino Americanas", *Série Culturas de Tuberosas Amiláceas Latino Americanas*, Vol. 3, Fundação Cargill, São Paulo.

Cereda, M.P. (2001), "Manejo, usos e tratamento de subprodutos da industrialização da mandioca", *Série Culturas de Tuberosas Amiláceas Latino Americanas*, Vol. 4, Fundação Cargill, São Paulo.

Chacón, J. et al. (2008), "Phylogenetic patterns in the genus *Manihot* (Euphorbiaceae) inferred from analyses of nuclear and chloroplast DNA regions", *Molecular Phylogenetics and Evolution*, Vol. 49, No. 1, pp. 260-267, http://dx.doi.org/10.1016/j.ympev.2008.07.015.

Chamsart, A. et al. (2007), "Two-stage sequential continuous enzymatic hydrolysis of cassava starch through in-line static mixer reactor", *Starch Update 2007*, The 4th International Conference on Starch Technology, Kasetsart University, Thailand.

Charles, A.L. et al. (2005), "Proximate composition mineral contents, hydrogen cyanide and phytic acid of five cassava genotypes", *Food Chemistry*, Vol. 92, No. 4, pp. 615-620, http://dx.doi.org/10.1016/j.foodchem.2004.08.024.

Chauynarong, N. et al. (2009), "The potential of cassava products in diets for poultry", *World's Poultry Science Journal*, Vol. 65, Issue 1, March, pp. 23-35, http://dx.doi.org/10.1017/S0043933909000026.

Chávez, A.L. et al. (2005), "Variation in quality traits in cassava roots evaluated in landraces and improved clones", *Euphytica*, No. 143, Issue 2, pp. 125-133, http://dx.doi.org/10.1007/s10681-005-3057-2.

Chiwona-Karltun, L. et al. (2004), "Bitter taste in cassava roots correlates with cyanogenic glucoside levels", *Journal of the Science of Food and Agriculture*, Vol. 84, Issue 6, pp. 581-590, http://dx.doi.org/10.1002/jsfa.1699.

CIAT (International Center for Tropical Agriculture) (2009), *CIAT Cassava Database*, www.ciat.cgiar.org, accessed in 2009

CIAT (2005), *Annual Report 2005 from IP3 Project, Improved Cassava for the Developing World*, CIAT Cali, Colombia.

Cock, J.H. (1985), *Cassava. New Potential for a Neglected Crop*, Westview Press, Boulder, Colorado.

Cock, J.H. (1982), "Cassava: A basic energy source in the tropics", *Science*, Vol. 218, No. 4574, pp. 755-762.

Codex (Codex Alimentarius Commission) (2003), *Guideline for the Conduct of Food Safety Assessment of Foods Derived from Recombinant DNA Plants*, CAC/GL 45/2003, available at: www.codexalimentarius.net/download/standards/10021/CXG_045e.pdf.

Conn, E.E. (1979), "Cyanide and cyanogenic glycosides", in Rosenthal, G.A. and D.H. Janzen (eds.), *Herbivores: Their Interaction with Secondary Plant Metabolites*, Academic Press, Inc., New York-London, pp. 387-412.

Conn, E.E. (1973), "Cyanogenic glycosides", in: *Toxicants Occurring Naturally in Foods*, National Academy of Science, pp. 299-308.

Devendra, C. (1977), "Cassava as a feed source for ruminants", in Nestle, B. and M. Graham (eds.), *Proceedings, Cassava as Animal Feed Workshop*, U. Guelph 18-20 April 1977, IDRC-095e. Ottawa, Ontario, Canada.

Du, T.H. and T.R. Preston (2005), "The effects of simple processing methods of cassava leaves on HCN content and intake by growing pigs", *Livestock Research for Rural Development*, Vol. 17, No. 9, available at: www.lrrd.org/lrrd17/9/hang17099.htm.

Eggum, B.O. (1970), "The protein quality of cassava leaves", *British Journal of Nutrition*, Vol. 24, pp. 761-768.

El-Sharkawy, M.A. (2004), "Cassava biology and physiology", *Plant Molecular Biology*, Vol. 56, Issue 4, pp. 481-501.

Embrapa (Empresa Brasileira de Pesquisa Agropecuária) (2005), *Embrapa Mandioca e Fruticultura Tropical, Processamento*, Da Silva Souza, L. et al., A.R. Faria et al., p. 547.

Ezeala, D.O. (1985), "Effect of fermentation on the fatty acid content and composition of cassava tuber meal", *Journal of Food Biochemistry*, Vol. 9, Issue 3, September, pp. 249-254, http://dx.doi.org/10.1111/j.1745-4514.1985.tb00352.x.

FAO (2009) website, www.fao.org (accessed January 2009).

FAO (2008), *Cassava for food and energy security: Investing in cassava research and development could boost yields and industrial uses*, Food and Agriculture Organization, Rome, available at: www.fao.org/newsroom/en/news/2008/1000899/index.html.

FAO (2001), *Improving Nutrition through Home Gardening, A Training Package for Preparing Field Workers in Africa*, Nutrition Programmes Service, Food and Nutrition Division, Food and Agriculture Organization, Rome, available at: www.fao.org/docrep/003/X3996E/X3996E00.HTM.

FAOSTAT (2009), FAO Statistics online website, http://faostat.fao.org (accessed July 2009).

Fasuyi, A.O. (2005), "Nutrient composition and processing effects on cassava (*Manihot esculenta*, Crantz) leave antinutrients", *Pakistan Journal of Nutrition*, Vol. 4, No. 1, pp. 37-42, Asian Network for Scientific Information, available at: www.pjbs.org/pjnonline/fin256.pdf.

Fasuyi, A.O. and V.A. Aletor (2005), "Varietal composition and functional properties of cassava (*Manihot esculenta* Crantz) leaf meal and leaf protein concentrates", *Pakistan Journal of Nutrition*, Vol. 4, No. 1, pp. 43-49, Asian Network for Scientific Information, available at: http://scialert.net/qredirect.php?doi=pjn.2005.43.49&linkid=pdf.

Favaro, S.P. et al. (2008), "The roles of cell wall polymers and intracellular components in the thermal softening of cassava roots", *Food Chemistry*, Vol. 109, No. 1, pp. 220-227, http://dx.doi.org/10.1016/j.foodchem.2007.10.070.

Forbes, R.M. and J.W. Erdman (1983), "Bioavailability of tree mineral elements", *American Review of Nutrition*, Vol. 3, pp. 213-231.

Galvao, C.E.S. et al. (2004), "Latex allergy and cross-reactivity to manioc: Report of 2 cases", *Journal of Allergy and Clinical Immunology*, Vol. 113, No. 2, Supplement, p. S61, http://dx.doi.org/10.1016/j.jaci.2003.12.188.

Gaspar, A. et al. (2004), "Allergy to manioc: Cross-reactivity to latex", *Food Allergy and Intolerance*, Vol. 5, No. 2, pp. 101-113.

Gaspar, A. et al. (2003), "Anaphylactic reaction to manioc: Cross-reactivity to latex", *Allergy*, Vol. 58, No. 7, pp. 683-684.

Gomez, G. and M. Valdivieso (1984), "Cassava for animal feeding: Effect of variety and plant age on production of leaves and roots", *Animal Feed Science and Technology*, Vol. 11, Issue 1, pp. 49-55, http://dx.doi.org/10.1016/0377-8401(84)90053-1.

Grace, M.R., (1977), *Cassava Processing*, FAO Plant Production and Protection Series No. 3, Food and Agriculture Organization, Rome, available at: www.fao.org/docrep/x5032e/x5032e00.htm.

Hahn, S.K. (1992), "An overview of African traditional cassava processing and utilization", in: Hahn, S.K. et al. (eds.), *Cassava as Livestock Feed in Africa: Proceedings of the IITA/ILCA/University of*

*Ibadan Workshop on the Potential Utilization of Cassava as Livestock Feed in Africa*, IITA, Ibadan, Nigeria, pp. 16-27.

Hahn, S.K. (1989), "An overview of African traditional cassava processing and utilization", *Outlook on Agriculture*, Vol. 18, No. 3, pp. 110-118.

Hang, D.T. and T.R. Preston (2005), "The effect of simple processing methods of cassava leaves on HCN content and intake by growing pigs", MEKARN-CelAgrid, Workshop on Forages for Pigs and Rabbits, 21-24 August 2006.

Haque, M.R. and J. H. Bradbury (2004), "Preparation of linamarin from cassava leaves for use in a cassava cyanide kit", *Food Chemistry*, Vol. 85, No. 1, pp. 27-29, http://dx.doi.org/10.1016/j.foodchem.2003.06.001.

Hock-Hin, Y. and T. Van-Den (1996), "Protein contents, amino acid compositions and nitrogen-to-protein conversion factors for cassava roots", *Journal of the Science of Food and Agriculture*, Vol. 70, Issue 1, pp. 51-54, http://dx.doi.org/10.1002/(SICI)1097-0010(199601)70:1<51::AID-JSFA463>3.0.CO;2-W.

Howeler, R. (2004), "Cassava in Asia: Present situation and its future potential in agro-industry", *Journal of Root Crops*, Vol. 30, No. 2, pp. 81-92.

Hung Nguyen, J.J. et al. (2002), "Effects of long-term nitrogen, phosphorus and potassium fertilization on cassava yield and plant nutrient composition in North Vietnam", *Journal of Plant Nutrition*, Vol. 25, No. 3, pp. 425-442.

Ibero, M. et al. (2007), "Allergy to Cassava: A new allergenic food with cross-reactivity to latex", *Journal of Investigational Allergology and Clinical Immunology*, Vol. 17, No. 6, pp. 409-12.

Ibero, M. et al. (2004), "Allergy to cassava: A new allergen to consider in the latex-fruit syndrome?" in: Anklam, E. and F. Morlin (eds.), *9th International Symposium on Immunological, Chemical and Clinical Problems of Food Allergy*, Budapest, Hungary.

Iglesias, C. et al. (1997), "Genetic potential and stability of carotene content in cassava roots", *Euphytica*, Vol. 94, Issue 3, pp. 367-373.

IITA (International Institute of Tropical Agriculture)-ICP (Integrated Cassava Project) (n.d.), website, www.cassavabiz.org (last accessed December 2008).

Izuagie, A.A. et al. (2007), "The relative glucose yields of some carbohydrate foods: Need for caution", *Journal of Applied Sciences Research*, Vol. 3, No. 10, pp. 961-963.

JIRCAS (2006), "Research highlights 2006: Development of effective ethanol production technology using cassava pulp", A. Kosugi, Japan International Research Center for Agricultural Sciences, available at: www.jircas.affrc.go.jp/english/publication/highlights/2006/2006_16.html.

Jobling, S. (2004), "Improving starch for food and industrial applications", *Current Opinion in Plant Biology*, Vol. 7, pp. 210-218.

Johnson, R.N. and W.D. Raymond (1965), "The chemical composition of some tropical food plants. IV. Manioc", *Tropical Science*, Vol. 7, No. 3, pp. 109-115.

Kakes, P. (1990), "Properties and functions of the cyanogenic system in higher plants", *Euphytica*, Vol. 48, Issue 1, pp. 25-43.

Khor, H.T. and H.L. Tan (2006), "The lipids of young cassava (*Manihot esculenta* Crantz) leaves", *Journal of the Science of Food and Agriculture*, Vol. 32, No. 4, pp. 399-402, http://dx.doi.org/10.1002/jsfa.2740320414.

Kiyothong, K. and M. Wanapat (2003), "Cassava hay and Stylo 184 hay to replace concentrates in diets for lactating dairy cows", *Livestock Research for Rural Development*, Vol. 15, No. 11, available at: www.lrrd.org/lrrd15/11/krai1511.htm.

Kozloski, G.V. et al. (2006), "Intake and digestion by lambs fed a low-quality grass hay supplemented or not with urea, casein or cassava meal", *Animal Feed Science and Technology*, Vol. 136, Issues 3-4, pp. 191-202, http://dx.doi.org/10.1016/j.anifeedsci.2006.09.002.

Kurup, V.P. et al. (2005), "Specific IgE response to purified and recombinant allergens in latex allergy", *Clinical and Molecular Allergy*, Vol. 3, No. 11, http://dx.doi.org/10.1186/1476-7961-3-11.

Lalaguna, F. and M. Agudo (1988), "Lipid classes of fresh cassava roots (*Manihot esculenta* Crantz): Identification and quantification", *Journal of the American Oil Chemists' Society*, Vol. 65, Issue 11, November, pp. 1 808-1 811.

Lancaster, P.A. and J.E. Brooks (1983), "Cassava leaves as human food", *Economic Botany*, Vol. 37, Issue 3, pp. 331-348.

Lehman, U. and F. Robin (2007), "Slowly digestible starch – Its structure and health implications: A review", *Trends in Food Science & Technology*, Vol. 18, No. 7, pp. 346-355.

Luyken, R. et al. (1961), *Nutritional Value of Foods from New Guinea – I. "Net Protein Utilization, Digestibility and Biological Value of Sweet Potatoes and Cassava Leaves from New Guinea"*, Utrecht Central Institute for Nutrition and Food Research.

Maynard, L.A. et al. (1979), *Animal Nutrition 7th Edition*, McGraw-Hill Book Company, New York, NY.

Mkpong, O.E. et al. (1990), "Purification, characterization, and localization of linamarase in cassava", *Plant Physiology*, Vol. 93, pp. 176-181.

Mkumbira, J. et al. (2003), "Classification of cassava into 'bitter' and 'cool' in Malawi: From farmers' perception to characterisation by molecular markers", *Euphytica*, Vol. 132, Issue 1, pp. 7-22.

National Academy of Sciences (2005), *Dietary Reference Intakes for Energy, Carbohydrate, Fiber, Fat, Fatty Acids, Cholesterol, Protein, and Amino Acids (Macronutrients)*, Food and Nutrition Board, Institute of Medicine of the National Academies, National Academy of Sciences, Washington, DC, available at: www.nap.edu/catalog.php?record_id=10490.

Nassar, N.M.A. and M. Vale de Sousa (2007), "Amino acids profile in cassava, its interspecific hybrid and progenies", *Genetics and Molecular Research*, Vol/ 6, No. 2, pp. 292-297, available at: www.geneticsmr.com/year2007/vol6-2/pdf/gmr0280.pdf.

Nhu Phuc, B.H. et al. (2000), "Effect of replacing soybean protein with cassava leaf protein in cassava root meal based diets for growing pigs on digestibility and nitrogen retention", *Animal Feed Science and Technology*, Vol. 83, No. 3, pp. 223-235, http://dx.doi.org/10.1016/S0377-8401(99)00136-4.

Nweke, F. (1990), "Processing potential for cassava production growth in Africa", *COSCA Working Paper No. 11*, Collaborative Study of Cassava in Africa, Proceedings of the 8th Symposium of the International Society for Tropical Root Crops, IITA, Ibadan, Nigeria.

Nweke, F. et al. (2002), *The Cassava Transformation*, Michigan State University Press, East Lansing, Michigan.

Ogunsua, A. (1989), "Total cyanide levels in bread made from wheat/cassava composite flours", *International Journal of Food Science and Technology*, Vol. 24, No. 4, pp. 361-365, http://dx.doi.org/10.1111/j.1365-2621.1989.tb00655.x.

Oguntimein, G.B. (1988), "Processing cassava for animal feeds", in: Hahn, S.K. et al., *Cassava as Livestock Feed in Africa: Proceedings of the IITA/ILCA/University of Ibadan Workshop on the Potential Utilization of Cassava as Livestock Feed in Africa*, IITA and ILRI, available at: www.fao.org/wairdocs/ILRI/x5458E/x5458e0d.htm (accessed 22 September 2005).

Oke, O.L. (1978), "Problems in the use of cassava as animal feed", *Animal Feed Science and Technology*, Vol. 3, Issue 4, pp. 345-380.

Oke, O.L. (1968), "Cassava as a food in Nigeria", *World Review of Nutrition and Dietetics*, Vol. 9, pp. 272-293.

Olsen, K.M. and B.A. Schaal (2001), "Microsatellite variation in cassava and its wild relatives: Further evidence for a southern Amazonian origin of domestication", *American Journal of Botany*, Vol. 88, No. 1, pp. 131-142.

Osei, S.A. et al. (1990), "Effects of fermented cassava peel meal on the performance of layers", *Animal Feed Science and Technology*, Vol. 24, Issues 3-4, pp. 295-301.

Ospina, B. et al. (2002), "Research on cassava foliage production in Colombia", *Proceedings of the 7th Regional Cassava Workshop*, Bangkok, Thailand, available at: http://ciat-library.ciat.cgiar.org/Articulos_Ciat/proceedings_workshop_02/481.pdf.

Padmaja, G. (2000), "Cassava ensiling", *Technical Bulletin No. 27*, Central Tuber Crops Research Institute, Trivandrum, India, p. 26.

Pandey, A. et al. (2000), "Biotechnology potential of agro-industrial residues II: Cassava bagasse", *Bioresource Technology*, Vol. 74, No. 1, pp. 81-87.

Promkot, C. and M. Wanapat (2003), "Ruminal degradation and intestinal digestion of crude protein of tropical protein resources using nylon bag technique and three-step *in vitro* procedure in dairy cattle", *Livestock Research for Rural Development,* Vol. 15, No. 11, available at: www.lrrd.org/lrrd15/11/prom1511.htm.

Purdue University, Center for New Crops & Plant Products (1995), "Cassava", contributed by S.K. O'Hair, Tropical Research and Education Centre, University of Florida, *New Crop FactSheet*, available at: www.hort.purdue.edu/newcrop/cropfactsheets/cassava.html.

Ravindran, V. (1991), "Preparation of cassava leaf product and their use as animal feed", *Proceedings of the FAO expert consultation*, CIAT, Cali, Columbia, No. 21-25, pp. 81-95, referred to in the *Pakistan Journal of Nutrition*, Vol. 7, No. 1, pp. 13-16.

Ravindran, V. et al. (1987), "Influence of processing methods and storage time on the cyanide potential of cassava leaf meal", *Animal Feed Science and Technology*, Vol. 17, No. 4, pp. 227-234, http://dx.doi.org/10.1016/0377-8401(87)90054-X.

Reed, W.R. et al. (1982), "Phytate in legumes and cereals", *Advanced Food Research*, Vol. 28, pp. 1-9.

Rickard, J.E. (2006), "Tannin levels in cassava, a comparison of methods of analysis", *Journal of the Science of Food and Agriculture*, Vol. 37, Issue 1, January, pp. 37-42, http://dx.doi.org/10.1002/jsfa.2740370106.

Rickard , J.E. et al. (1991), "The physico-chemical properties of cassava starch", *Tropical Science*, Vol. 31, pp. 189-207.

Sable, M. et al. (1992), "Nutritional value of cassava meal in diets for geese", *Animal Feed Science and Technology*, Vol. 36, Issues 1-2, pp. 29-40, http://dx.doi.org/10.1016/0377-8401(92)90083-I.

Sackey, I.S. and R.J. Bani (2007), "Survey of waste management practices in cassava processing to gari in selected districts of Ghana", *Food, Agriculture and Environment*, Vol. 5, No. 2, pp. 325-328.

Salami, R.I. (2000), "Preliminary studies on the use of parboiled Ccassava peel meal as a substitute for maize in layers' diets", *Tropical Agriculture*, Trinidad and Tobago, Vol. 77, No. 3, pp. 199-204.

Sánchez, T. et al. (2009), "Screening of starch quality traits in cassava *(Manihot esculenta* Crantz)", *Starch/Starke*, Vol. 61, No. 5, May, pp. 12-19, http://dx.doi.org/10.1002/star.200990027.

Sautter, C. et al. (2007), "Biofortification of essential nutritional compounds and trace elements in rice and cassava", *Proceedings of the Nutrition Society*, Vol. 65, No. 2, May, pp. 153-159.

Scott, G.J.M. et al. (2000), "Roots and tubers for the 21st century: Trends, projections and policy options", *Food, Agriculture and Environment Discussion Papers*, No. 31, International Food Policy Research Institute (IFPRI), Washington, DC, at: www.ifpri.org/sites/default/files/pubs/2020/dp/2020dp31.pdf.

Siritunga, D. and R.T Sayre (2003), "Generation of cyanogens free transgenic cassava", *Planta*, Vol. 217, No. 3, pp. 367-373.

Smith, O.B. (1988), "A review of ruminant responses to cassava-based diets", in: Hahn, S.K. et al. (eds.), *Cassava as Livestock Feed in Africa: Proceedings of the IITA/ILCA/University of Ibadan Workshop on the Potential Utilization of Cassava as Livestock Feed in Africa*, IITA, Ibadan, Nigeria, available at: www.fao.org/wairdocs/ILRI/x5458E/x5458e07.htm.

Soccol, C.R. (1996), "Biotechnological products from cassava roots by solid state fermentation", *Journal of Scientific & Industrial Research*, Vol. 55, pp. 358-364.

Suthsmma, Y. and C. Sorapipatana (2007), "A study of ethanol production cost for gasoline substitution in Thailand and its competitiveness", *Thammasat International Journal of Science and Technology*, Vol. 12, No. 1, January-March, available at: http://tujournals.tu.ac.th/tijsat/detailart.aspx?ArticleID=286.

Szylit, O. et al. (1978), "Raw and steam-pelleted cassava, sweet potato and yam *Cayenensis* as starch sources for ruminant and chicken diets", *Animal Feed Science and Technology*, Vol. 3, Issue 1, pp. 73-87, http://dx.doi.org/10.1016/0377-8401(78)90025-1.

Taylor, N. et al. (2004), "Development and application of transgenic technologies in cassava", *Plant Molecular Biology*, Vol. 56, Issue 4, pp. 671-688.

Tien Dung, N. et al. (2005), "Effect of replacing a commercial concentrate with cassava hay (*Manihot esculenta* Crantz) on the performance of growing goats", *Animal Feed Science and Technology*, Vol. 119, Issues 3-4, pp. 271-281.

TIME IS (Technology Innovation Management & Entrepreneurship Information Service) (2005.), "Cassava starch, alcohol from cassava", Joint Project of NSTEDB, Department of Science & Technology, Government of India and FICCI, www.techno-preneur.net (accessed July 2009).

Tonukari, N.J. (2004), "Cassava and the future of starch", *Electronic Journal of Biotechnology*, Vol. 7, No. 1, 15 April 2004, available at: www.ejbiotechnology.info/content/vol7/issue1/issues/2.

Ubalua, A.O. (2007), "Cassava wastes: Treatment options and value addition alternatives", *African Journal of Biotechnology*, Vol. 6, No. 18, pp. 2 065-2 073.

USDA Agricultural Research Service (2008), *USDA National Nutrient Database for Standard Reference*, Release 21, Nutrient Data Laboratory, available at: http://ndb.nal.usda.gov.

USDA Agricultural Research Service (2005), *USDA National Nutrient Database for Standard Reference*, Release 18, Nutrient Data Laboratory, available at: http://ndb.nal.usda.gov.

Vongsamphanh, P. and M. Wanapat (2004), "Comparison of cassava hay yield and chemical composition of local and introduced varieties and effects of levels of cassava hay supplementation in native beef cattle fed on rice straw", *Livestock Research for Rural Development*, Vol. 16, No. 8, available at: www.lrrd.org/lrrd16/8/vong16055.htm.

Wanapat, M. (1995), "The use of local feed resources for livestock production in Thailand", in: Tingshuang, G. (ed.), *Proceedings of the International Conference on Increasing Animal Production with Local Resources*, China Forestry Publishing House, Ministry of Agriculture, China.

Wanapat, M. et al. (1997), "Cassava hay: A new strategic feed for ruminants during the dry season", *Livestock Research for Rural Development*, Vol. 9, No. 99, available at: www.lrrd.org/lrrd9/2/metha92.htm.

WHO (2007), "Protein and amino acid requirements in human nutrition", Report of a Joint WHO/FAO/UNU Expert Consultation, *WHO Technical Report Series No. 935*, World Health Organization, Geneva.

Wobeto, C. et al. (2007), "Antinutrients in the cassava (*Manhihot esculenta* Crantz) leaf powder at three ages of the plant", *Food Science and Technology (Campinas)*, Vol. 27, No. 1, January/March, pp. 108-112, http://dx.doi.org/10.1590/S0101-20612007000100019.

Yeoh, H.H. and M.Y. Chew (1976), "Protein content and amino acid composition of cassava leaf", *Phytochemistry*, Vol. 15, No. 11, pp. 1 597-1 599.

Yeoh, H.H. and V.D. Truong (1996), "Protein contents, amino acid compositions and nitrogen-to-protein conversion factors for cassava roots", *Journal of the Science of Food and Agriculture*, Vol. 70, Issue 1, January, pp. 51-54, http://dx.doi.org/10.1002/(SICI)1097-0010(199601)70:1<51::AID-JSFA463>3.0.CO;2-W.

Yousuf, M.B. et al. (2007), "Protein supplementary values of cassava-, leucaena- and gliricidia- leaf meals in goats fed low quality *Panicum maximum* hay", *Livestock Research for Rural Development*, Vol. 19, No. 2, available at: http://lrrd.cipav.org.co/lrrd19/2/yous19023.htm.

Zhang, P. et al. (2003), "Transfer and expression of an artificial storage protein (ASP1) gene in cassava (*Manihot exculenta* Crantz)", *Transgenic Research*, Vol. 12, No. 2, pp. 243-250.

# Chapter 4

# Grain sorghum (*Sorghum bicolor*)

*This chapter, prepared by the OECD Task Force for the Safety of Novel Foods and Feeds with the United States and South Africa as lead countries, deals with the composition of grain sorghum (*Sorghum bicolor*). It contains elements that can be used in a comparative approach as part of a safety assessment of foods and feeds derived from new varieties. Background is given on sorghum production, uses and processing, followed by appropriate varietal comparators and characteristics screened by breeders. Nutrients in sorghum grain, silage and ethanol production by-products, as well as anti-nutrients, are then detailed. The final sections suggest key products and constituents for analysis of new varieties for food use and for feed use.*

## Background

### *Production of sorghum varieties for food and feed*

Sorghum includes a wide variety of related plant species used for a variety of purposes. The major species of grain sorghum is *Sorghum bicolor* (L.) Moench, an annual cereal crop of African origin (De Alencar Figueiredo et al., 2008). Numerous varieties, including hybrid ones, have been developed by companies and institutions to serve different end-uses (Kriegshauser et al., 2006; Salinas et al., 2006). Grain varieties of sorghum may be characterised depending on their starch content, structure and functional properties for cooking (Sang et al., 2008). Some cultivars of *Sorghum bicolor* (L.) Moench are referred to as "sweet sorghums" due to the high sucrose content in their stalks (Ali et al., 2008). Broom sorghum (broomcorn, *Sorghum vulgare*) is also grown in some regions of the world; however, the taxonomic designation of *S. vulgare* is now considered to be a subspecies of *S. bicolor* in the main worldwide reference databases (Integrated Taxonomic Information System [North America; ITIS] and Germplasm Resources Information Network [GRIN] websites, 2009). Sudangrass (*Sorghum sudanense* [Piper] Stapf), classified as a nothosubspecies of *S. bicolor* in the *GRIN database*, is used as a forage source for livestock in many countries and may be crossed with *S. bicolor* to increase yield. Johnsongrass (*Sorghum halepense* [L] Pers), a perennial crop, is closely related to Sudangrass but is regarded as a noxious weed in many countries.

Sorghum has many other common names including great millet, guinea corn, aura, mtama, jowar, cholam, kaoliang, milo and milo-maize (FAO, 1995). Sorghum grain consists of three distinct anatomical parts: the outer layer, or pericarp; the storage tissue endosperm; and the germ or embryo. The relative proportions of these parts within the grain depend on the cultivar and environmental conditions. The outermost layer of the pericarp, the epicarp, is usually covered with a thin layer of wax, and two or three cell layers of pigmented cells. Below the epicarp lies the mesocarp, which in sorghum, unlike other cereals, contains starch granules. Most of the starch and protein (including enzymes) is stored in the endosperm of sorghum grain, whereas the germ contains most of the oil and minerals, to support initial growth of the embryonic plant (Serna-Saldivar and Rooney, 1995; Waniska and Rooney, 2000).

Approximately 75% of the weight of *S. bicolor* grain is starch, comprised of amylose and amylopectin arranged radially in spherical granules in a pseudo crystalline matrix (having both crystalline and amorphous regions). These granules are insoluble in cold water, and relatively inaccessible to hydrolysis by amylase. Sorghum starch has properties and uses similar to maize starch and the procedures for milling sorghum are similar to that for milling maize. Pigmented sorghum pericarp will sometimes yield starch with a pinkish colour, which can be bleached with $NaClO_2$, or rinsed with NaOH or methanol during wet milling to produce a more acceptable colour (Waniska and Rooney, 2000).

Sorghum is considered the fifth most important cereal crop in the world behind wheat, rice, maize and barley (CGIAR, 2009). Sorghum is grown on approximately 44 million hectares in 99 countries (ICRISAT, 2009). An estimation of the worldwide tonnage produced in 2007-08 is shown in Table 4.1.

Table 4.1. **World sorghum production 2007-08**

| Country | Production ('000 tonnes) | % of total |
|---|---|---|
| United States | 12 827 | 20 |
| Nigeria | 10 000 | 16 |
| India | 7 780 | 12 |
| Mexico | 6 100 | 10 |
| Sudan | 4 500 | 7 |
| Ethiopia | 3 230 | 5 |
| Argentina | 2 900 | 5 |
| Australia | 2 691 | 4 |
| China (People's Republic of) | 1 900 | 3 |
| Burkina Faso | 1 800 | 3 |
| Brazil | 1 700 | 3 |
| Other countries | 6 880 | 12 |
| Total | 62 308 | 100 |

*Source:* US Grains Council (2008).

## *Uses*

According to the US National Sorghum Producers Association (2006), approximately 50% of the world production of sorghum grain is used as human food, while the Food and Agriculture Organization (FAO) estimates that 95% of its total food use occurs in Africa and Asia (FAO, 1995). Sorghum grain is a staple diet in Africa, the Middle East, Asia and Central America, where its processed grain may be consumed in many forms including porridge, steam-cooked product, tortillas, baked goods or as a beverage (CGIAR, 2009). Sorghum represents a large portion of the total calorie intake in many African countries (FAO, 1995). The People's Republic of China and India account for almost all of the food use of sorghum in Asia.

Sorghum is genetically more closely related to maize than it is to wheat, rye or barley, and as such is considered a safe food for patients with celiac disease (Ciacci et al., 2007; US Grains Council, 2008).

Several million tonnes of sorghum are used across Africa for traditional beer brewing, and in west, east and central Africa for lager/stout production. Research from Mexico suggests that waxy sorghum (a mutant variety that is nearly 100% amylopectin) may be advantageous for brewing; however, normal sorghum (approximately 75% amylopectin and 25% amylose) is more commonly used for beer production (Del Pozo-Insfran et al., 2004; Figueroa et al., 1995).

In other parts of the world, sorghum grain is used mainly as an animal feed. Such use is concentrated in Mexico, many South American countries, the United States, Japan and the Commonwealth of Independent States. The stover of sorghum also is used as fodder for animals. Brown midrib (BMR) varieties of *Sorghum bicolor* have been developed for use as forage sources for livestock because of their reduced lignin content and higher digestibility of the stover (Aydin et al., 1999; Oliver et al., 2004). Broom sorghum (broomcorn, *S. vulgare*) is also used as a source of animal feed in some regions, although it is less digestible than *S. bicolor* (Nikkhah et al., 2004). Sudangrass and sudangrass hybrids may be used as pasture, hay, green-chop or silage for livestock. According to FAO (1995), the use of sorghum for feed has been the driving force behind increasing its global production and trade.

Sweet sorghums are used for the production of syrup or molasses, and are being considered as potential sources for fuel ethanol (Gibbons et al., 1986; ICRISAT, 2009). Production of ethanol from sorghum grain or sweet sorghum biomass (stalks) has gained increasing interest in recent years (Ali et al., 2008; Gibbons et al., 1986; Wang et al., 2008; Zhao et al., 2008). To produce ethanol from sorghum grain, the whole grain is ground, gelatinized and converted to fermentable carbohydrates using enzymes. The by-product, distillers' grains, contains approximately 30% protein, and is commonly used as feed for livestock in either wet or dry form (Al-Suwaiegh et al., 2002; Lodge et al., 1997; Rooney and Waniska, 2000).

## *Processing*

Sorghum grain and biomass processing depends on the intended final product. Dry milling of grain is used for production of ethanol, preparation of flour for baking or porridge, or for use as animal feed. Malting is used for production of beverages, porridges or baked goods. Sorghum stalks and leaves may be crushed to extract juice, or fed green or dried to livestock. Whole sorghum grain used for feeding non-ruminant livestock is processed mainly by hammer milling. The ground meal may be pelleted for use in poultry and swine feeds. Steam flaking is widely used on cattle feedlots to improve palatability and rumen fermentation (Rooney and Waniska 2000; Zinn et al., 2008).

The process for elaborating diverse types of animal feed from sorghum is depicted in Figure 4.1.

Figure 4.1. **Processing sorghum for animal feed**

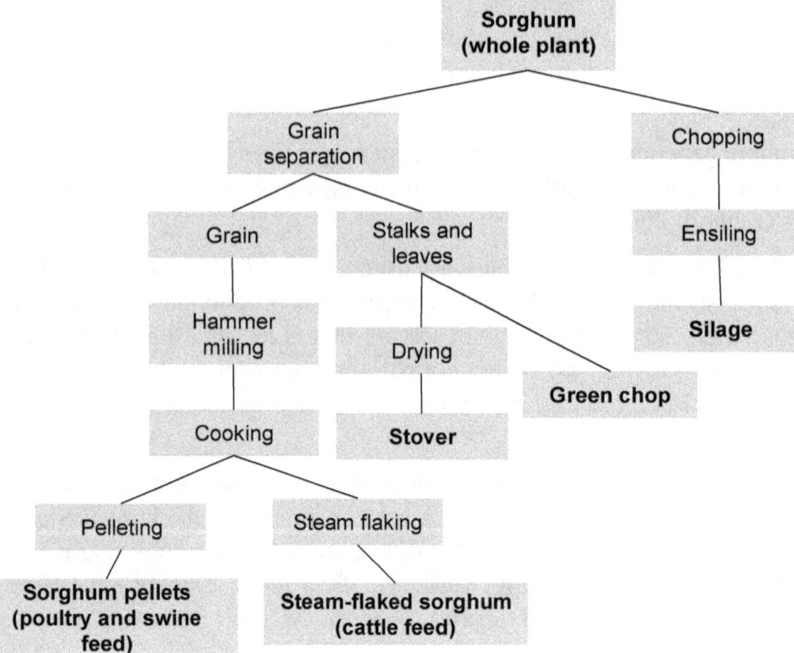


*Source*: Rooney and Waniska (2000).

There are various methods included in the broad topic of "dry milling" which includes cracking, decortication (dehulling, degermination), hammer milling, disc milling and roller milling, and may involve more than one of these methods depending on the type of sorghum and the desired end-products.

Roller milling works best for sorghums with soft, floury endosperm that is easily crushed and removed from the pericarp. For roller milling, tempering the grain to 15-16% moisture just before milling improves the separation of bran from flour. Dehulling (degermination) produces highly refined fractions of flour, bran, germ, meal and grits.

The milling properties of sorghum are affected by both genetics and environmental conditions (Rooney and Waniska, 2000).

The process steps for sorghum grain dry milling are schematised in Figure 4.2.

Figure 4.2. **Dry milling sorghum grain**

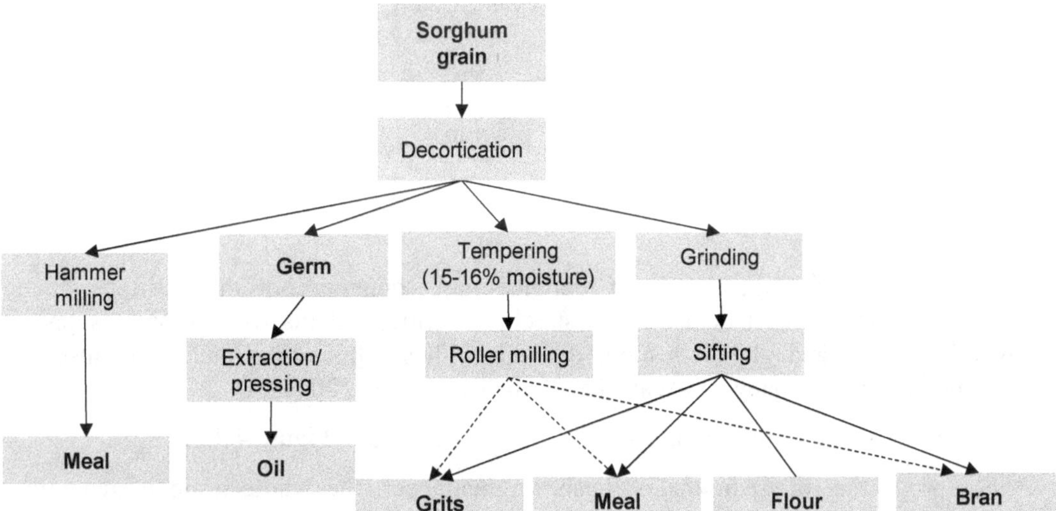


*Source*: Adapted from Rooney and Waniska (2000).

Sorghum syrup, molasses and sugar are produced from juice extracted from sweet sorghum stalks which are high in sucrose. A roller-type mill is used to extract the juice from the stalks shortly after harvest. The juice is then clarified, typically by heating, and solids are concentrated by evaporating water from the juice to produce syrup.

Fermentable carbohydrates in sweet sorghum stalks comprise approximately 80% soluble sugars and 20% starch. To optimise production of ethanol from sweet sorghum biomass requires both liquefying and saccharifying enzymes (Rooney and Waniska, 2000). Figure 4.3 sketches the sweet sorghum processing.

Figure 4.3. **Processing sweet sorghum**

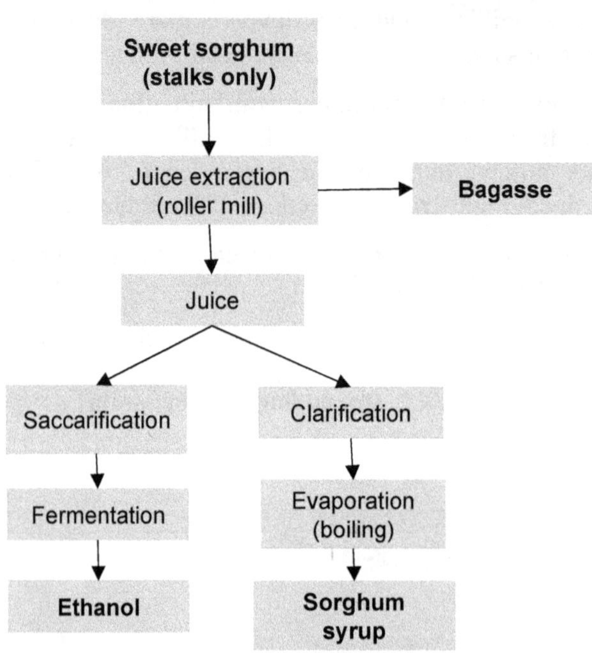


*Source*: Adapted from Rooney and Waniska (2000).

Wet milling sorghum to extract starch is not common, but may still be done in some countries to meet demands for starch, particularly if maize is in short supply. Separating starch and gluten is more difficult with sorghum than maize, because of its fragile pericarp (Rooney and Waniska, 2000; Taylor et al., 2006).

The process of wet milling of sorghum grain is outlined in Figure 4.4.

For production of traditional African sorghum beer, the whole grain is generally malted by steeping, allowing the grain to germinate, and drying. The resulting malt is then ground, mixed with water, mashed (saccharified), filtered, boiled and inoculated with yeast and allowed to ferment to produce a cloudy beer. Alternatively, ground malt is mixed with water, allowed to sour, then boiled with an adjunct (maize or sorghum grits), cooled and saccharified with additional sorghum malt before inoculating with yeast to produce an opaque beer (Rooney and Waniska, 2000).

Figure 4.5 summarises the whole process.

Figure 4.4. **Wet milling of sorghum grain**

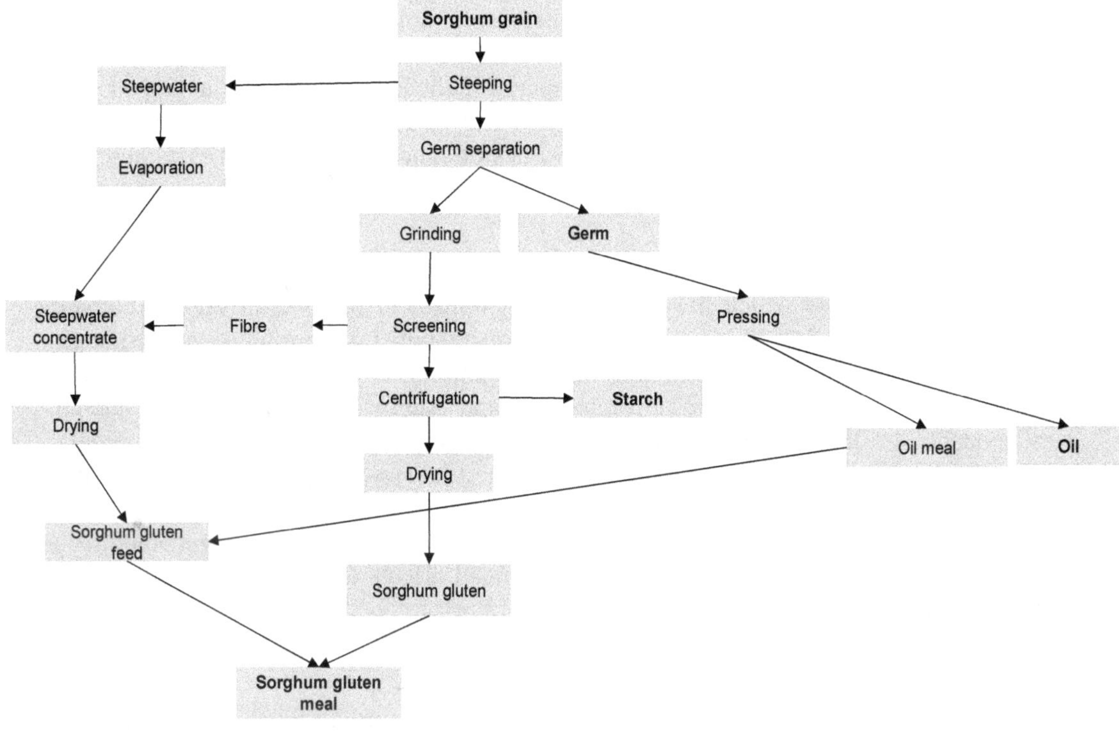


*Source*: Adapted from FAO and AFRIS (1993) and Kent and Evers (1994).

Figure 4.5. **Traditional African sorghum beer production**

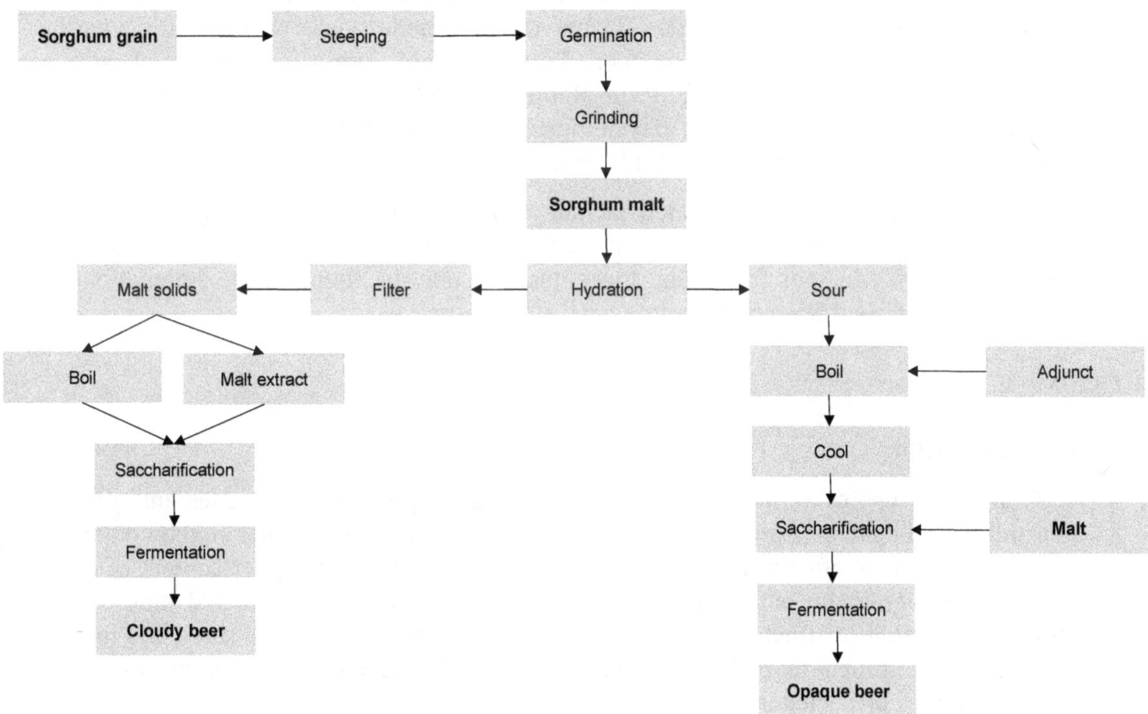


*Source*: Adapted from Rooney and Waniska (2000).

### *Appropriate comparators for testing new varieties*

This chapter suggests parameters that sorghum breeders should measure when developing new modified varieties. The data obtained in the analysis of a new sorghum variety should ideally be compared to those obtained from an appropriate near isogenic non-modified variety, grown and harvested under the same conditions.[1] The comparison can also be made between values obtained from new varieties and data available in the literature, or chemical analytical data generated from other commercial sorghum varieties.

Components to be analysed include key nutrients, toxicants and allergens. Key nutrients are those which have a substantial impact in the overall diet of humans (food) and animals (feed). These may be major constituents (fats, proteins, and structural and non-structural carbohydrates) or minor compounds (vitamins and minerals). Similarly, the levels of known anti-nutrients and allergens should be considered. Key toxicants are those toxicologically significant compounds known to be inherently present in the species, whose toxic potency and levels may impact human and animal health. Standardised analytical methods and appropriate types of material should be used, adequately adapted to the use of each product and by-product. The key components analysed are used as indicators of whether unintended effects of the genetic modification influencing plant metabolism has occurred or not.

### *Breeding characteristics screened by developers*

In the early stages of growth, breeders evaluate germination percentage, emergence, seedling vigour (measured as leaf area, number, length and width assessed at 15-20 days post emergence) and cold tolerance. As the plant matures, it is evaluated for plant height, standability, stalk diameter, half-blooming (Liang and Walter, 1968; Rattunde, 1998), drought-tolerance, and pest- and disease-resistance (Krausz et al., 1994; Teetes and Pendelton, 1999; Partridge, 2008). The harvested sorghum is evaluated for yield, head type, head weight, kernel number, kernel weight, grain colour, grain size, and threshability (Liang and Walter, 1968; Rattunde, 1998). For forage and dual-purpose sorghum varieties, days to flower, physiological maturity, plant height, lodging score, stover yield and biomass yield may also be evaluated (Rattunde, 1998).

End-use criteria that are evaluated include tannin concentration, endosperm texture and malting quality. These criteria will vary depending on the intended market for the final product. Methods for testing these qualities may be found at the International Association for Cereal Science Technology (ICC) website.

## Nutrients

### *Sorghum grain*

Grain size, type of pericarp and seed coat vary among sorghum varieties and affect their nutritional content. Larger grain varieties are associated with higher starch content, while smaller grains often have a proportionally larger germ, and a higher content of oil. Smaller grains typically have a higher seed coat to seed content ratio, and thus have a higher proportion of structural carbohydrate (fibre). Sorghum seeds are small and must be cracked or ground to make the nutrients available. The nutrient composition of sorghum grain is presented in Tables 4.2-4.6.

Table 4.2. **Proximate analysis of *S. bicolor* grain (dry matter basis)**

| Component | USDA | Ensminger et al. | NRC | Kriegshauser et al. | Preston | Ragaee et al. | FAO and AFRIS | Range of mean values |
|---|---|---|---|---|---|---|---|---|
| Moisture (%) | 9.2 | 10.0 | 10.0-12.5 | | 11.0 | | 9.3-12.3 | 9.2-12.5 |
| | | | % of dry matter | | | | | |
| Crude protein[1] | 12.4 | 12.8 | 10.1-12.6 | 12.1-14.1 | 11.0 | 12.1 | 10.8-15.6 | 10.1-15.6 |
| Total fat[2] | 3.6 | 2.9 | 3.0-3.3 | 3.1-3.8 | 3.1 | 3.32 | 0.8-4.3 | 0.8-4.3 |
| Ash | 1.7 | 1.9 | 1.9-2.0 | 1.5-1.6 | 2.0 | 1.87 | 1.5-3.3 | 1.5-3.3 |
| Nitrogen-free extract[3] | 82.7 | | | 70.8-73.3 | | | 74.6-84.9 | 70.8-84.9 |
| Crude fibre | 6.9 | 2.8 | 2.6-3.0 | 2.1-2.7 | 3.0 | | 1.7-2.1 | 1.7-6.9 |
| Neutral detergent fibre | | 18.0 | 10.9-23.0 | | 15.0 | | | 10.9-23.0 |
| Acid detergent fibre | | 9.0 | 5.0-9.3 | | 6.0 | | | 5.0-9.0 |

*Notes:* 1. Crude protein = nitrogen x 6.25. 2. Total fat as measured by ether extract. 3. NFE = 100 – (ash + ether extract + crude protein + crude fibre).

*Sources:* USDA Agricultural Research Service (2009); Ensminger et al. (1990); NRC (1994, 1998, 2000, 2001); Kriegshauser et al. (2006); Preston (2010); Ragaee et al. (2006); FAO and AFRIS (1993).

Table 4.3. **Mineral concentrations in *S. bicolor* grain (dry matter basis)**

| Minerals | Unit | Preston | Ragaee et al. | USDA | Ensminger et al. | NRC | Range of mean values |
|---|---|---|---|---|---|---|---|
| Calcium (Ca) | % | 0.04 | 0.03 | 0.03 | 0.06 | 0.03-0.07 | 0.03-0.07 |
| Sodium (Na) | % | | 0.005 | 0.01 | 0.03 | 0.01 | 0.005-0.03 |
| Potassium (K) | % | 0.40 | 0.24 | 0.39 | 0.38 | 0.38-0.47 | 0.24-0.47 |
| Phosphorus (P) | % | 0.32 | 0.35 | 0.32 | 0.35 | 0.32-0.36 | 0.32-0.36 |
| Magnesium (Mg) | % | | 0.19 | | 0.16 | 0.14-0.17 | 0.14-0.19 |
| Chlorine (Cl) | % | 0.10 | | | 0.09 | 0.06-0.10 | 0.06-0.10 |
| Sulfur (S) | % | 0.14 | | | 0.17 | 0.09-0.14 | 0.09-0.17 |
| Iron (Fe) | mg/kg | | 10.6 | 48.8 | 70.0 | 48.7-89.0 | 10.6-89.0 |
| Cobalt (Co) | mg/kg | | | | 0.31 | | 0.31 |
| Copper (Cu) | mg/kg | | 0.2 | | 10.8 | 4.7-11.5 | 0.2-11.5 |
| Manganese (Mn) | mg/kg | | | | 10.9 | 16.9-21.0 | 10.9-21.0 |
| Selenium (Se) | mg/kg | | | | | 0.23-0.46 | 0.23-0.46 |
| Zinc (Zn) | mg/kg | 18.0 | | | 47.1 | 16.9-25 | 16.9-47.1 |
| Molybdenum (Mo) | mg/kg | | | | | 1.0 | 1.0 |
| Chromium (Cr) | mg/kg | | 0.8 | | | | 0.8 |

*Sources:* Preston (2010); Ragaee et al. (2006); USDA Agricultural Research Service (2009); Ensminger et al. (1990); NRC (1994, 1998, 2000, 2001).

Table 4.4. **Vitamin concentrations in *S. bicolor* grain**

Dry matter basis

| Vitamins | Unit | USDA | Ensminger et al. | NRC | Range of mean values |
|---|---|---|---|---|---|
| A[1] | IU | 0.0 | 2.2 | 1.0 | 0.0-2.2 |
| D | IU | | | 29.0 | 29.0 |
| E | IU | | | 5.6-12.0 | 5.6-12.0 |
| Thiamin | mg/kg | 2.6 | 5.0 | 3.4-3.7 | 2.6-5.0 |
| Riboflavin | mg/kg | 1.6 | 1.4 | 1.5 | 1.4-1.6 |
| Niacin | mg/kg | 32.2 | 52.0 | 39.0-47.1 | 32.2-52.0 |
| Biotin | mg/kg | | 0.29 | 0.29-0.42 | 0.29-0.42 |
| Choline | mg/kg | | 762 | 737.0-767.8 | 737.0-767.8 |
| Folacin | mg/kg | | 0.24 | 0.19-0.23 | 0.19-0.24 |
| Pantothenic acid | mg/kg | | 11.3 | 12.5-14.3 | 11.3-14.3 |
| Pyridoxine | mg/kg | | 6.0 | 5.0-6.0 | 5.0-6.0 |

*Note:* 1. Measured as β-carotene.

*Sources:* USDA Agricultural Research Service (2009); Ensminger et al. (1990); NRC (1982, 1994, 1998, 2000, 2001).

Table 4.5. **Amino acid composition of *S. bicolor* grain**

% dry matter basis

| Amino acid | USDA | NRC[1] | Kriegshauser et al. | FAO and AFRIS | Range of mean values |
|---|---|---|---|---|---|
| Methionine | 0.19 | 0.15-0.21 | | 0.04-0.06 | 0.04-0.21 |
| Cystine | 0.14 | 0.12-0.22 | | 0.12-0.14 | 0.12-0.22 |
| Lysine | 0.25 | 0.24-0.91 | 0.25-0.26 | 0.23-0.29 | 0.23-0.91 |
| Tryptophan | 0.14 | 0.1-0.9 | 0.09-0.10 | | 0.09-0.9 |
| Threonine | 0.38 | 0.33-0.57 | 0.36-0.4 | 0.41-0.53 | 0.33-0.57 |
| Isoleucine | 0.48 | 0.4-0.7 | 0.43-0.49 | 0.41-0.53 | 0.4-0.7 |
| Histidine | 0.27 | 0.25-0.34 | 0.28-0.31 | 0.27-0.35 | 0.25-0.35 |
| Valine | 0.62 | 0.5-0.7 | 0.57-0.65 | 0.51-0.65 | 0.5-0.7 |
| Leucine | 1.64 | 1.1-1.6 | 1.47-1.75 | 1.5-1.9 | 1.1-1.9 |
| Arginine | 0.39 | 0.4-1.8 | 0.43-0.47 | 0.38-0.49 | 0.36-1.8 |
| Phenylalanine | 0.60 | 0.54-0.7 | | 0.59-0.75 | 0.54-0.75 |
| Glycine | 0.38 | 0.35-0.36 | | | 0.35-0.38 |
| Alanine | 1.14 | | | | 1.14 |
| Aspartic acid | 0.82 | | | | 0.82 |
| Glutamic acid | 2.69 | | | | 2.69 |
| Proline | 0.94 | | | | 0.94 |
| Serine | 0.51 | 0.46-0.55 | | | 0.46-0.55 |
| Tyrosine | 0.35 | 0.19-0.46 | | 0.45-0.57 | 0.19-0.57 |

*Notes:* 1. Range of values based on NRC as referenced below in the sources.

*Sources:* USDA Agricultural Research Service (2009); NRC (1982, 1994, 1998, 2000, 2001); Kriegshauser et al. (2006); FAO and AFRIS (1993).

Table 4.6. **Fatty acid composition of *S. bicolor* and *S. vulgare* grain**

% of total fatty acids

| Fatty acid | Sorghum vulgare NRC[1] | Sorghum bicolor USDA | Sorghum bicolor Rooney and Serna-Saldivar | Sorghum bicolor Osman et al.[2] | Range of mean values |
|---|---|---|---|---|---|
| 12:0 | | 0.19 | | 0.22-1.17 | 0.19-1.17 |
| 14:0 | | 0.25 | | 0.10-2.27 | 0.10-2.27 |
| 16:0 | 17.80 | 11.19 | 14.3 | 14.21-17.92 | 11.19-17.92 |
| 16:1 | 4.77 | 0.80 | 1.0 | 1.56-2.83 | 0.80-2.83 |
| 18:0 | 0.96 | 0.96 | 2.1 | 1.51-2.89 | 0.96-2.89 |
| 18:1 | 28.29 | 26.52 | 31.0 | 32.16-37.34 | 26.52-37.34 |
| 18:2 | 35.92 | 35.91 | 38.29 | 38.29-44.29 | 35.91-44.29 |
| 18:3 | 1.91 | 1.79 | | 1.04-1.65 | 1.04-1.79 |

*Notes:* 1. Based on *S. vulgare*. 2. Analysis of six cultivars of *S. bicolor*.

*Sources:* NRC (1994); USDA Agricultural Research Service (2009); Rooney and Serna-Saldivar (2000); Osman et al. (2000).

## *Forage sorghum*

The leaves and stalks of *Sorghum bicolor* may be harvested and fed as hay or straw, or animals may be grazed on stover after the grain has been harvested. Broom sorghum (*S. vulgare*) and sudangrass (*S. sudanense*) may be grazed, or whole plants may be harvested for forage as green chop, silage or hay. The leaves and stalks of brown midrib mutant (BMR) varieties of *S. bicolor* are reported to have lower lignin concentrations compared to cultivars lacking this trait, making them a more desirable roughage source for livestock. However, the data on fibre fractions, particularly lignin, vary among the available published sources, and methods used to conduct analyses are not always reported. Table 4.7 compares the nutrient composition of silage from several sorghum species.

Table 4.7. **Nutrient concentrations of silages produced by *S. bicolor*, BMR mutant S. *bicolor*, Sudangrass (*S. sudanense*) and the hybrid *S. bicolor* BMR x *S. sudanense***

| Nutrient | Sorghum bicolor: Sorghum bicolor[2,3] | Sorghum bicolor: *S. bicolor* BMR mutant[3] | Sorghum bicolor: Range of mean values | Sorghum sudanense: *Sorghum sudanense*[2] | Sorghum sudanense: BMR x *S. sudanense*[4] | Sorghum sudanense: Range of mean values |
|---|---|---|---|---|---|---|
| Dry matter (%) | 28.8-30.0 | 30.0 | 28.8-30.0 | 28.8 | 28.2 | 28.2-28.8 |
| | % of dry matter | | | % of dry matter | | |
| Crude protein | 7.3-9.1 | 7.9 | 7.3-9.1 | 10.8 | 10.8 | 10.8 |
| Ash | 7.5 | | 7.5 | 10.9 | 7.64 | 7.64-10.9 |
| Total fat | 2.9 | | 2.9 | 3.6 | 3.9 | 3.6-3.9 |
| Acid detergent fibre | 36.6-38.7 | 39.8 | 36.6-39.8 | 40.7 | 41.6 | 40.7-41.6 |
| Neutral detergent fibre | 59.0-60.7 | 60.4 | 59.0-60.7 | 63.3 | 66.2 | 63.3-66.2 |
| Lignin | 6.5-10.3 | 7.5 | 6.5-10.3 | 5.9 | 4.6 | 4.6-5.9 |
| NSC[1] | | | | | 5.8 | 5.8 |

*Notes and sources:* 1. Non-structural carbohydrate = 100 – (ash + total fat + crude protein + neutral detergent fibre). 2. NRC (2001). 3. BMR (brown midrib mutant) of *S. bicolor:* Grant et al. (1995).4. Dann et al. (1988).

Very little information is available on mineral composition of forage sorghums. Macro- and micro-mineral contents of silages produced from *S. bicolor*, *S. sudanense* and the BMR mutant of *S. bicolor* x *S. sudanense* hybrid are presented in Table 4.8.

Table 4.8. **Mineral composition of sorghum silages from *S. bicolor*, Sudangrass (*S. sudanense*) and the hybrid *S. bicolor* BMR x *S. sudanense***

| Mineral | Sorghum bicolor[1] | Sorghum sudanense[1] | BMR x S. sudanense[2] |
|---|---|---|---|
| | | % of dry matter | |
| Calcium (Ca) | 0.5 | 0.64 | 0.66 |
| Phosphorus (P) | 0.21 | 0.24 | 0.20 |
| Magnesium (Mg) | 0.27 | 0.31 | 0.39 |
| Potassium (K) | 1.75 | 2.57 | 1.87 |
| Sulfur (S) | 0.12 | 0.15 | 0.12 |
| Sodium (Na) | 0.02 | 0.03 | 0.01 |
| Chlorine (Cl) | 0.6 | 0.56 | 1.07 |
| | | mg/kg | |
| Iron (Fe) | 392 | 990 | 518 |
| Zinc (Zn) | 31 | 33 | 24.0 |
| Copper (Cu) | 9 | 11 | 9.0 |
| Manganese (Mn) | 65 | 79 | 79.0 |
| Molybdenum (Mo) | 1.9 | 2.7 | 0.8 |
| Selenium (Se) | 0.03 | | |

*Sources:* 1. From NRC (2001). 2. From Dann et al. (1988).

## *By-products of sorghum processing*

The by-product of sorghum ethanol production is distillers' grains. Data on distillers' grains are available (Lodge et al., 1997; Al-Suwaiegh et al., 2002), but are somewhat limited in scope. Table 4.9 presents the available nutritional information for wet and dry sorghum distillers' grains, and dry grains plus solubles. Distiller's dried grains with solubles (DDGS) contain all fermentation residues, including yeast, remaining after ethanol is removed by distillation (Shurson, 2009).

Table 4.9. **Nutrient composition of sorghum distillers' grains**

| | Wet distillers' grains | Dry distillers' grains | Dry distillers' grains + solubles |
|---|---|---|---|
| Dry matter (%) | 23.5-35.3 | 91.4 | 91.4 |
| | | % dry matter basis | |
| Crude protein | 31.2-31.6 | 32.9 | 31.4 |
| Ash | 2.5 | | |
| Total fat | 11.3-13.3 | 13.0 | 11.8 |
| Acid detergent fibre | 28.5 | 28.4 | |
| Neutral detergent fibre | 41.3-45.4 | 45.8 | 51.1 |
| NSC[1] | 9.2 | 3.3 | |
| Starch[2] | 10.2 | | 7.4 |

*Notes:* 1. Non-structural carbohydrate, expressed as 100 – (ash + ether extract + crude protein + neutral detergent fibre), from Al-Suwaiegh et al. (2002). 2. Expressed as starch, although method not provided, from Lodge et al. (1997).

By-products of sorghum starch extraction include bran, sorghum hominy, sorghum gluten feed, sorghum gluten meal, and oil meal. Very little nutritional information is published on these by-products. FAO and AFRIS published nutrient compositions for several of these feedstuffs, but most of the values are based on data from 1970.

In the absence of more recent data, Table 4.10 provides available information as a general guideline. As mentioned in the previous section, sorghum starch production is limited, and so it is expected that by-products of sorghum starch production will not be widely available.

Table 4.10. **Nutrient and essential amino acid composition of by-products of sorghum (*S. bicolor*) starch extraction**

| Component | Bran | Hominy | Gluten feed | Gluten meal | Oil meal |
|---|---|---|---|---|---|
| Moisture (%) | 12.0 | 11.0 | 10.5 | 10.7 | 0.9 |
| | | | **% of dry matter basis** | | |
| Crude protein | 8.9 | 11.2 | 24.6 | 46.9 | 16.6 |
| Ash | 2.4 | 2.7 | 8.2 | 3.8 | 1.6 |
| Total fat | 5.5 | 6.5 | 4.9 | 7.2 | 7.8 |
| Crude fibre | 8.6 | 3.8 | 9.5 | 5.3 | 13.2 |
| Nitrogen-free extract[1] | 74.6 | 75.8 | 52.8 | 36.8 | 60.8 |
| Arginine | | | 0.111 | 0.145 | 0.108 |
| Cystine | | | 0.054 | 0.080 | 0.055 |
| Glycine | | | 0.138 | 0.145 | 0.123 |
| Histidine | | | 0.069 | 0.103 | 0.066 |
| Isoleucine | | | 0.096 | 0.235 | 0.081 |
| Leucine | | | 0.273 | 0.835 | 0.176 |
| Lysine | | | 0.054 | 0.066 | 0.055 |
| Methionine | | | 0.054 | 0.103 | 0.066 |
| Phenylalanine | | | 0.111 | 0.314 | 0.095 |
| Threonine | | | 0.111 | 0.145 | 0.081 |
| Tryptophan | | | 0.015 | 0.038 | 0.136 |
| Tyrosine | | | 0.081 | 0.235 | 0.055 |
| Valine | | | 0.192 | 0.286 | 0.123 |

*Note:* 1. Expressed as 100 – (ash + ether extract + crude protein + crude fibre).

## Anti-nutrients

### *Cyanogenic glycosides*

Cyanogenic glycosides are mainly present in germinating seeds, sprouts and the leaves of immature sorghum plants. Traore et al. (2004) showed that malted red sorghum that had been dried contained on average 320 ppm cyanogens. The most abundant of cyanogen is dhurrin, which may comprise 3-4% of the leaves of germinating seeds (Waniska and Rooney, 2000). Stressors such as drought, frost, heavy insect infestation or overgrazing can result in increased levels of these compounds, which, along with tannins, are part of the plants' defence mechanisms. The use of potassium nitrate fertiliser was also shown to increase cyanogen production in sorghum (Busk and Moller, 2002). In the stomach of livestock, cyanogenic glycosides may be converted into hydrogen cyanide, which is very toxic, and at a low level chronic exposure may result in poor growth or reduced milk production.

Although sprouted sorghum can contain high levels of cyanogens, typical methods of processing sprouted sorghum grain for human consumption, such as manual degermination (removal of roots and shoots), removes most the toxin (Traore et al., 2004; Dada and Dendy, 1987). Therefore, cyanogenic glycosides are not generally a concern for humans. Malted sorghum is commonly used in the production of beer and other beverages and baked goods in Africa, which are consumed without resulting health problems associated with the formation of low levels of hydrogen cyanide (Waniska and Rooney, 2000).

Processing of germinating seeds for feed may result in the release of cyanide. It is generally recommended not to graze animals on young plants or cut them for green chop until they are at least 18-51 cm tall (Undersander and Lane, 2001). However, traditional curing processes such as drying for hay, and malting processes of sprouts such as heating and drying, reduces the concentration of this toxin below a level of concern (Dada and Dendy, 1987; Waniska and Rooney, 2000). With proper management, such as waiting until the plants have reached an appropriate height before grazing or harvesting, appropriate stocking rates and good growing conditions, the levels of these compounds are low and do not pose a risk to livestock. Sorghum varieties developed specifically for grazing (e.g. Sudangrass) have low or non-detectable levels of cyanogenic glycosides (Waniska and Rooney, 2000).

### *Tannins*

Early literature identified tannic acid as an anti-nutritional factor in sorghum grain. However, more recent research indicates that tannic acid is not a sorghum component (Dykes and Rooney, 2006). Some, but not all, sorghum varieties have pigmented testa containing condensed tannins, polyphenolic compounds that possibly give the seed a bitter taste and have been known to reduce intake, digestibility (particularly of protein), growth and feed efficiency of livestock (Gilani et al., 2005; Waniska and Rooney, 2000). Sorghums are classified based on their tannin content: type I, no detectable tannin; type II, tannins in pigmented testa; type III, tannins in pigmented testa and pericarp (Waniska and Rooney, 2000).

Digestibility and utilisation of absorbed nutrients may be reduced 3-15% by tannins (Waniska and Rooney, 2000). Tannins act as a plant defence against consumption by birds, and also provide some resistance to mould. In livestock production, tannins reduce the availability of key nutrients such as protein, energy, vitamins and minerals. Tannins are associated with the outer layers of the pericarp and testa of the sorghum kernel. White sorghum varieties without a pigmented testa are free of tannins. Red, brown or black varieties may contain significant amounts of tannins, but only if they have a pigmented testa.

The preferred method to determine if sorghum grains have a pigmented testa and hence contain tannins is to perform a "Clorox" bleach test as described by the ICC (2008). This test, a standard analysis method of the ICC, is used by the Federal Grain Inspection Service of the United States (USDA, Federal Grain Inspection Service-Grain Inspection, Packers and Stockyards Administration) in classifying sorghum. Decortication of sorghum grain is sometimes made to remove or reduce tannin content (Waniska and Rooney, 2000).

### *Phytic acid*

Like all grain species, sorghum contains phytic acid which binds minerals and reduces their availability to the consumer. Its phytic acid levels are similar to those reported for wheat, barley and maize, but lower than that of soybeans and other oilseeds. Since sorghum grain is usually low in mineral content (with phytin and mineral contents equivalent to maize), and the presence of phytic acid likely rendering its low mineral content unavailable, supplementation with other mineral sources is necessary where sorghum is a major component of the diet. As with tannin content, phytic acid content (and mineral content) may be reduced by abrasive decortication of the grain to remove the pericarp and aleurone layers (Waniska and Rooney, 2000).

### *Enzyme inhibitors*

Sorghum contains protease inhibitors that specifically inhibit serine proteases such as trypsin and chymotrypsin, and most varieties also contain α-amylase inhibitors. These inhibitors are potent antifungal agents and are inactivated by germination and heat treatments (Waniska and Rooney, 2000).

Concentrations of anti-nutrients in sorghum sprouts and grain as reported in available literature are summarised in Table 4.11.

Table 4.11. **Concentrations of anti-nutrients in sorghum sprouts and grain**

| Anti-nutrient | Unit | Grain | | | | Sprouts | |
|---|---|---|---|---|---|---|---|
| | | Waniska and Rooney | Salinas et al. | Kayode et al. | Range of mean values | Waniska and Rooney | Traore et al. |
| Cyanogens | ppm | | 0 | 0 | | 613 | 320[1] |
| Tannins | g/100 g DM | | 0.55-1.05 | 0.22 | 0.22-1.05 | | |
| Phytic acid | g/100 g DM | 0.17-0.38 | | 0.80 | 0.17-0.80 | | |

*Note:* DM: dry matter. 1. Based on dried sprouted seed.

*Sources:* Waniska and Rooney (2000); Salinas et al. (2006); Kayode et al. (2007); Traore et al. (2004).

## Suggested constituents to be analysed related to food use

### *Key products consumed by humans*

Consumption of sorghum as food has been increasing since the early 1980s, particularly in the more arid regions of developing countries in Africa, Asia, the Caribbean, and Central and South America. Eaten in a variety of forms depending on the region, sorghum may be consumed as whole grain, popped as a snack or boiled into porridge, processed into flour for baking, or fermented to produce beer or other baked goods. According to the Encyclopedia of Life Support Systems (UNESCO-EOLSS website), the main sorghum-based foods are: flatbread (unleavened and prepared from fermented or unfermented dough); fermented or unfermented porridges, couscous, grits; fermented or unfermented beverages; deep-fried preparations, and many others. Sorghum may also be used alone or in combination with maize to produce tortillas.

Sorghum flour used for baking does not contain viscoelastic gluten, such as that found in wheat, barley and rye doughs. Although this makes sorghum flour acceptable for use in products for patients with celiac disease (Ciaci et al., 2007), yeast-leavened products from 100% sorghum flour are difficult to obtain (Waniska and Rooney, 2000), and may have undesirable characteristics such as poor rising, coarse crumb and brittleness (Taylor et al., 2006). Addition of gums, starch, enzymes, emulsifiers and fat sources improve the quality and texture of sorghum breads, and using a soft batter rather than firm dough also improves quality of leavened breads (Taylor et al., 2006). Unleavened breads, tortillas and snacks are successfully produced with 100% sorghum flour and mixtures of sorghum and maize flour (Waniska and Rooney, 2000; Taylor et al., 2006).

Malting and brewing of sorghum has been used to produce lager, stout (referred to as "clear beers") as well as traditional opaque beers in parts of Africa (Taylor et al., 2006). The basic process for producing beer involves making gruel of cooked, gelatinized starchy adjunct which is then liquefied and saccharified by enzymes in a malted cereal (Daiber and Taylor, 1995).

The malting process involves soaking viable grain and allowing it to germinate under conditions that permit activation of enzyme systems while minimising respiration losses. Sorghum starch has a higher gelatinization temperature and lower β-amylase activity in the malt compared to barley. If tannins are present in the sorghum, they can inactivate amylases and the sorghum has to be chemically treated to inactivate the tannins (Taylor et al., 2006). Thus, various modifications in brewing clear beer have been developed to overcome these limitations, including the use of sorghum only as a starchy adjunct, use of barley malt or commercial enzymes to hydrolyze sorghum starch, and potentially the use of waxy (high amylopectin) sorghums rather than normal sorghums in the malting and brewing process (Taylor et al., 2006).

### *Suggested analyses for food use*

Sorghum's primary contribution to the human diet is energy in the form of starch and proteins. The protein content of sorghum varies across varieties, and, like many grains, is low in essential amino acids, particularly lysine and threonine. It has slightly more tryptophan than maize. However, complementary proteins from legumes can meet dietary requirements for these amino acids (Klopfenstein and Hoseney, 1995).

Sorghum is not an important source of fatty acids, minerals or fat-soluble vitamins; however, it does contain reasonably high amounts of choline and vitamin B6. Although sorghum appears to have relatively high levels of niacin, the availability of this vitamin is questionable. The availability of these nutrients for absorption depends on the processing of the kernel and the concentration of tannin in the grain. Traditional forms of processing (steeping, parboiling, fermentation, malting, popping, roasting, drying, alkali or acid treatment, and milling) may make starch and protein more available, but some of these methods will destroy vitamins, thereby reducing the concentration of certain vitamins. Tannin content should be estimated qualitatively in whole grain and sorghum bran. Methods for estimating tannin content are described by the ICC (2008).

Constituents suggested for analysis in grain sorghum for food and beverage use are listed in Table 4.12. When one considers all of the sorghum products that might be used as human food, their nutrient content should not be expected to change if the content of the whole seed is not changed. Hence, only the whole grain sorghum seed are suggested to be analysed.

Table 4.12. **Suggested constituents to be analysed in grain sorghum (*S. bicolor*) for food use**

| Parameter | Whole grain |
|---|---|
| Moisture | X |
| Crude protein | X |
| Crude fat (ether extract) | X |
| Ash | X |
| Total dietary fibre | X |
| Starch | X |
| Fatty acids | X |
| Amino acids | X |
| Tannins | X |
| Phytic acid | X |
| Pyridoxine (vitamin B6) | X |

## Suggested constituents to be analysed related to feed use

### *Key products consumed by animals*

Most parts of the sorghum plant are used as animal feed. Growing sorghum may be grazed, or the aerial parts of the plant may be ensiled or dried and fed as stover or silage for ruminant animals. Whole sorghum grain is cracked, ground or steam flaked and fed to poultry, swine, dairy and beef cattle as a source of energy. Although not common, by-products of sorghum starch extraction such as hominy and gluten feed or gluten meal may also be fed to livestock. The Association of Animal Feed Control Officials in the United States (AAFCO, 2009) defines sorghum gluten feed and gluten meal as follows:

- Grain sorghum gluten feed is that part of the grain of grain sorghums that remains after the extraction of the larger part of the starch and germ, by the processes employed in the wet milling manufacture of starch or syrup.
- Grain sorghum gluten meal is that part of the grain of grain sorghums that remains after the extraction of the larger part of the starch and germ, and the separation of the bran by the processes employed in the wet milling manufacture of starch or syrup.

  *Note*: Milo, Hegari, Kaffir or Feterita may substitute for the words "grain sorghum" in the above-mentioned definitions. If the name of the type is given, it must correspond thereto.

### *Suggested analyses for feed use*

As mentioned previously, nutritional composition of sorghum grain and stover varies with environmental conditions. However, some standard analyses for nutrient composition may be warranted.

Proximate analysis is typical for feed ingredients for non-ruminants. This analysis typically includes moisture, crude protein (N x 6.25), crude fibre (composed of cellulose, hemicellulose and lignin), fat (expressed as ether extract) and ash. Nitrogen-free extract (dry matter basis) includes starch, sugars and the soluble fraction of hemicellulose, and is derived by difference [100 – (crude P + crude fibre + ether extract + ash = NFE)].

Starch may also be analysed directly, and this may be preferred for whole grain and bran, as nitrogen-free extract may also include hemicellulose, cellulose and lignin, which are indigestible for non-ruminants. Another means of expressing soluble carbohydrates is NSC, also derived by difference [100 – (crude protein + ether extract + ash + neutral detergent fibre)]. For proximate analysis of animal feeds, acid detergent fibre and neutral detergent fibre are preferred to crude fibre analysis, particularly for ruminant feeds. These give an improved indication of the digestibility and the energetic feeding value of the feed, which is particularly important. Amino acids and fatty acids should be individually quantified. Among the fatty acids, linoleic is of key importance for sorghum grain.

Tannin is the major anti-nutrient of concern in sorghum grain products, particularly bran, in varieties that contain tannins. As mentioned above, analyses to estimate tannin content are available and not difficult to perform (ICC, 2008).

Phytic acid is common to all grains. With the use of the enzyme phytase, it is possible to break down part of the phytic acid and release bound phosphorus and calcium. Hence, the phytic acid content of the grain is beneficial to know.

As hydrogen cyanide poisonings have been reported in livestock grazing sorghum stubble (Waniska and Rooney, 2000), cyanogenic glycosides should be quantified.

Constituents suggested for analysis in grain sorghum for feed use are listed in Table 4.13. When one considers all of the sorghum products that might be used as animal feed, their nutrient content should not be expected to change if the content of the whole seed and the whole plant is not changed. Hence, only the whole grain sorghum seed or the whole sorghum plant are suggested to be analysed.

Table 4.13. **Suggested constituents to be analysed in grain sorghum for feed use**

| Parameter | Whole grain | Whole plant |
|---|---|---|
| Moisture | X | X |
| Crude protein | X | X |
| Crude fat (ether extract) | X | X |
| Ash | X | X |
| Acid detergent fibre | X | X |
| Neutral detergent fibre | X | X |
| Amino acids | X | |
| Fatty acids | X | |
| Calcium | X | X |
| Phosphorus | X | X |
| Tannins | X | |
| Phytic acid | X | |
| Cyanogenic glycosides | | X |

# Note

1. For additional discussion of appropriate comparators, see Codex Alimentarius Commission (2003; paragraphs 44 and 45).

# *References*


AAFCO (Association of American Feed Control Officials), Inc. (2009), Official Publication, Perdue University, West Lafayette, Indiana, United States.

Ali, M.L. et al. (2008), "Assessment of genetic diversity and relationship among a collection of US sweet sorghum germplasm by SSR markers", *Molecular Breeding*, Vol. 21, Issue 4, pp. 497-509, http://dx.doi.org/10.1007/s11032-007-9149-z.

Al-Suwaiegh, S. et al. (2002), "Utilization of distillers grains from the fermentation of sorghum or corn in diets for finishing beef and lactating dairy cattle", *Journal of Animal Science*, Vol. 80, No. 4, pp. 1 105-1 111.

Aydin, G. et al. (1999), "Brown midrib sorghum in diets for lactating dairy cows", *Journal of Dairy Science*, Vol. 82, No. 10, pp. 2 127-2 135.

Busk, P.K. and B.L. Moller (2002), "Dhurrin synthesis in sorghum is regulated at the transcriptional level and induced by nitrogen fertilization in older plants", *Plant Physiology*, Vol. 129, No. 3, pp. 1 222-1 231.

CGIAR (Consultative Group on International Agricultural Research) (2009), www.cgiar.org.

Ciacci, C. et al. (2007), "Celiac disease: *In vitro* and *in vivo* safety and palatability of wheat-free sorghum food products", *Clinical Nutrition*, Vol. 26, No. 6, pp. 799-805.

Codex (Codex Alimentarius Commission) (2003; with Annexes II and III adopted in 2008), *Guideline for the Conduct of Food Safety Assessment of Foods Derived from Recombinant DNA Plants*, CAC/GL 45/2003, available at: www.codexalimentarius.net/download/standards/10021/CXG_045e.pdf.

Dada, L.O. and D.A.V. Dendy (1987), "The effect of various processing techniques on the cyanide content of processed germinated sorghum", *Tropical Science*, Vol. 27, pp. 101-104.

Daiber, K.H. and J.R.N. Taylor (1995), "Opaque beers", Chapter 10 in: Dendy, D.A.V. (ed.), *Sorghum and the Millets: Chemistry and Technology*, American Association of Cereal Chemists, St. Paul, Minnesota, pp. 298-322.

Dann, H.M. et al. (1998), "Comparison of brown midrib sorghum-sudangrass with corn silage on lactational performance and nutrient digestibility in holstein dairy cows", *Joural of Dairy Science*, Vol. 91, No. 2, pp. 663-672.

De Alencar Figueiredo, L.F. et al. (2008), "Phylogenetic of crop neodiversity of sorghum", *Genetics*, Vol. 179, No. 2, pp. 997-1 008, http://dx.doi.org/10.1534/genetics.108.087312.

Del Pozo-Insfran, D. et al. (2004), "Effect of amyloglucosidase on wort composition and fermentable carbohydrate depletion in lager beers", *Journal of the Institute of Brewing*, Vol. 110, Issue 2, pp. 124-132, http://dx.doi.org/10.1002/j.2050-0416.2004.tb00191.x.

Dykes, L. and L.W. Rooney (2006), "Sorghum and millet phenols and antioxidants", *Journal of Cereal Science*, Vol. 44, No. 3, November, pp. 236-251.

Ensminger, M.E. et al. (1990), *Feeds & Nutrition, Second Edition*, The Ensminger Publishing Company, Clovis.

FAO (1995), "Nutritional inhibitors and toxic factors", Chapter 6, in: *Sorghum and Millets in Human Nutrition*, FAO Agriculture and Consumer Protection, Corporate Document Repository, available at: www.fao.org/docrep/t0818e/t0818e0j.htm.

FAO and AFRIS (Animal Feed Resources Information System) (1993), "Sorghum bicolor – Sorghum (grain)", available at: www.fao.org/ag/aga/agap/frg/afris/default.htm.

Figueroa, J.D.C. et al. (1995), "Effect of sorghum endosperm type on the quality of adjuncts for the brewing industry", *Journal of the American Society of Brewing Chemists*, Vol. 53, No. 1, pp. 5-9.

Gibbons, W.R. et al. (1986), "Intermediate-scale, semi-continuous solid-phase fermentation process for production of fuel ethanol from sweet sorghum", *Applied and Environmental Microbiology*, Vol. 51, No. 1, pp. 115-122.

Gilani, G.S. et al. (2005), "Effects of antinutritional factors on protein digestibility and amino acid availability in foods", *Journal of the AOAC International*, Vol. 88, No. 3, pp. 967-987.

Grant, R.J. et al. (1995), "Brown midrib sorghum silage for midlactation dairy cows", *Journal of Dairy Science*, Vol. 78, No. 9, pp. 1970-1 980.

ICC (International Association for Cereal Science and Technology) (n.d.), www.icc.or.at/index.php.

ICC (2008), *Draft Standard Methods: No. 174 Determination of Total Defects in Sorghum Grain* (approved 2007); *No. 175 Determination of Germinative Energy of Sorghum Grain* (approved 2007); *No. 176 Estimation of Sorghum Grain Endosperm Texture (Rev. Jan. 07)* (approved 2008); *No. 177 Detection of Tannin Sorghum Grain by the Bleach Test* (approved 2008).

ICRISAT (International Crops Research Institute for the Semi-Arid Tropics) (2009), www.icrisat.org.

ITIS (Integrated Taxonomic Information System, North America) (2009), *ITIS Standard Report* Sorghum vulgare *Pers.*, available at: http://www.itis.gov/servlet/SingleRpt/SingleRpt?search_topic=TSN &search_value=521838.

Kayode, A.P.P. et al. (2007), "Impact of sorghum processing on phytate, phenolic compounds and *in vitro* solubility of iron and zinc in thick porridges", *Journal of the Science of Food and Agriculture*, Vol. 87, Issue 5, April, pp. 832-838, http://dx.doi.org/10.1002/jsfa.2782.

Kent, N.L. and A.D. Evers (1994), "Wet milling: Starch and gluten", in: *Kents Technology of Cereals, Fourth Edition*, Elsevier, Oxford, p.261.

Klopfenstein, C.F. and R.C. Hoseney (1995), "Nutritional properties of sorghum and the millets", in: Dendy, D.A.V. (ed.), *Sorghum and the Millets: Chemistry and Technology*, American Association of Cereal Chemists, St. Paul, Minnesota, pp. 125-168.

Krausz, J.P. et al. (1994), *Disease Response of Grain Sorghum Hybrids*, Texas A&M University, available at: http://lubbock.tamu.edu/files/2011/10/b6004.pdf.

Kriegshauser, T.D. et al. (2006), "Variation in nutritional value of sorghum hybrids with contrasting seed weight characteristics and comparisons with maize in broiler chicks", *Crop Science*, Vol. 46, No. 2, pp. 695-699, http://dx.doi.org/10.2135/cropsci2005.07.0225.

Liang, G.H.L. and T.L. Walter (1968), "Heritability estimates and gene effects for agronomic traits", *Crop Science*, Vol. 8, No. 1, pp. 77-81, http://dx.doi.org/10.2135/cropsci1968.0011183X000800010022x.

Lodge, S.L. et al. (1997), "Evaluation of corn and sorghum distillers by-products", *Journal of Animal Science*, Vol. 75, No. 1, pp. 37-43.

Nikkhah, A. et al. (2004), "Evaluation of feeding ground or steam-flaked broom sorghum and ground barley on performance of dairy cows in midlactation", *Journal of Dairy Science*, Vol. 87, pp. 122-130.

NRC (2001), *Nutrient Requirements of Dairy Cattle*, Seventh Revised Edition, National Research Council, National Academies Press, Washington, DC, available at: www.nap.edu/catalog/9825/nutrient-requirements-of-dairy-cattle-seventh-revised-edition-2001.

NRC (2000), *Nutrient Requirements of Beef Cattle*, Seventh Revised Edition: Update 2000, National Research Council, National Academies Press, Washington, DC, available at: www.nap.edu/catalog/9791/nutrient-requirements-of-beef-cattle-seventh-revised-edition-update-2000.

NRC (1998), *Nutrient Requirements of Swine*, Tenth Revised Edition, National Research Council, National Academies Press, Washington, DC, available at: www.nap.edu/catalog/6016/nutrient-requirements-of-swine-10th-revised-edition.

NRC (1994), *Nutrient Requirements of Poultry*, Ninth Revised Edition, National Research Council, National Academies Press, Washington, DC, available at: http://books.nap.edu/catalog/2114.html.

NRC (1982), *United States – Canadian Tables of Feed Composition: Nutritional Data for United States and Canadian Feeds*, Third Revision, National Academies Press, Washington, DC, available at: www.nap.edu/catalog/1713/united-states-canadian-tables-of-feed-composition-nutritional-data-for.

Oliver, A.L. et al. (2004), "Comparison of brown midrib -6 and -18 forage sorghum with conventional sorghum and corn silage in diets of lactating dairy cows", *Journal of Dairy Science*, Vol. 87, No. 3, pp. 637-644.

Osman, R.O. et al. (2000), "Oil content and fatty acid composition of barley and sorghum grains", *Grasas y Aceites*, Vol. 51, pp. 157-162.

Partridge, J.E. (2008), *Common Diseases of Sorghum*, University of Nebraska-Lincoln, Department of Plant Pathology, Lincoln, Nebraska.

Preston, R.L. (2010), "2010 feed composition tables: Beef", in: *Beef Magazine*, Minneapolis, Minnesota, available at: http://beefmagazine.com/nutrition/feed-composition-tables/feed-composition-value-cattle--0301/index1.html.

Ragaee, S. et al. (2006), "Antioxidant activity and nutrient composition of selected cereals for food use", *Food Chemistry*, Vol. 98, Issue 1, pp. 32-38.

Rattunde, H.F.W. (1998), "Early-maturing dual-purpose sorghums: Agronomic trait variation and covariation among landraces", *Plant Breeding*, Vol. 117, No. 1, pp. 33-36, http://dx.doi.org/10.1111/j.1439-0523.1998.tb01444.x.

Rooney, L.W. and S.O. Serna-Saldivar (2000), "Sorghum", Chapter 5 in: Kulp, K. and J.G. Ponts, Jr (eds.), *Handbook of Cereal Science and Technology, Second Edition*, Marcel Dekker, Inc. New York, New York, pp. 149-176.

Rooney, L.W. and R.D. Waniska (2000), "Chapter 4.2: Sorghum food and industrial utilization", in: Smith, C.W. and R.A. Frederiksen (eds.), *Sorghum Origin, History, Technology, and Production*, J. Wiley & Sons, New York, New York, pp 689-729.

Salinas, I. et al. (2006), "Compositional variation amongst sorghum hybrids: Effect of kafirin concentration on metabolizable energy", *Journal of Cereal Science*, Vol. 44, Issue 3, pp. 342-346.

Sang, Y. et al. (2008), "Structure and functional properties of sorghum starches differing in amylose content", *Journal of Agricultural and Food Chemistry*, Vol. 56, No. 15, pp. 6 680-6 685, http://dx.doi.org/10.1021/jf800577x.

Serna-Saldivar, S. and L.W. Rooney (1995), "Structure and chemistry of sorghum and millets", in: Dendy, D.A.V. (ed.), *Sorghum and Millets: Chemistry and Technology*, American Association of Cereal Chemists, St. Paul, Minnesota, p.75.

Shurson, J. (2009), *Distillers Grains By-products in Livestock and Poultry Feeds, Overview: Distiller's Dried Grains with Solubles*, Department of Animal Science, College of Food, Agriculture and Natural Resource Science, University of Minnesota, St. Paul, Minnesota.

Taylor, J.R.N. et al. (2006), "Novel food and non-food uses for sorghum and millets", *Journal of Cereal Science*, Vol. 44, No. 3, November, pp. 252-271.

Teetes, G. and B.B. Pendelton (1999), *Insect Pests of Sorghum*, Texas A&M University, Department of Entomology.

Traore, T. et al. (2004), "Changes in nutrient composition, phytate and cyanide contents and alpha-amylase activity during cereal malting in small production units in Ouagadougou (Burkina Faso)", *Food Chemistry*, Vol. 88, pp. 105-114, http://dx.doi.org/10.1016/j.foodchem.2004.01.032.

Undersander, D. and W. Lane (2001), *Sorghums, Sudangrasses, and Sorghum-sudangrass Hybrids for Forage*, University of Wisconsin, Cooperative Extension, Madison, Wisconsin, available at: www.uwex.edu/ces/forage/pubs/sorghum.htm.

UNESCO (United Nations Educational, Scientific and Cultural Organization*), EOLSS (Encyclopedia of Life Support Systems)*, virtual dynamic library, http://www.eolss.net/ (accessed 2009).

USDA Agricultural Research Service (2009), *USDA National Nutrient Database for Standard Reference*, Release 22, Nutrient Data Laboratory, available at: http://ndb.nal.usda.gov.

USDA Agricultural Research Service, (2009), "Taxon Name = Sorghum", *USDA GRIN Taxonomy for Plants Database*, available at: www.ars-grin.gov/cgi-bin/npgs/html/tax_search.pl (accessed 2009)

US Grains Council (2008), "Sorghum", www.grains.org.

US National Sorghum Producers Association (2006), www.sorghumgrowers.com.

Wang, D. et al. (2008), "Grain sorghum is a viable feedstock for ethanol production", *Journal of Industrial Microbiology and Biotechnology – Bioenergy Special Issue*, Vol. 35, No. 5, pp. 313-320, http://dx.doi.org/10.1007/s10295-008-0313-1.

Waniska, R.D. and L.W. Rooney (2000), "Chapter 4.1: Structure and chemistry of the sorghum caryopsis", in: Smith, C.W. and R.A. Frederiksen (eds.), *Sorghum Origin, History, Technology, and Production*, J. Wiley & Sons, New York, New York, pp 649-688.

Zhao, R. et al. (2008), "Impact of mashing on sorghum proteins and its relationship to ethanol fermentation", *Journal of Agricultural and Food Chemistry*, Vol. 56, No. 3, pp. 946-953, http://dx.doi.org/10.1021/jf072590r.

Zinn, R.A. et al. (2008), "Influence of dry-rolling and tempering agent addition during the steam-flaking of sorghum grain on its feeding value for feedlot cattle", *Journal of Animal Science*, Vol. 86, No. 4, pp. 916-922.

# *Chapter 5*

# Sweet potato (*Ipomoea batatas*)

*This chapter, prepared by the OECD Task Force for the Safety of Novel Foods and Feeds with South Africa as lead country and Japan as co-lead, deals with the composition of sweet potato (*Ipomea batatas*). It contains elements that can be used in a comparative approach as part of a safety assessment of foods and feeds derived from new varieties. Background is given on sweet potato production, processing and uses, followed by appropriate varietal comparators and characteristics screened by breeders. Nutrients in storage roots and leaves of sweet potato, anti-nutrients, toxicants, allergens and other components are then detailed. The final sections suggest key constituents in storage roots and leaves for analysis of new varieties for food use and for feed use.*

## Background

### *General description of sweet potato*

The sweet potato (*Ipomoea batatas* [L.] Lam.) belongs to the *Convolvulaceae* or morning glory family (Jones, 1965; Austin, 1977). It is considered as the only major economically important species of the *Ipomoea* genus (Hall and Phatak, 1993). *Ipomoea batatas* is thought to have originated in Mexico and possibly Central America (Zhang and Corke, 2001). Common names in Latin America are "*batata*", "*camote*" and "*boniato*" (Spanish); "*batata doce*" (Portuguese); "*apichu*" and "*kumara*" in some Andean regions (Martin and Jones, 1986; Woolfe, 1992). Indigenous South Americans have probably cultivated sweet potatoes for thousands of years. The crop was spread to other parts of the world such as Polynesia and New Zealand during the 8th century. Introduction into the People's Republic of China (hereafter "China") occurred during the 14th century, probably from the Philippines and into Japan during the 17th century first from England, which was unsuccessful and then from China (Woolfe, 1992). Sweet potato was introduced to the tropical areas of Africa, Europe, China, India and Indonesia during the 16th century (Janssens, 2001). Along its long-standing domestication process, the crop has developed secondary centres of genetic diversity; many types of sweet potato that are genetically distinct from those found in their area of origin can be found in Papua New Guinea and in other parts of Asia (CIP, 2009).

Sweet potato is a perennial plant, although it is typically cultivated as an annual crop (Janssens, 2001). Despite its name, the sweet potato is not related to the potato (which belongs to the *Solanaceae* family). This herbaceous plant does not develop a tuber (thickened stem), but certain of its roots produce edible storage roots (Jones et al., 1986). Nearly half of the sweet potato produced in Asia (the world's largest producing region) is used for animal feed, while the remainder is primarily used for human consumption. In Africa in contrast, most of the crop is cultivated for human consumption. Sweet potato is high in carbohydrates and vitamin A and can produce more edible energy per hectare per day than wheat, rice or cassava. It has an abundance of uses ranging from consumption of fresh roots or leaves to processing into animal feed, starch, flour, candy and alcohol (CIP, 2009). Various publications review the crop characteristics, agronomy, and food and feed applications of sweet potato, which is indicative of the interest of the scientific community in this particular crop (Bovell-Benjamin, 2007; Chassy et al., 2008; Lebot, 2009; Woolfe, 1992).

### *Production*

Sweet potato is an important crop in many parts of the world, being cultivated in more than 100 countries. As a world crop, it ranks seventh from the viewpoint of total production after wheat, rice, maize, potato, barley and cassava (Kays, 2005). In monetary terms, it ranks 13th globally in the production value of commodities, and is 5th on the list of the developing countries' most valuable food crops (Woolfe, 1992). The annual world production was 110.1 million tonnes (Mt) in 2008, with 84% produced in Asia (92.5 Mt), 12.7% in Africa (14 Mt), 2.6% in Americas (2.6 Mt), 0.7 Mt in Oceania and less than 0.1 Mt in Europe (FAOSTAT, 2008; see Table 5.1). Furthermore, the crop accounts for about one-third of the production of root and tuber crops in developing countries. China is by far the largest producer, accounting in 2008 for more than 77% of the world supply, followed by Nigeria, Uganda, Indonesia and Viet Nam (Table 5.1). The global land area under production in 2008 was estimated to reach 8.2 million hectares, with an average yield of 13.5 t/ha (Table 5.1).

Table 5.1. **Sweet potato production in selected countries, 2008**

| Country | **Production** ('000 tonnes) | **Area** ('000 hectares) | **Average yield** (tonnes/hectare) |
|---|---|---|---|
| Asia total | 92 490 | 4 433 | 20.9 |
| China (People's Republic of) | 85 213* | 3 685* | 23.1 |
| Indonesia | 1 877 | 174 | 10.8 |
| Viet Nam | 1 324 | 162 | 8.2 |
| India | 1 146 | 126 | 9.1 |
| Japan | 968* | 41* | 23.8 |
| Philippines | 572 | 116 | 4.9 |
| Korea, Democratic People's Republic | 380 | 28* | 13.6 |
| Korea | 329 | 19 | 16.9 |
| Bangladesh | 307 | 32 | 9.7 |
| Africa total | 14 013 | 3 312 | 4.2 |
| Nigeria | 3 318 | 1 106 | 3.0 |
| Uganda | 2 707 | 599 | 4.5 |
| Tanzania | 1 322* | 505* | 2.6 |
| Kenya | 895 | 63 | 14.3 |
| Madagascar | 890* | 127 | 7.0 |
| Burundi | 874* | 131* | 6.7 |
| Rwanda | 800* | 140* | 5.7 |
| Angola | 710* | 145* | 4.9 |
| Ethiopia | 526 | 62 | 8.4 |
| Americas total | 2 852 | 301 | 9.5 |
| United States | 837 | 39 | 21.2 |
| Brazil | 519* | 47* | 10.9 |
| Cuba | 375 | 59 | 6.4 |
| Argentina | 340* | 24* | 14.2 |
| Oceania | 706 | 125 | 5.6 |
| Papua New Guinea | 580* | 115* | 5.0 |
| Europe | 67 | 6 | 12 |
| World total | 110 128 | 8 178 | 13.5 |

*Note:* * FAO estimate.

*Source:* FAOSTAT (2009).

Since China contributes the largest portion of sweet potato production in the world, production in one of the Chinese provinces should be mentioned. The Shandong province has an approximate annual production of about 17 million tonnes, which is produced on 600 000 ha (Fuglie et al., 1999). Compared with the 2008 FAO data, this would account for approximately 20% of China's production and 15% of the world sweet potato production.

It is important to note that when compiling a review on the worldwide production of sweet potato, discrepancies in data might occur. Since sweet potato is mainly produced by small farmers on non-contiguous plots, harvested several times a year and not sold through regulated domestic markets, estimating the exact production and trade of this crop is difficult.

A very small amount of the world production of sweet potato is traded internationally (as unprocessed roots, 0.17% in 2007). The main 2007 exporters were the United States, China, Israel and France, while the main importers were the United Kingdom, Canada, France and Japan (FAOSTAT, 2007; Table 5.2).

Table 5.2. **Sweet potato import and export figures for selected countries, 2007**

| Importing country | Import ('000 tonnes) | Exporting country | Export ('000 tonnes) |
|---|---|---|---|
| Asia | 39.5 | Asia | 43.4 |
| Hong Kong, China | 3.2 | China (People's Republic of) | 16.0 |
| Japan | 14.6 | Israel | 12.3 |
| Singapore | 6.4 | Indonesia | 8.4 |
| Malaysia | 5.6 | | |
| Saudi Arabia | 3.5 | | |
| Africa | 1.2 | Africa | 9.7 |
| | | Egypt | 7.1 |
| Americas | 44.7 | Americas | 67.1 |
| United States | 7.8 | United States | 38.9 |
| Canada | 24.9 | Dominican Republic | 8.2 |
| Argentina | 7.9 | Brazil | 5.9 |
| | | Honduras | 5.4 |
| | | Paraguay | 3.4 |
| Europe | 93.9 | Europe | 25.7 |
| United Kingdom | 37.1 | France | 10.1 |
| Albania | 12.7 | Italy | 6.8 |
| France | 15.6 | Netherlands | 5.8 |
| Italy | 6.0 | | |
| Netherlands | 12.0 | | |
| Oceania | 0.3 | Oceania | 0.2 |
| World total | 179.6 | World total | 146.0 |

*Source:* FAOSTAT (2007).

## *Processing and uses*

Sweet potato is an important root crop and, besides human consumption, the roots, stems and leaves are readily eaten by cattle, goats, pigs and poultry as forage for animals.

Sweet potato roots can be sliced, dried and ground in order to produce flour that remains in good condition for a long time. Dried root slices are a suitable means of storage in humid areas. In Indonesia, sweet potato is soaked in salt water for about an hour to inhibit microbial growth before drying. The flour is used as a dough conditioner for bread, biscuit and cake processing (it may substitute for up to 20% of wheat flour), as well as in gluten-free pancake preparation (Shih et al., 2006). Sweet potato flour is used as a stabiliser in the ice-cream industry, and powder made from dehydrated sweet potato is used in instant soups.

Mashed sweet potato is used as an ingredient of ice cream, tarts, baking products and desserts as a substitute for more expensive ingredients. As puree, it is used in pie fillings, sauces (e.g. tomato sauce in Uganda), frozen patties, baby foods and in fruit-flavoured sweet potato jams together with pineapple, mango, guava and orange.

In the United States, sweet potato in whole, halved, chunks or pureed form is canned. Sweet potato can be frozen as cubes, slices, french fries, mash, halves, quarters or whole roots. In Japan, sweet potato slices are steamed and dried to produce *hoshiimo* or *mushikiri* (Woolfe, 1992).

Sweet potato is further processed as sugar-coated or salted crisps for snack foods (Woolfe, 1992). Sweet potato crisps are produced in much the same way as potato and the product is now popular in Asia. The sugar-coated chips are popular in China and the salted variety is popular in the United States.

The major industrial use of sweet potato is for the production of starch (Figure 5.1). Sweet potato starch is produced under alkaline (pH 8.6) conditions by using lime, which helps to flocculate impurities and dissolve the pigments. The uncooked starch of the sweet potatoes is very resistant to the hydrolysis by amylase. When cooked, its susceptibility to the enzyme increases. Thus, after cooking, the easily hydrolysable starch fraction of sweet potato increases from 4-55% (Cerning-Beroard and Le Dividich, 1976). The starch shows properties intermediate between potato starch and maize/cassava starch, e.g. in terms of viscosity. In Japan about 90% of the starch produced from sweet potato is used to manufacture starch syrup, glucose and isomerised glucose syrup (high-fructose syrup), lactic acid beverages, bread, as well as other products in the food industry such as distilled spirits called *shochu*. In China, the starch is used for making pasta (Singh et al., 2004) and for producing alcoholic beverages. Non-alcoholic juices are also made in African countries such as Uganda.

Figure 5.1. **Sweet potato starch production, generalised process scheme**

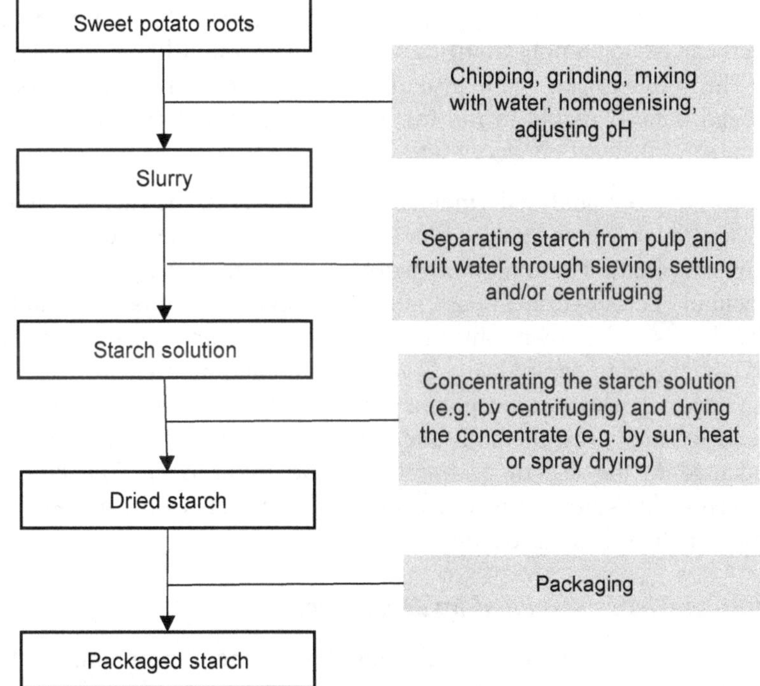


*Source*: Adapted from Woolfe (1992) and Bovell-Benjamin (2007).

In Africa, sweet potato is mainly consumed as food in an array of consumption patterns across production areas (Manrique, 1998; CIP, 2009). In eastern and southern

Africa, roots are eaten either just boiled or cooked together with beans, vegetables and other foods, and sometimes fried as chips. In a few areas the roots are peeled, sliced and dried. In South Africa, sweet potato is mainly consumed fresh and the largest portion of the production is sold on the commercial fresh produce markets. Many resource-poor farmers in South Africa grow sweet potato in home gardens. In these cases, sweet potato is an important subsistence food crop where resource-poor farmers boil and eat sweet potatoes as part of a hot meal, or cold with tea, and to a lesser extent in stews. At this level, sweet potato processing is limited to occasional drying, freezing, and baking crisps and bread.

Sweet potato has potential use in the bio-fuel industry. Some sweet potato varieties have a carbohydrate content that approaches the lower limits of those of sugarcane, the highest yielding ethanol crop. At this point, sweet potato is not an economically competitive fuel source. It costs more to grow and process sweet potato than many other fuel sources. In addition to ethanol, the ethanol manufacturing process also produces by-products as marketable products. Selling these by-products (e.g. residual mash) from this industry to the feed industry is an important economic outlet for ethanol manufacturers (USDA Agricultural Research Service, 2008).

In South America, the juice of red sweet potatoes is combined with lime juice to make a fabric dye.

### *Appropriate comparators for testing new varieties*

This chapter suggests parameters that sweet potato breeders should measure when developing new modified varieties. The data obtained in the analysis of a new sweet potato variety should ideally be compared to those obtained from an appropriate near isogenic non-modified variety, grown and harvested under the same conditions.[1] The comparison can also be made between values obtained from new varieties and data available in the literature, or chemical analytical data generated from other commercial sweet potato varieties.

Components to be analysed include key nutrients, toxicants and allergens. Key nutrients are those which have a substantial impact in the overall diet of humans (food) and animals (feed). These may be major constituents (fats, proteins, and structural and non-structural carbohydrates) or minor compounds (vitamins and minerals). Similarly, the levels of known anti-nutrients and allergens should be considered. Key toxicants are those toxicologically significant compounds known to be inherently present in the species, whose toxic potency and levels may impact human and animal health. Standardised analytical methods and appropriate types of material should be used, adequately adapted to the use of each product and by-product. The key components analysed are used as indicators of whether unintended effects of the genetic modification influencing plant metabolism has occurred or not.

### *Breeding characteristics screened by developers*

The major goals of current research and development programmes focusing on the improvement of sweet potato include:

- Improved root yield, dry matter yield, foliage yield and maturity period.
- Disease and pest resistance, e.g. resistance to sweet potato feathery mottle virus (SPFMV) disease (Okada et al., 2001), resistance to the sweet potato weevil (*Cylas formicarius*) and tolerance to sweet potato virus disease (SPVD).

- Increased nutritional value focusing on the improvement of the beta-carotene content of sweet potato as part of the HarvestPlus Crop Biofortification Program (HarvestPlus, 2004; Nestel et al., 2006). Focusing on cultivars with purple or orange flesh with high levels of anthocyanins (e.g. CIP-VITAA Partnership, 2004).
- Taste attributes.
- Processing attributes such as starch quality. New sweet potato lines are being developed having low gelatinization and altered starch structures (Katayama et al., 2004, 2006) as a convenient-cooking cultivar.
- Tolerance to abiotic stress including salinity, drought and acid soils.

## Nutrients

Sweet potato is a staple food source for many indigenous populations in China, Central and South Americas, Ryukyu Islands, Africa, the Caribbean, the Māori people, Hawaiians and Papua New Guineans. It serves as an important protein source for many world populations (Bovell-Benjamin, 2007) and is an important source of starch and other carbohydrates. The energy value of sweet potato exceeds that of potato, cassava and other known tubers (Janssens, 2001). The carbohydrate content of the storage roots varies from 25-30%, while the rest is composed of water (58-72%). Leaves contain about 3% protein, approximately twice the amount of storage roots (Woolfe, 1992).

Sweet potato contains various micro-nutrients. Substantial quantities of vitamin C, moderate quantities of thiamin (vitamin B1), riboflavin (vitamin B2) and niacin, some quantities of panthothenic acid (vitamin B5), pyridoxine (vitamin B6), folic acid and satisfactory quantities of vitamin E are present. Sweet potato also contains some essential minerals and trace elements having especially high quantities of iron. Two other important minerals present are potassium and calcium (Woolfe, 1992). Moderate quantities of zinc, sodium, magnesium and manganese are also present (Antia et al., 2006; Suda et al., 1999).

The major contribution which sweet potato makes to human nutrition is the beta-carotene present in orange-fleshed varieties. Beta-carotene is converted to vitamin A in the human body. Dark orange varieties can contain up to 20 000 µg beta-carotene per 100 g fresh storage root weight (Woolfe, 1992; Takahata et al., 1993; Bovell-Benjamin, 2007; Teow et al., 2007). Other crops such as maize, rice and wheat contain very little beta-carotene. Orange-fleshed sweet potato is used in food diversification programmes for the alleviation of vitamin A deficiency.

A 100-gramme portion of sweet potato root may supply the following nutrients required daily for an adult African male (Woolfe, 1992):

- 100% of beta-carotene (dark orange sweet potato)
- 57% of vitamin C
- 6% of thiamin
- 8% of riboflavin
- 3% of niacin
- 6% of folate

- 2-7% of iron.

Sweet potato leaves have a high value as feed for farm animals, with 3 kg of green leaves equivalent to 1 kg of maize with a nutritional value rated at 95-100% that of maize. Dry leaves have a higher nutritional value when compared with alfalfa hay as forage (Reed, 1976). Dried sweet potato leaves consist of 22% crude protein, 46% crude fibre and 9% total ash. The digestible crude protein is 9% and the total digestible nutrients are 22.4% (Satapathy et al., 2006).

The nutrient compositions of the sweet potato leaves, roots and processed products as related to proximates, minerals, vitamins, lipids, proteins and some secondary metabolites are presented in Tables 5.3-5.10. The refuse values indicated in the tables does not affect the values presented.

Table 5.3. **Proximate composition of raw sweet potato**

| | **Storage roots** | | | | | **Leaves** | | |
|---|---|---|---|---|---|---|---|---|
| | **With skins** | **Without skin** | | | | | | |
| **Nutrient** | Raw unprepared[1] | Raw, frozen unprepared[2] | Raw[3] | Raw[4] | Range of mean values – raw without skin | Raw[5] | Raw[4] | Range of mean values – raw |
| | **g/100 g fresh weight** | | | | | **g/100 g fresh weight** | | |
| Water | 77.3 | 74.9 | 66.1 | 68.8-73.3 | 66.1-74.9 | 88.0 | 86.7 | 86.7-88.0 |
| | **g/100 g dry weight** | | | | | **g/100 g dry weight** | | |
| Protein | 6.9 | 6.8 | 3.5 | 3.7-6.1 | 3.5-6.8 | 33.2 | 24.1 | 24.1-33.2 |
| Total fat | 0.2 | 0.7 | 0.6 | | 0.6-0.7 | 2.5 | | |
| Ash | 4.4 | 4.0 | 2.9 | 2.6-3.2 | 2.6-4.0 | 11.3 | 10.5 | 10.5-11.3 |
| Carbohydrate | 88.6 | 88.5 | 92.9 | 91.0-95.0 | 88.5-95.0 | 53.0 | 60.2 | 53.0-60.2 |
| Crude fibre | | | | 3.0-3.2 | 3.0-3.2 | | 12.0 | 12.0 |
| Dietary fibre | 13.2 | 6.8 | 6.8 | | 6.8 | 16.6 | | 16.6 |
| Sugars, total | 18.4 | | | | | | | |
| Sucrose | 11.1 | | | | | | | |
| Glucose | 4.2 | | | | | | | |
| Fructose | 3.1 | | | | | | | |
| Starch | 55.7 | | | | | | | |

*Notes:* 1. USDA *National Nutrient Database for Standard Reference*, Release #22 (2009), NBD No. 11507, Refuse 28% (non-edible parings and trimmings).

2. USDA *National Nutrient Database for Standard Reference*, Release #22 (2009), NBD No. 11516, Refuse 0%.

3. Ministry of Education, Culture, Sports, Science, and Technology, The Council for Science and Technology, Subdivision on Resources (2005).

4. O'Hair (1984).

5. USDA *National Nutrient Database for Standard Reference*, Release #22 (2009), NBD No. 11505, Refuse 6% (tough stems and bruised leaves).

For references 1, 2 and 5, dry weight values were calculated using reported values for water.

Table 5.4. **Proximate composition of processed sweet potato**

| Nutrient | Storage roots | | | | | Leaves |
|---|---|---|---|---|---|---|
| | Baked, frozen, without skin[1] | Boiled, without skin[2] | Cooked[3] | Steamed[4] | Baked[4] | Cooked[5] |
| | g/100 g fresh weight | | | | | g/100 g fresh weight |
| Moisture | 73.7 | 80.1 | 72.9 | 66.4 | 58.1 | 88.7 |
| | g/100 g dry weight[6] | | | | | g/100 g dry weight[6] |
| Protein | 6.5 | 6.9 | 6.3 | 3.6 | 3.4 | 20.5 |
| Total fat | 0.5 | 0.7 | 0.4 | 0.6 | 0.5 | 2.7 |
| Ash | 4.1 | 3.2 | 0.0 | 3.0 | 3.1 | 12.0 |
| Carbohydrate | 89.0 | 89.2 | 78.6 | 92.9 | 94.0 | 64.8 |
| Dietary fibre | 6.8 | 12.6 | 11.1 | 11.3 | 8.4 | 16.81 |
| Sugars total | 34.9 | 28.9 | | | | 48.0 |
| Sucrose | | 7.2 | | | | |
| Glucose | | 2.7 | | | | |
| Fructose | | 2.2 | | | | |
| Maltose | | 16.8 | | | | |
| Starch | | 26.3 | | | | |

*Notes:* 1. USDA *National Nutrient Database for Standard Reference*, Release #22 (2009), NBD No. 11507, Refuse 0%. 2. USDA *National Nutrient Database for Standard Reference*, Release #22 (2009), NBD No. 11510, Refuse 0%. 3.Medical Research Council, South Africa (1998). 4. Ministry of Education, Culture, Sports, Science, and Technology, The Council for Science and Technology, Subdivision on Resources (2005). 5. USDA *National Nutrient Database for Standard Reference*, Release #22 (2009), NBD No. 11506, Refuse 0%. 6. Dry weight values.

Table 5.5. **Mineral composition of raw sweet potato (per 100 g dry weight)[1]**

| Mineral | Unit | Storage roots | | | | | Leaves raw[1] | | |
|---|---|---|---|---|---|---|---|---|---|
| | | *With skin* | *Without skin* | | | | | | |
| | | Raw unprepared[2] | Raw, frozen unprepared[3] | Raw[4] | Raw[5] | Range of mean values – raw without skin | Raw[6] | Raw[5] | Range of mean values |
| Calcium (Ca) | mg | 132.0 | 147.4 | 118.0 | 79.0-106.0 | 79.0-147.4 | 307.3 | 647.0 | 307.3-647 |
| Iron (Fe) | mg | 2.7 | 2.1 | 2.1 | 3.4-6.4 | 2.1-6.4 | 8.4 | 33.8 | 8.4-33.8 |
| Magnesium (Mg) | mg | 110.0 | 87.6 | 73.7 | | 73.7-87.6 | 506.6 | | 506.6 |
| Phosphorus (P) | mg | 206.9 | 179.2 | 135.7 | 142.0-160.0 | 135.7-179.2 | 780.7 | 609.0 | 609.0-780.7 |
| Potassium (K) | mg | 1 483.3 | 1 453.6 | 1 386.4 | 724.0 | 724.0-1453.6 | 4 302.3 | | |
| Sodium (Na) | mg | 242.1 | 23.9 | 11.8 | 107.0 | 11.8-107.0 | 74.8 | | |
| Zinc (Zn) | mg | 1.3 | 1.2 | 0.6 | | 0.6-1.2 | 2.4 | | |
| Manganese (Mn) | mg | 1.1 | 2.7 | 1.3 | | 1.3-2.6 | 2.1 | | |
| Copper (Cu) | mg | 0.7 | 0.7 | 0.5 | | 0.5-0.7 | 0.3 | | |
| Selenium (Se) | mcg | 2.6 | 2.4 | 0 | | 2.4 | 7.5 | | |

*Notes:* 1. Dry weight values were calculated using reported values for water from the respective reference in Table 5.3. 2. USDA *National Nutrient Database for Standard Reference*, Release #22 (2009), NBD No. 11507, Refuse 28% (non-edible parings and trimmings). 3. USDA *National Nutrient Database for Standard Reference, Release* #22 (2009), NBD No. 11516, Refuse 0%. 4. Ministry of Education, Culture, Sports, Science, and Technology, The Council for Science and Technology, Subdivision on Resources (2005). 5. O'Hair (1984). 6. USDA *National Nutrient Database for Standard Reference*, Release #22 (2009), NBD No. 11505, Refuse 6% (tough stems and bruised leaves).

Table 5.6. **Mineral content of processed sweet potato (per 100 g dry weight)**[1]

| Nutrient | Unit | Storage roots | | | | | Leaves |
|---|---|---|---|---|---|---|---|
| | | Baked frozen without skin[2] | Boiled without skin[2] | Cooked[3] | Steamed[4] | Baked[4] | Cooked[5] |
| Calcium (Ca) | mg | 133.1 | 135.9 | 103.3 | 139.9 | 81.1 | 212.6 |
| Iron (Fe) | mg | 2.1 | 3.6 | 1.9 | 1.8 | 1.7 | 5.3 |
| Magnesium (Mg) | mg | 79.8 | 90.6 | 73.8 | 56.6 | 54.9 | 540.3 |
| Phosphorus (P) | mg | 167.3 | 161.0 | 203.0 | 125.0 | 131.3 | 531.4 |
| Potassium (K) | mg | 1 433.5 | 1 157.5 | 1 284.1 | 1 458.3 | 1 288.8 | 4 225.0 |
| Sodium (Na) | mg | 30.4 | 135.9 | 36.9 | 11.9 | 31.0 | 115.1 |
| Zinc (Zn) | mg | 1.1 | 1.0 | 1.1 | 0.6 | 0.5 | 2.3 |
| Manganese (Mn) | mg | 2.5 | 1.3 | 0 | 1.5 | 0.8 | 2.0 |
| Copper (Cu) | mg | 0.7 | 0.5 | 0 | 0.5 | 0.5 | 0.3 |
| Selenium (Se) | mcg | 2.3 | 1.0 | 0 | 0 | 0 | 8.0 |

*Notes:* 1. Dry weight values were calculated using reported values for water from Table 5.5 for the respective references. 2. USDA *National Nutrient Database for Standard Reference*, Release #22 (2009), NBD No. 11510, Refuse 0%. 3. Medical Research Council, South Africa (1998). 4. Ministry of Education, Culture, Sports, Science, and Technology, The Council for Science and Technology, Subdivision on Resources (2005). 5. USDA *National Nutrient Database for Standard Reference*, Release #22 (2009), NBD No. 11506, Refuse 0%.

Table 5.7. **Vitamin composition of raw sweet potato (per 100 g dry weight)**[1]

| Nutrient | Unit | Storage roots | | | | | Leaves raw[1] | | |
|---|---|---|---|---|---|---|---|---|---|
| | | *With skins* | *Without skins* | | | | | | |
| | | Raw unprepared[2] | Frozen unprepared[3] | Raw[4] | Raw[5] | Range | Raw[6] | Raw[5] | Range of mean values |
| Vitamin C, total | mg | 10.56 | 52.97 | 29.00 | 79.00-119.00 | 29.00-119.00 | 91.36 | 127.80 | 91.36-127.80 |
| Thiamin | mg | 0.34 | 0.27 | 0.11 | 0.35 | 0.11-0.35 | 1.30 | 0.80 | 0.80-1.30 |
| Riboflavin | mg | 0.27 | 0.20 | 0.03 | 0.16 | 0.03-0.20 | 2.87 | 1.60 | 1.60-2.87 |
| Niacin | mg | 2.45 | 2.38 | 0.80 | 2.40 | 0.8-2.38 | 9.39 | 5.30 | 5.30-9.39 |
| Pantothenic acid | mg | 3.52 | 2.05 | 0.96 | 0 | 0.96-2.05 | 1.87 | 0 | |
| Vitamin B6 | mg | 0.92 | 0.70 | 0.28 | 0 | 0.28-0.70 | 1.58 | 0 | |
| Folate, total | mcg | 48.42 | 83.63 | 49.00 | 0 | 49.0-83.63 | 664.45 | 0 | |
| Folate, DFE | mcg | 48.28 | 83.63 | | | | 664.45 | | |
| Choline | mg | 54.14 | | | | 0 | | | |
| Carotene, beta | mcg | 37 451.58 | 24 771.01 | | | | | | |
| Carotene, alpha | mcg | 30.81 | | | | 0 | | | |
| Vitamin A, IU[7] | IU | 62 442.78 | 41 286.34 | 0 | 0 | | 8 538.21 | 0 | |
| Vitamin A, RAE[8] | mcg | 3 120.60 | 2 062.92 | 0 | 0 | | 423.59 | 0 | |
| Vitamin E | mcg | 1.14 | 0 | 1.60 | 0 | | 0 | 0 | 0 |
| Vitamin K | mcg | 7.92 | 0 | 0 | 0 | 0 | 0 | 0 | 0 |

*Notes:* 1. Dry weight values were calculated using reported values for water from Table 5.3 for the respective references. 2. USDA *National Nutrient Database for Standard Reference*, Release #22 (2009), NBD No. 11507, Refuse 28% (non-edible parings and trimmings). 3. USDA *National Nutrient Database for Standard Reference*, Release #22 (2009), NBD No. 11516, Refuse 0%. 4. Ministry of Education, Culture, Sports, Science, and Technology, The Council for Science and Technology, Subdivision on Resources (2005). 5. O'Hair (1984). 6. USDA *National Nutrient Database for Standard Reference*, Release #22 (2009), NBD No. 11505, Refuse 6% (tough stems and bruised leaves). 7. IU: International units. 8. RAE: retinol activity equivalents.

Table 5.8. **Vitamin composition of processed sweet potato (per 100 g dry weight)**[1]

| Nutrient | Unit | Storage roots | | | | | Leaves |
|---|---|---|---|---|---|---|---|
| | | Baked frozen without skin[2] | Boiled without skin[3] | Cooked[4] | Steamed[5] | Baked[5] | Cooked[6] |
| Vitamin C, total | mg | 34.60 | 64.42 | 92.25 | 59.52 | 23 | 13.29 |
| Thiamin | mg | 0.25 | 0.28 | 0.26 | 0.30 | 0.12 | 0.99 |
| Riboflavin | mg | 0.21 | 0.24 | 0.48 | 0.09 | 0.06 | 2.36 |
| Niacin | mg | 2.11 | 2.71 | 2.21 | 2.08 | 1.0 | 8.88 |
| Pantothenic acid | mg | 2.13 | 2.92 | | 2.89 | 1.30 | 1.77 |
| Vitamin B6 | mg | 0.71 | 0.83 | 0.89 | 0.68 | 0.33 | 1.42 |
| Folate, total | mcg | 83.65 | 30.20 | | 136.90 | 47 | 434.01 |
| Folic acid | mcg | 0.00 | 0.00 | 84.87 | | 0 | 0.00 |
| Folate, total | mcg | 83.65 | 30.20 | | | 0 | 434.01 |
| Folate, DFE[7] | mcg | 83.65 | 30.20 | | | 0 | 434.01 |
| Choline | mg | 0 | 54.35 | | | | 86.01 |
| Carotene, beta | mcg | 47 520.91 | 47 528.94 | | | | 4 871.57 |
| Carotene, alpha | mcg | 178.71 | 0.00 | | | | |
| Vitamin A, IU[8] | IU | 79 353.61 | 79 214.90 | 8 051.66 | | 0 | 8 113.37 |
| Vitamin A, RAE[9] | mcg | 3 965.78 | 3 960.74 | | | 0 | 407.44 |
| Vitamin E | mcg | 2.93 | 4.73 | | 4.46 | 1.3 | 8.50 |
| Vitamin K | mcg | 9.51 | 10.57 | | 0.00 | 0.0 | 961.91 |

*Notes:* 1. Dry weight values were calculated using respective reported values for water in Table 5.4. 2. USDA *National Nutrient Database for Standard Reference*, Release #22 (2009), NBD No. 11517, Refuse 0%. 3. USDA *National Nutrient Database for Standard Reference*, Release #22 (2009), NBD No. 11510, Refuse 0%. 4. Medical Research Council, South Africa (1998). 5. Ministry of Education, Culture, Sports, Science, and Technology, The Council for Science and Technology, Subdivision on Resources (2005). 6. USDA *National Nutrient Database for Standard Reference*, Release #22 (2009), NBD No. 11505, Refuse 6% (tough stems and bruised leaves). 7. DFE: dietary folate equivalents. 8. IU: International units. 9. RAE: retinol activity equivalents.

Table 5.9. **Fatty acid composition of sweet potato (per 100 g dry weight)**[1]

| Fatty acid | Unit | Storage roots | | | | | | Leaves | |
|---|---|---|---|---|---|---|---|---|---|
| | | With skins | Without skins | | | | | | |
| | | Raw[2] | Frozen, raw[3] | Raw[4] | Range of mean values | Boiled[5] | Frozen/baked[6] | Raw[7] | Cooked[8] |
| 16:0 | g | 0.079 | 0.139 | 0.068 | 0.069-0.139 | 0.031 | 0.084 | 0.049 | 0.523 |
| 18:0 | g | 0.004 | 0.016 | 0.012 | 0.012-0.0162 | 0.0 | 0.011 | 0.050 | 0.053 |
| 18:1 undifferentiated | g | 0.004 | 0.028 | 0.006 | 0.006-0.028 | 0.0 | 0.019 | 0.100 | 0.106 |
| 18:2 undifferentiated | g | 0.057 | 0.267 | 0.150 | 0.150-0.267 | 0.061 | 0.160 | .0.939 | 1.001 |
| 18:3 undifferentiated | g | 0.004 | 0.052 | 0.021 | 0.021-0.052 | 0.0 | 0.027 | 0.174 | 0.186 |

*Notes:* 1. Dry weight values were calculated using reported values for water from Tables 5.3 and 5.4 for the respective references. 2. USDA *National Nutrient Database for Standard Reference*, Release #22 (2009), NBD No. 11507, Refuse 28% (non-edible parings and trimmings). 3. USDA *National Nutrient Database for Standard Reference*, Release #22 (2009), NBD No. 11516, Refuse 0%. 4. Ministry of Education, Culture, Sports, Science, and Technology, The Council for Science and Technology, Subdivision on Resources (2005). 5. USDA *National Nutrient Database for Standard Reference*, Release #22 (2009), NBD No. 11510, Refuse 0%. 6. USDA *National Nutrient Database for Standard Reference, Release* #22 (2009), NBD No. 11517, Refuse 0%. 7. USDA *National Nutrient Database for Standard Reference*, Release #22 (2009), NBD No. 11505, Refuse 6% (tough stems and bruised leaves). 8. USDA *National Nutrient Database for Standard Reference*, Release # 22 (2009), NBD No. 11506, Refuse 0%.

Table 5.10. **Amino acid composition of sweet potato (per 100 g dry weight)**[1]

| Amino acid | Unit | Storage roots | | | | Leaves | |
|---|---|---|---|---|---|---|---|
| | | With skins | Without skins | | | | |
| | | Raw[2] | Raw, frozen[3] | Boiled | Baked/frozen[5] | Raw[6] | Cooked[7] |
| Tryptophan | g | 0.136 | 0.084 | 0.141 | 0.08 | 0.291 | 0.177 |
| Threonine | g | 0.365 | 0.339 | 0.367 | 0.323 | | |
| Isoleucine | g | 0.242 | 0.339 | 0.242 | 0.327 | | |
| Leucine | g | 0.405 | 0.498 | 0.408 | 0.479 | | |
| Lysine | g | 0.290 | 0.335 | 0.292 | 0.319 | 1.894 | 1.169 |
| Methionine | g | 0.128 | 0.167 | 0.126 | 0.160 | 0.714 | 0.443 |
| Cystine | g | 0.097 | 0.056 | 0.096 | 0.053 | 0.390 | 0.239 |
| Phenylalanine | g | 0.392 | 0.406 | 0.393 | 0.393 | 0 | 0 |
| Tyrosine | g | 0.150 | 0.279 | 0.151 | 0.266 | 0 | 0 |
| Valine | g | 0.379 | 0.446 | 0.377 | 0.426 | 0 | 0 |
| Arginine | g | 0.242 | 0.315 | 0.242 | 0.304 | 0 | 0 |
| Histidine | g | 0.136 | 0.127 | 0.136 | 0.122 | 0 | 0 |
| Alanine | g | 0.339 | 0.370 | 0.337 | 0.357 | 0 | 0 |
| Aspartic acid | g | 1.681 | 1.163 | 1.686 | 1.114 | 0 | 0 |
| Glutamic acid | g | 0.682 | 0.665 | 0.679 | 0.639 | 0 | 0 |
| Glycine | g | 0.277 | 0.307 | 0.277 | 0.297 | 0 | 0 |
| Proline | g | 0.229 | 0.299 | 0.232 | 0.285 | 0 | 0 |
| Serine | g | 0.387 | 0.350 | 0.388 | 0.338 | 0 | 0 |

*Notes:* 1. Dry weight values were calculated using reported values for water from Tables 5.3 and 5.4 for the respective references. 2. USDA *National Nutrient Database for Standard Reference*, Release #22 (2009), NBD No. 11507, Refuse 28% (non-edible parings and trimmings). 3. USDA *National Nutrient Database for Standard Reference*, Release #22 (2009), NBD No. 11516, Refuse 0%. 4. USDA *National Nutrient Database for Standard Reference*, Release #22 (2009), NBD No. 11510, Refuse 0%. 5. USDA *National Nutrient Database for Standard Reference*, Release #22 (2009), NBD No. 11517, Refuse 0%. 6. USDA *National Nutrient Database for Standard Reference*, Release #22 (2009, NBD No. 11505, Refuse 6% (tough stems and bruised leaves). 7. USDA *National Nutrient Database for Standard Reference*, Release #22 (2009), NBD No. 11506, Refuse 0%.

## Other constituents

### *Anti-nutrients*

#### *Oxalate*

Oxalate is found in uncooked sweet potato leaves at levels of 73 mg/100 g (Ravindran et al., 1995) to 89 mg/100 g (Lebot, 2009) and levels up to 308 mg/100 g are found in dry matter (Antia et al., 2006). A high intake of oxalate reduces the calcium availability, as indicated by the intermediate calcium absorption index of sweet potato roots (0.423 ± 0.0255; Weaver et al., 1997). Proper boiling of sweet potato leaves before consumption significantly reduces the total oxalate content (Antia et al., 2006) since more than 60% of oxalates are present in water soluble form which leaches out into the water (Holloway et al., 1989). Oxalates, both free and as calcium oxalate, are present in the roots and the total levels are generally similar to those in other root crops (Lebot, 2009; Woolfe, 1992).

### *Trypsin inhibitors*

The first non-leguminous plant reported to contain a trypsin inhibitor was sweet potato (Sohonie and Bhandarkar, 1954). Strong inhibition of trypsin has been demonstrated *in vitro* and this could indicate interference with protein digestion *in vivo*, thus having nutritional implications in humans (Woolfe, 1992).

Since trypsin inhibitors are present in uncooked sweet potato roots, this is especially true for those snacking on raw sweet potato. Sporamins, which are the major storage proteins in sweet potato roots, have an inhibiting effect on protein degradation by trypsin and belong to the Kunitz type of trypsin inhibitors, which are also found in various other crops (Shewry, 2003).

Trypsin inhibitor activity (TIA) varies considerably between different cultivars. It is suggested that genotypes with high protein content and low TIA, as well as appropriate methods of processing, can improve the utilisation of sweet potato for food as well as feed (Zhang and Corke, 2001). Analysis of 8 sweet potato cultivars and 199 breeding lines indicated the average TIA of the cultivars is 197 U/mg DW (range of 65-392 U/mg DW) and the average TIA of the breeding lines is 273 U/mg DW (range of 38-944 U/mg DW; Jun et al., 2005). Some sweet potato lines of low trypsin inhibitor activity have been bred (Toyama et al., 2006). The levels of TIA in sweet potatoes can generally be regarded as low and cooking/microwaving sweet potato tubers at high temperatures (100°C) destroys most TIA (Sasi Kiran and Padmaja, 2003; Ravindran et al., 1995).

### *Polyphenols*

Sweet potato is a source of polyphenols. These include phenolic acids, such as chlorogenic-, caffeic- and dicaffeoylquinic acids (Padda and Picha, 2007), anthocyanins (cyanidin and peonidin which cause the purple colour found in some sweet potato varieties) (Oki et al., 2002) and flavonols such as quercetin and rutin (Guan et al., 2006). Polyphenols are known to chelate metals such as iron and zinc and reduce their absorption. Polyphenols may also inhibit some digestive and cellular enzymes (Halliwell, 2007).

### *Phytic acid*

Phytic acid (myo-inositol 1,2,3,4,5,6-hexakis [dihydrogen phosphate]) is present in sweet potato. Phytic acid is estimated to bind 60-75% of the phosphorus in the form of phytate (National Academies of Sciences, 2005). Dilworth et al. (2005) found 4.98 X $10^3$ mmole/g phytate in uncooked sweet potato roots oven dried at 65°C, and 2.238 X $10^3$ mmole/g in cooked roots.

## *Toxicants*

Sweet potato produces certain metabolites in response to injury and on exposure to infectious agents such as fungi. Some of these metabolites, known as phytoalexins, especially the furano-terpenoids, are known to be toxic (Woolfe, 1992). Fungal contamination of sweet potato roots by *Ceratocystis fimbriata* and several *Fusarium* species, especially *F. solani*, and damage caused by weevils leads sweet potato plants to produce of phytoalexins (Schneider et al., 1984), which can also be further converted biochemically by fungi growing on the diseased potato. Phytoalexins produced upon insect damage are produced by the storage roots (Wilson et al., 1971), and small amounts can be found in the leaves and stems upon injury (Clark et al., 1981).

Furano-terpenoids are formed by sweet potato roots in response to stress (e.g. mould infection) and then converted to toxic forms by the moulds growing on the roots (Chassy et al., 2008). The furano-terpenoid, 4-ipomeanol (exert cytochrome-P450-mediated toxicity) is produced specifically upon *F. solani* infection and it has been shown to be a major component of "lung oedema factor" (Wilson et al., 1971; Boyd, 1976).

The differences among different species of husbandry animals and humans in toxic effects – and target organs – of phytoalexins such as 4-ipomeanol appear to be linked with the differential induction of, and metabolism by, cytochrome P450 isozymes (Lakhanpal et al., 2001). However, the levels of furano-terpenoids are decreased by baking or cooking and because of the bitter taste of furano-terpenoids, infected sweet potatoes are usually discarded. Levels of furano-terpenoids in non-diseased sweet potatoes may be negligible (Woolfe, 1992).

### *Allergens*

There have been three reported cases of allergy to sweet potato, with symptoms including generalised urticaria, hypotension, nausea, vomiting and loss of consciousness (Velloso et al., 2004). No other cases have been reported in scientific literature.

### *Other components*

#### *Raffinose*

Raffinose is a sugar that is not digested in the upper digestive tract and that is fermented by colon bacteria to yield flatus gases (e.g. hydrogen and carbon dioxide; Palmer, 1982). This process is not known to occur in low-sugar cultivars. Therefore, the higher the raffinose content and the sweeter the taste, the higher the probability for flatulence to occur (Martin and Deshpande, 1985).

It has been reported that 0.5% of the fresh weight of baked sweet potato consists of raffinose (Palmer, 1982). The level of raffinose depends on the sweet potato cultivar; a study on cultivars from Chinese Taipei indicated that raffinose levels ranged from 0.102% to 1.08% dry weight (Tsou and Yang, 1984).

#### *Lutein*

Sweet potato leaves contain relatively high levels of lutein, a carotenoid. Lutein levels in sweet potato leaves have been found ranging from 0.37 mg/g FW (Ishiguro and Yoshimoto, 2006) to 0.58 mg/g FW (Menelaou et al., 2006).

## Suggested constituents to be analysed related to food use

Sweet potato is an important staple food for large sectors of the world population in the tropics. In many places, sweet potato is a key security food especially during periods when other foods are in short supply (Manrique, 1998). For instance, in Papua New Guinea, sweet potato is the main staple food of the highlands and often supplies 90% of the caloric intake.

Although sweet potato is considered to be a low protein food in regions where food is abundant, it serves as an important source of protein in other countries, e.g. in East Africa. Protein is evaluated in relationship to its biological value, which is markedly influenced by the relative amounts of indispensable (essential) and dispensable (non-essential) amino acids and the form of nitrogen in the diet (WHO, 2007).

WHO (2007) and the National Academies of Sciences (2005) list the following nine amino acids as indispensable, i.e. those having carbon skeletons that cannot be synthesised to meet the body's needs from simpler molecules: histidine, isoleucine, leucine, lysine, methionine, phenylalanine, threonine, tryptophan and valine. Additionally, the National Academies of Sciences (2005) lists six amino acids as "conditionally indispensable", i.e. those amino acids requiring a dietary source when endogenous synthesis cannot meet metabolic needs: arginine, cysteine, glutamine, glysine, proline and tyrosine. However, WHO (2007) indicated that the requirement for indispensable amino acids is not an absolute value, and one must consider the total nitrogen content of the diet, including the dispensable amino acids particularly at lower levels of nitrogen consumption.

Worldwide, sweet potato provides significant amounts of carbohydrates, macro-nutrients, as well as substantial quantities of micro-nutrients (Woolfe, 1992). Also, potassium and calcium are important minerals to consider for both tubers and leaves. Leaves are a fair source of iron. The vitamins beta-carotene and C are also important. Raw storage roots and leaves also contain trypsin inhibitor and raffinose, two anti-nutrients.

Table 5.11 shows suggested nutritional and compositional parameters to be analysed in sweet potato for food use.

Table 5.11. **Suggested nutritional and compositional parameters to be analysed in sweet potato matrices for food use**

| Parameter | Storage roots raw | Leaves raw |
|---|---|---|
| Moisture[1] | X | X |
| Crude protein[1] | X | X |
| Crude fat (ether extractable)[1] | X | X |
| Ash[1] | X | X |
| Carbohydrates[2] | X | X |
| Dietary fibre | X | X |
| Potassium | X | X |
| Calcium | X | X |
| Iron | | X |
| Beta-carotene | X | X |
| Vitamin C | X | X |
| Amino acids | X | X |
| Trypsin inhibitor | X | X |
| Raffinose | X | X |

*Notes:* 1. These components should be measured using a method suitable for the measurement of proximates. 2. Carbohydrates are calculated as follows: 100 – (water + crude protein + total fat + ash) g/100 g fresh weight.

## Suggested constituents to be analysed related to feed use

Sweet potato can be fed to all domestic animals, including ruminants and non-ruminants. For instance, in China, 40% of sweet potato produced is used as animal feed, in Brazil 35% and in Madagascar 30% (Woolfe, 1992). Both the roots and leaves can be used in either a fresh or dried form or as silage and fed to cattle. In countries such as China, Japan and Chinese Taipei, where sweet potato is processed into starch and alcohol, its by-products are also used as animal feed.

The storage roots serve as a source of energy in animal diets. Peeled sweet potato storage roots can replace up to 75% of maize in the diets of layer chickens without influencing their performance (Agwunobi, 1993). It has been shown that dried sweet potato can replace up to 50% of the maize in pig diets (Dominguez, 1992).

Nwokolo (1990) has reviewed the use of sweet potatoes for swine feed and concluded that leaves, stems and roots can be safely fed to swine. Fresh roots contain trypsin inhibitor, which adversely affects swine and thus it is best if sweet potatoes roots are cooked prior to feeding to inactivate the trypsin inhibitor. Cutting and drying fresh roots into chips also improves digestibility and utilisation. It has been found that supplementing swine diets containing sweet potato roots with lysine and sulphur amino acid improves the utilisation of sweet potatoes.

Table 5.12 shows suggested nutritional and compositional parameters to be analysed in sweet potato for feed use. The constituents of key importance are crude protein, crude fat (ether extractable), ash, carbohydrates, dietary fibre, calcium and phosphorus.[2] Although there are 20 primary amino acids that occur in proteins, only 10 or 11 are recognised as essential, i.e. a need has been shown to be supplied by the diet (National Academies of Science, 2005). According to the National Academies of Sciences (2005), the essential amino acids for swine include arginine, histidine, isoleucine, leucine, lysine, methionine, phenylalanine, tryptophan, valine and threonine. There is also a requirement for cystine and tyrosine, but these amino acids can be synthesised from methionine and phenylalanine, respectively. Amino acid content is also important, especially in swine and poultry diets. The National Academies of Sciences (2005) lists the same amino acids as essential for poultry, with the addition of lysine.

In cattle and sheep, where microbial protein from the rumen has been considered the primary protein source for the animal, there is increased interest in proteins that escape rumen fermentation, particularly in high producing dairy cattle. Thus, nutritionists are taking a closer look at the potential for cattle to also have certain limiting amino acids. Methionine, lysine, phenylalanine and threonine have been suggested as being limiting amino acids for cattle.

Calcium and phosphorus are major minerals in animal feed and should be measured. For swine and poultry, trypsin inhibitor and phytic acid are also important. Oxalate may also be a compound of interest as it also binds minerals making them unavailable for digestion by the animal (Almazan, 1995).

Table 5.12. **Suggested nutritional and compositional parameters to be analysed in sweet potato matrices for feed use**

| Parameter | Storage root raw | Leaves raw |
|---|---|---|
| Moisture[1] | X | X |
| Crude protein[1] | X | X |
| Crude fat (ether extractable)[1] | X | X |
| Ash[1] | X | X |
| Carbohydrates[2] | X | X |
| Dietary fibre | X | X |
| Calcium | X | X |
| Phosphorus | X | X |
| Amino acids | X | X |
| Oxalate | | X |
| Trypsin inhibitor | X | |

*Notes:* 1. These components should be measured using a method suitable for the measurement of proximates.
2. Carbohydrates are calculated as follows: 100 – (water + crude protein + total fat + ash) g/100 g fresh weight.

# Notes

1. For additional discussion of appropriate comparators, see Codex Alimentarius Commission (2003: paragraphs 44 and 45).
2. Analysis of acid detergent fibre (ADF) and neutral detergent fibre (NDF) in sweet potato may be relevant to ruminant nutrition.

# *References*


Agwunobi, L.N. (1993), "Sweet potato [*Ipomoea batatas* (L.) Lam.] meal as a partial substitute for maize in layers ration", *Tropical Agriculture*, Vol. 70, No. 3, pp. 291-293.

Almazan, A.M. (1995), "Antinutritional factors in sweetpotato greens", *Journal of Food Composition and Analysis*, Vol. 8, Issue 4, pp. 363-368.

Antia, B.S. et al. (2006), "Nutritive and anti-nutritive evaluation of sweet potato (*Ipomoea batatas*) leaves", *Pakistan Journal of Nutrition*, Vol. 5, No. 2, pp. 166-168, Asian Network for Scientific Information, available at: www.pjbs.org/pjnonline/fin379.pdf.

Austin, D.F. (1977), "Hybrid polyploids in *Ipomoea* section *batatas*", *Journal of Heredity*, Vol. 68, No. 4, pp. 259-260.

Bovell-Benjamin, A. (2007), "Sweet potato: A review of its past, present, and future role in human nutrition", *Advances in Food and Nutrition Research*, Vol. 52, pp. 1-59.

Boyd, M.R. (1976), "Role of metabolic inactivation in the pathogenesis of chemically induced pulmonary disease: Mechanisms of action of the lung-toxic furan, 4-ipomeanol", *Environmental Health Perspectives*, Vol. 16, pp. 127-138.

Cerning-Beroard, J. and J. Le Dividich (1976), "Feeding value of some starchy tropical products *in -vitro* and *in-vivo* study of sweet-potato (*Ipomoea batatas*), yam (*Dioscorea alata*), malanga (*Xanthosoma sagitifolium*), breadfruit (*Artocarpus communis*) and banana (*Musa* Sp.)", *Annales de Zootechnie (Paris)*, Vol. 25, pp. 155-168.

Chassy, B. et al. (2008), "Nutritionally improved sweetpotato", Chapter 4 in: *Nutritional and Safety Assessments of Foods and Feeds Nutritionally Improved through Biotechnology: Case Studies*, prepared by a Task Force of the ILSI International Food Biotechnology Committee, Comprehensive Reviews in Food Science and Food Safety, Vol. 7, International Life Sciences Institute, available at: www.ilsi.org/foodbiotech/publications/10_ilsi2008_casestudies_crfsfs.pdf.

CIP (Centro Internacional de la Papa – International Potato Center) (2009), "Sweet potato", CIP Lima, Peru, www.cipotato.org/sweetpotato (last accessed July 2009).

CIP (2004), *The VITAA Partnership, A Food-Based Approach to Vitamin A Deficiency in Sub-Saharan Africa*, International Potato Center.

Clark, C.A. et al. (1981), "Accumulation of furaterpenoids in sweet potato tissue following inoculation with different pathogens", *Phytopathology*, Vol. 71, No. 7, pp. 708-711.

Codex (Codex Alimentarius Commission) (2003; with Annexes II and III adopted in 2008), *Guideline for the Conduct of Food Safety Assessment of Foods Derived from Recombinant DNA Plants*, CAC/GL 45/2003, available at: www.codexalimentarius.net/download/standards/10021/CXG_045e.pdf.

Dilworth, L.L. et al. (2005), "Digestive and absorptive enzymes in rats fed phytic acid extract from sweet potato (*Ipomoea batatas*), *Diabetologia Croatia*, Vol. 34, No. 2, at: www.idb.hr/diabetologia/05no2-3.pdf.

Dominguez, P.L. (1992), "The use of sweet potato (*Ipomoea batatas*) for feeding pigs", *Desarrollo de productos de raices y tuberculos*, CIP, Vol. 2, pp. 223-231.

FAOSTAT (2007, 2008, 2009), FAO Statistics online website, http://faostat.fao.org.

Fuglie K.O. et al. (1999), *Economic Impact of Virus-Free Sweet Potato Seed in Shandong Province, China*, International Potato Center, Lima, Peru, available at: http://pdf.usaid.gov/pdf_docs/PNACK412.pdf.

Guan, Y. et al. (2006), "Determination of pharmacologically active ingredients in sweet potato (*Ipomoea batatas* L.) by capillary electrophoresis with electrochemical detection", *Journal of Agricultural and Food Chemistry*, Vol. 54, No. 1, pp. 24-28.

Hall, M.R. and S.C. Phatak (1993), "Sweet potato *Ipomoea batatas* (L.) Lam.", in: Kalloo, G. and B.D. Bergh (eds.), *Genetic Improvement of Vegetable Crops*, Pergamon Press, Oxford, pp. 693-708.

Halliwell, B. (2007), "Dietary polyphenols: Good, bad, or indifferent for your health?", *Cardiovascular Research*, Vol. 73, No. 2, pp. 341-347.

HarvestPlus (2004), "Breeding crops for better nutrition", *HarvestPlus Brief*, International Food Policy Research Institute, Washington, DC.

Holloway, W.D. et al. (1989), "Organic acids and calcium oxalate in tropical root crops", *Journal of Agricultural and Food Chemistry*, Vol. 37, No. 2, pp. 337-341, http://dx.doi.org/10.1021/jf00086a014.

Ishiguro, K. and M. Yoshimoto (2006), "Lutein content of sweet potato leaves", *Sweet Potato Research Front*, Vol. 20, No. 4.

Janssens, M. (2001), "Sweet potato", in: Raemaekers, R.H. (ed.), *Crop Production in Tropical Africa*, CIP Royal Library Albert I, Brussels, pp. 205-221.

Jones, A. (1965), "Cytological observations and fertility measurements of sweet potato [*Ipomoea batatas* (L.) Lam]", *Proceedings of the American Society of Horticultural Science*, Vol. 86, pp. 527-537.

Jones, A. et al. (1986), "Sweet potato breeding", in: Basset, M.J. (ed.), *Breeding Vegetable Crops*, AVI Publishing Co., Westport, Connecticut, pp. 1-35.

Jun, T. et al. (2005), "Varietal differences in trysin inhibitor activity of sweet potato roots", *Breeding Research*, Vol. 7, No. 1, pp. 17-23, http://dx.doi.org/10.1270/jsbbr.7.17.

Katayama, K. et al. (2006), "New sweet potato cultivar 'quick sweet' having low gelatinization temperature and altered starch structure", *ISHS Acta Horticulturae*, II International Symposium on Sweetpotato and Cassava: Innovative Technologies for Commercialization, Vol. 703, pp. 55-61.

Katayama, K. et al. (2004), "New sweet potato line having low gelatinization temperature and altered starch structure", *Starch-Starke*, Vol. 54 No.2, pp.51-57, http://dx.doi.org/10.1002/1521-379X(200202)54:23.0.CO;2-6.

Kays, S.J. (2005), "Sweet potato production worldwide: Assessment, trends and the future", *ISHS Acta Horticulturae*, International Symposium on Root and Tuber Crops: Food Down Under, Vol. 670, pp. 19-25.

Lakhanpal, S. et al. (2001), "Phase II study of 4-Ipomeanol, a naturally occurring alkylating furan, in patients with advanced hepatocellular carcinoma", *Investigational New Drugs,* Vol. 19, No. 1, pp. 69-76.

Lebot, V. (2009), *Tropical Root and Tuber Crops: Cassava, Sweet Potato, Yams and Aroids (Crop Production Science in Horticulture)*, Oxford University Press, United Kingdom.

Manrique, L.A. (1998), *Sweet Potato: Production Principles and Practices*, Manrique International Agrotech, Honolulu, Hawaii.

Martin, F.W. and S.N. Deshpande (1985), "Sugars and starches in a non-sweet sweet potato compared to conventional varieties", *Journal of Agriculture of the University of Puerto Rico*, Vol. 69, pp. 99-106.

Martin, F.W. and A. Jones (1986), "Breeding sweet potatoes", *Plant Breeding Reviews*, Vol. 4, pp. 313-345.

Medical Research Council, South Africa (1998), Kruger M., N. Sayed, M. Langenhoven and F. Holing, "Composition of South African foods". *Vegetables and Fruit, Supplement to the MRC Food Composition Tables*, Tygerberg.

Menelaou, E. et al. (2006), "Lutein content in sweet potato leaves", *Horticultural Science*, Vol. 41, No. 5, pp. 1 269-1 271.

Ministry of Education, Culture, Sports, Science, and Technology, The Council for Science and Technology, Subdivision on Resources (2005), *Standard Tables of Food Composition in Japan,* Japan.

National Academy of Sciences (2005), *Dietary Reference Intakes for Energy, Carbohydrate, Fiber, Fat, Fatty Acids, Cholesterol, Protein, and Amino Acids (Macronutrients)*, Food and Nutrition Board, Institute of Medicine of the National Academies, National Academy of Sciences, Washington, DC, available at: www.nap.edu/catalog.php?record_id=10490.

Nestel, P. et al. (2006), "Biofortification of staple food crops – Symposium: Food fortification in developing countries", *Journal of Nutrition,* Vol. 136, No. 4, pp. 1 064-1 067, available at: http://jn.nutrition.org/content/136/4/1064.full.pdf+html.

Nwokolo, E. (1990), "Sweet potato", in: Thacker, P.A. and R.N. Kirkwood (eds.), *Non-traditional Feed Sources for Use in Swine Production*, Butterworths, London, pp. 481-491.

O'Hair, S.K. (1984), "Farinaceous crops", in: Martin F.W. (ed.), *CRC Handbook of Tropical Food Crops*, CRC Press, Boca Raton, Florida, pp. 109-136.

Okada, Y. et al. (2001), "Virus resistance in transgenic sweet potato [*Ipomoea batatas* L. (Lam)] expressing the coat protein gene of sweet potato feathery mottle virus", *Theoretical and Applied Genetics*, Vol. 103, Issue 5, pp. 743-751.

Oki, T. et al. (2002), "Involvement of anthocyanins and other phenolic compounds in radical-scavenging activity of purple-fleshed sweet potato cultivars", *Journal of Food Science*, Vol. 67, No. 5, pp. 1 752-1 756, http://dx.doi.org/10.1111/j.1365-2621.2002.tb08718.x.

Padda, M.S. and D.H. Picha (2007), "Methodology optimization for quantification of total phenolics and individual phenolic acids in sweet potato (*Ipomoea batatas* L.) roots", *Journal of Food Science*, Vol. 72, No. 7, pp. 412-416.

Palmer, J.K. (1982), "Carbohydrates in sweet potato", in: Villareal, R.L. and T.D. Griggs (eds.), *Sweet Potato*, Proceedings of the First International Symposium AVRDC, Shanhua, T'ainan, pp. 137-137.

Ravindran, V. et al. (1995), "Biochemical and nutritional assessment of tubers from 16 cultivars of sweet potato (*Ipomoea batatas* L)", *Journal of Agricultural and Food Chemistry*, Vol. 43, pp. 2 646-2 651.

Reed, C.F. (1976), *Information Summaries on 1000 Economic Plants*, Typescripts submitted to the USDA.

Sasi Kiran, K. and G. Padmaja (2003), "Inactivation of trypsin inhibitors in sweet potato and taro tubers during processing", *Plant Foods for Human Nutrition*, Vol. 58, No. 2, pp. 153-163.

Satapathy, M.R. et al. (2006), "Nutritional evaluation of dried sweet potato haulms in goats", *Journal of Interacademia*, Vol. 10, pp. 254-256.

Schneider, J.A. et al. (1984), "The fate of the phytoalexins ipomoearone: Furaterpenes and butenolides from *Ceratocystis fimbriata* infected sweet potatoes", *Phytochemistry*, Vol. 23, No. 4, pp. 759-764.

Shewry, P.R. (2003), "Tuber storage proteins", *Annals of Botany*, Vol. 91, pp. 755-769, http://dx.doi.org/10.1093/aob/mcg084.

Shih, F.F. et al. (2006), "Physicochemical properties of gluten-free pancakes from rice and sweet potato flours", *Journal of Food Quality*, Vol. 29, No. 1, pp. 97-107.

Singh, S. et al. (2004), "Sweet potato-based pasta product: Optimization of ingredient levels using response surface methodology", *International Journal of Food Science and Technology*, Vol. 39, No. 2, pp. 191-200, http://dx.doi.org/10.1046/j.0950-5423.2003.00764.x.

Sohonie, K. and A.P. Bhandarkar (1954), "Trypsin inhibitors in Indian food-stuffs: Part 1 – Inhibitors in vegetables", *Journal of Science on Indian Research*, Vol. 13B, pp. 500-593.

Suda, I. et al. (1999), "Sweet potato potentiality: Prevention for life style-related disease induced by recent food habits in Japan", *Food and Food Ingredients Journal of Japan*, Vol. 181, pp. 59-68.

Takahata, Y. et al. (1993), "Varietal differences in chemical composition of the sweet potato storage root", *ISHS Acta Horticulturae*, Physiological Basis of Postharvest Technologies, Vol. 343, pp. 77-80.

Teow, C.C. et al. (2007), "Antioxidant activities, phenolic and β-carotene contents of sweet potato genotypes with varying flesh colours", *Food Chemistry*, Vol. 103, Issue 3, pp. 829-838.

Toyama, J. et al. (2006), "Selection of sweet potato lines with high protein content and/or low trypsin inhibitor activity", *Breeding Science*, Vol. 56, No. 1, pp. 17-23.

Tsou, S.C.S. and M.H. Yang (1984), "Flatulence factors in sweet potato", *ISHS Acta Horticulturae*, Symposium on Quality Vegetables, Vol. 163, pp. 179-186.

USDA Agricultural Research Service (2009), *USDA National Nutrient Database for Standard Reference*, Release 22, Nutrient Data Laboratory, available at: http://ndb.nal.usda.gov.

USDA Agricultural Research Service (2008), "Sweet potato out-yields corn in ethanol production study", United States Department of Agriculture, ARS, available at: www.ars.usda.gov/is/pr/2008/080820.htm.

Velloso, A. et al. (2004), "Anaphylaxis caused by *Ipomoea batatas*", *Journal of Allergy and Clinical Immunology*, Vol. 113, No. 2, Supplement, p.S242, http://dx.doi.org/10.1016/j.jaci.2004.01.331.

Weaver, C.M. et al. (1997), "Calcium bioavailability from high oxalate vegetables: Chinese vegetables, sweet potatoes and rhubarb", *Journal of Food Science*, Vol. 62, No. 3, pp. 524-525, http://dx.doi.org/10.1111/j.1365-2621.1997.tb04421.x.

WHO (2007), "Protein and amino acid requirements in human nutrition", Report of a Joint WHO/FAO/UNU Expert Consultation, *WHO Technical Report Series No. 935*, World Health Organization, Geneva.

Wilson, B.J. et al. (1971), "A lung oedema factor from mouldy sweet potatoes (*Ipomoea batatas*)", *Nature*, Vol. 231, pp. 52-53, http://dx.doi.org/10.1038/231052a0.

Woolfe, J.A. (1992), "Sweet potato – Past and present", in: *Sweet Potato: An Untapped Food Resource*, Cambridge University Press, Cambridge, Great Britain.

Zhang, Z. and H. Corke (2001), "Trypsin inhibitor activity in vegetative tissue of sweet potato plants and its response to heat treatment", *Journal of the Science of Food and Agriculture*, Vol. 81, Issue 14, pp. 1 358-1 363, http://dx.doi.org/10.1002/jsfa.945.

# *Chapter 6*

# Papaya (*Carica papaya*)

*This chapter, prepared by the OECD Task Force for the Safety of Novel Foods and Feeds with Thailand as lead country and the United States as co-lead, deals with the composition of papaya (*Carica papaya*). Background is given on papaya production, processing and uses for human and animal consumption, followed by appropriate varietal comparators and characteristics screened by breeders. Nutrients in papaya fruit, chemical composition of processing by-products, as well as other constituents (anti-nutrients, toxicants and allergens), are then detailed. The final sections suggest key products and constituents for analysis of new varieties for food use and for feed use.*

## Background

Papaya (*Carica papaya* L.) belongs to the family *Caricaceae* and is the only species in the genus *Carica*. It is an herbaceous perennial plant with a green or purple hollow stem which is usually single, erect and bears a crown of palmately lobed leaves. The leaves are clustered, 40-60 cm in width, normally with 7-9 lobes. The petioles are long, hollow and pale green or purple tinged in colour (Campostrini and Yamanishi, 2001).

Papaya is a polygamous species. It has male, female and hermaphrodite plants that produce staminate, pistillate and perfect flowers under different seasonal or environmental conditions. The fruit of papaya is a fleshy berry, variable in weight from 200 g up to 9 kg (Yon, 1994). The fruit shape is a sex-linked character. The fruits from female flowers are spherical to ovoid in shape while the fruit from hermaphrodite flowers are long, cylindrical or pyriform. The skin of unripe fruit is smooth and green. When ripe, the skin turns yellow or orange. The flesh of ripe fruit is yellow, orange or red in colour. Papaya seeds are in the ovarian cavity, which is larger in female fruit than in hermaphrodite ones. The seeds are small and dark brown or black with translucent sarcotesta mucilaginous (Yon, 1994; Paull et al., 2008).

### *Production of papaya*

#### *World production*

The world production of papaya (*Carica papaya* L.) in 2008 was estimated to be approximately 9.1 million tonnes (FAOSTAT, 2008). The countries with the largest papaya production in 2008 were India and Brazil (about 2.7 and 1.9 million tonnes respectively) followed by Nigeria, Indonesia and Mexico (Table 6.1). Within the "top-15" papaya-producing countries of that year, Indonesia showed the highest yield, rising at 72.7 tonnes/hectare (t/ha) on average, and Nigeria the greatest area harvested with 92 500 ha. Brazil, Colombia, Guatemala and the Philippines showed significant increases in papaya production and yield between 2004 and 2008.

Table 6.1. **World production of papaya**

| Country | **Production** (tonnes) | | **Area harvested** (hectares) | | **Yield** (t/ha) | |
|---|---|---|---|---|---|---|
| | 2004 | 2008 | 2004 | 2008 | 2004 | 2008 |
| India | 2 535 100 | 2 685 900[2] | 73 800 | 80 300[2] | 34.4[3] | 33.4[3] |
| Brazil | 1 612 348 | 1 900 000[2] | 34 445 | 36 750[2] | 46.8[3] | 51.7[3] |
| Nigeria | 755 000[2] | 765 000[2] | 91 000[2] | 92 500[2] | 8.3[3] | 8.3[3] |
| Indonesia | 732 611 | 653 276 | 9 134 | 8 982 | 80.2[3] | 72.7[3] |
| Mexico | 787 663 | 638 237 | 20 610 | 16 084 | 38.2[3] | 39.7[3] |
| Ethiopia | 260 000[2] | 260 000[2] | 12 500[2] | 12 500[2] | 20.8[3] | 20.8[3] |
| Democratic Republic of the Congo | 214 070 | 223 770[2] | 12 712 | 13 500[2] | 16.8[3] | 16.7[3] |
| Colombia | 103 870 | 207 698 | 4 464 | 5 498 | 23.3[3] | 37.8[3] |
| Guatemala | 84 000[2] | 184 530[2] | 2 100[2] | 3 500[2] | 40.0[3] | 52.7[3] |
| Philippines | 133 876 | 182 907 | 8 969 | 9 175 | 14.9[3] | 19.9[3] |
| Peru | 193 923 | 157 771[2] | 13 449 | 11 043[2] | 14.4[3] | 14.3[3] |
| Venezuela | 131 753 | 132 013[2] | 7 103 | 7 107[2] | 18.5[3] | 18.6[3] |
| China (People's Republic of) | 157 620[1] | 120 359[2] | 5 743[1] | 5 826[2] | 27.4[3] | 20.7[3] |
| Thailand | 125 000[2] | 131 000[2] | 10 500[2] | 11 000[2] | 11.9[3] | 11.9[3] |
| Cuba | 119 000 | 89 400 | 6 088 | 4 006 | 19.5[3] | 15.0[3] |

*Notes:* 1. Unofficial figure. 2. FAO estimate. 3 Calculated data, from FAOSTAT.

Mexico had the greatest papaya export value in 2007 (approximately USD 55 million) followed by Brazil, the United States, the Netherlands and Belize (Table 6.2). Although Belize and Malaysia did not belong to the "top-15" producing countries, they were in 2007 the fifth and sixth leading exporters of papaya in terms of value. The Netherlands was the fourth exporter and second importer in terms of value (Tables 6.2 and 6.3). This can be explained by the fact that the country, with its location and port facilities, imports agricultural commodities including papaya and re-exports them to other countries in the European Union (Carter, 1997).

The United States was a major papaya importing country in 2007 with an import value of approximately USD 73 million followed by the Netherlands, the United Kingdom, Canada and Germany (Table 6.3). Papaya imported into the US market in 2006 came from Mexico (69%), Belize (25%), Brazil (2.8%), Jamaica (1%) and elsewhere (2.2%) (Pollack and Perez, 2008). Between 2004 and 2007, several countries, primarily Canada, Portugal, Spain, France and the United Arab Emirates, increased the quantity of papaya imported.

Table 6.2. **World papaya export**

| Country | **Export value** (USD thousands) | | **Export quantity** (tonnes) | |
|---|---|---|---|---|
| | 2004 | 2007 | 2004 | 2007 |
| Mexico | 72 722 | 55 327 | 96 525 | 101 306 |
| Brazil | 26 563 | 34 404 | 35 930 | 32 267 |
| United States | 15 917 | 17 715 | 9 789 | 9 604 |
| Netherlands | 17 242 | 16 907 | 9 554 | 8 625 |
| Belize | 17 429[1] | 13 101 | 28 751[1] | 33 341 |
| Malaysia | 21 893 | 8 407 | 58 149 | 26 938 |
| Philippines | 4 182 | 6 374[1] | 3 324 | 4 880[1] |
| France | 2 802 | 3 766 | 1 307 | 1 830 |
| Côte d'Ivoire | 671 | 3 203 | 1 048 | 5 296 |
| Spain | 1 269 | 2 749 | 1 464 | 1 637 |
| Jamaica | 2 124 | 2 748 | 1 229 | 1 340 |
| India | 1 119 | 2 721 | 3 475 | 10 880 |
| Costa Rica | 482 | 2 525 | 579 | 2 972 |
| Ecuador | 2 057 | 2 383 | 7 196 | 5 486 |
| China (People's Republic of) | 817 | 2 277 | 4 455 | 10 067 |
| Dominican Republic | 741[1] | 2 108[1] | 1 515[1] | 5 200[1] |
| Guatemala | 372 | 1 372 | 1 069 | 6 680 |
| Fiji | 644 | 1 254 | 303 | 470 |
| Germany | 1 881 | 1 029 | 1 084 | 442 |
| Belgium | 3 004 | 800 | 980 | 527 |

*Note:* 1. Estimated data using trading partners database.

*Source:* FAOSTAT.

## *Domestic and foreign markets*

Small papaya farmers and commercial farmers in many countries grow papaya for both local and foreign markets. The local markets prefer medium- and large-fruited varieties that have yellow and red flesh. Exported papaya fruit are usually small or of medium size (Codex Alimentarius Commission, 2005; Stice et al., 2010), with yellow or red flesh (Picha, 2006; Pesante, 2003).

Both hermaphrodite fruits (pear-shaped) and female fruits (round) are accepted by consumers in some countries, but the fruits have to be fresh, free from bruises and blemishes, and uniform in size and ripeness.

The latest Codex Alimentarius standard for papaya, amended in 2005, included standards regarding quality, size, uniformity, packaging, labelling, contaminants and hygiene (Codex Alimentarius Commission, 2005).

Table 6.3. **World papaya import**

| Country | **Import value** (USD thousands) | | **Import quantity** (tonnes) | |
|---|---|---|---|---|
| | 2004 | 2007 | 2004 | 2007 |
| United States | 95 844 | 73 125 | 126 024 | 138 115 |
| Netherlands | 19 305 | 19 208 | 15 432 | 12 569 |
| United Kingdom | 18 422 | 18 231 | 11 108 | 8 588 |
| Canada | 11 965 | 17 987 | 10 324 | 14 487 |
| Germany | 16 433 | 16 873 | 10 581 | 8 155 |
| Portugal | 8 909 | 12 932 | 5 682 | 5 992 |
| Spain | 5 849 | 11 695 | 3 541 | 6 686 |
| Japan | 12 547 | 9 497 | 4 763 | 3 996 |
| France | 4 906 | 8 533 | 2 048 | 3 414 |
| Hong Kong, China | 11 953 | 5 075 | 25 972 | 9 800 |
| Italy | 3 343 | 4 097 | 1 630 | 2 008 |
| Singapore | 4 224 | 4 040 | 24 606 | 19 086 |
| Switzerland | 3 118 | 3 628 | 1 345 | 1 339 |
| United Arab Emirates | 1 371[1] | 1 847[1] | 3 152[1] | 6 315[1] |
| New Zealand | 734 | 1 620 | 393 | 874 |
| Sweden | 1 350 | 1 460 | 603 | 580 |
| Belgium | 2 247 | 1 392 | 1 302 | 847 |
| Austria | 833 | 1 190 | 466 | 406 |
| China (People's Republic of) | 3 582 | 1 126 | 4 734 | 1 411 |
| Norway | 292 | 1 016 | 95 | 293 |

*Note:* 1. Estimated data using trading partners database.

*Source:* FAOSTAT.

## *Papaya for human and animal consumption*

Papaya fruit is consumed at both the unripe and ripe stages. Unripe fruits are cooked and utilised as vegetables, processed products and as a source of papain. Ripe papaya is consumed as a fresh fruit and is also used for processing (Figure 6.1).

Figure 6.1. **Papaya processing**

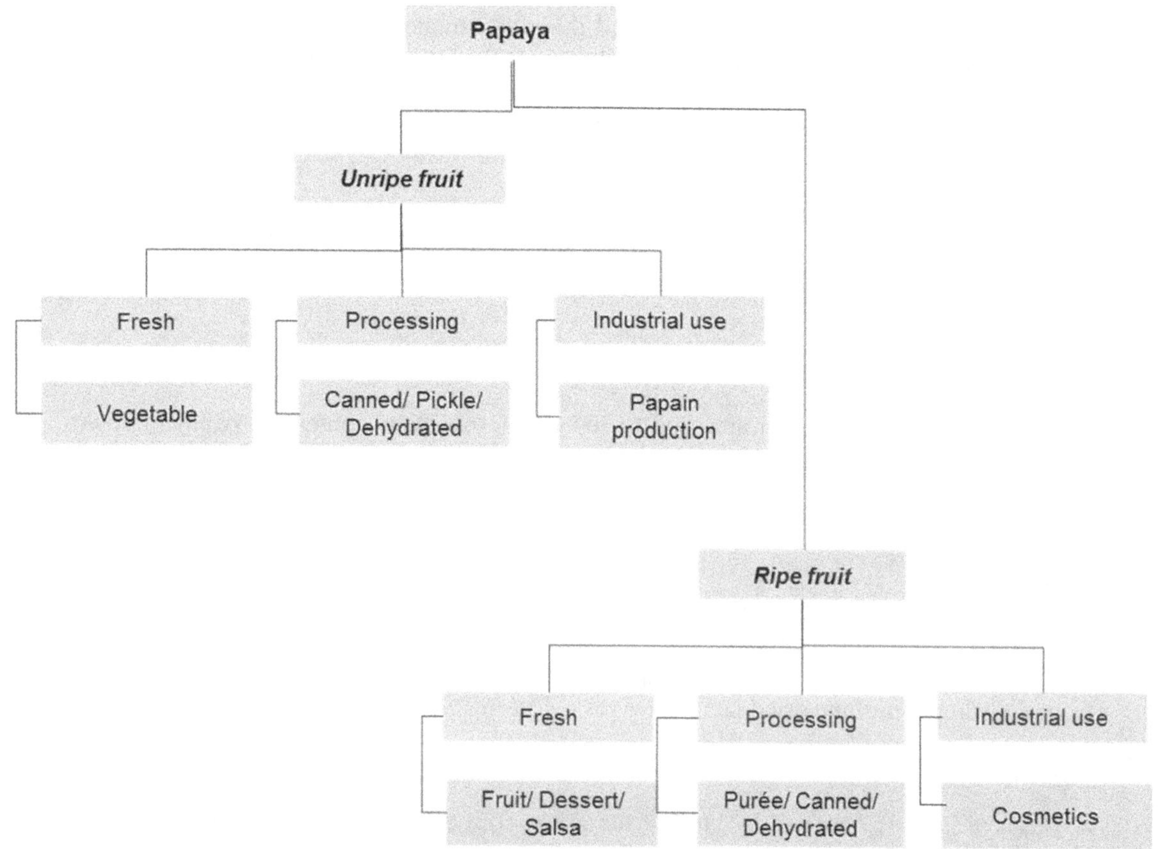


### *Unripe fruit or green fruit*

At the unripe stage, the fruit is consumed as a cooked vegetable in some Asian countries where papaya is widely grown (Mendoza, 2007; Mano et al., 2009). In Thailand, unripe fruits are used as ingredients in papaya salad and cooked dishes (Sone et al., 1998). In Puerto Rico, unripe fruits are canned in sugar syrup and sold either in local markets or exported (Morton, 1987). The preserved unripe papaya fruit, which contains high sugar content, is used as an additive in ice cream. Green or unripe papaya must be cooked (often boiled) prior to consumption to denature the papain in the latex (Odu et al., 2006; Morton, 1987).

### *Ripe fruit*

Ripe papaya fruit is consumed in many different ways. The most common way is to eat it like a melon. It can be peeled, the seeds removed, cut into pieces and served as a fresh fruit. It can also be cut into wedges and then served with lime or lemon. Ripe papaya is also used in jam, jelly, marmalade and other products containing added sugar. Other processed products include purée, nectar (a non-fermented beverage produced from fruit juice, sugar and water; Matsuura et al., 2004), juice, frozen slices or chunks, mixed beverages, papaya powder, baby food, concentrated and candied items (Mugula et al., 1994; OECD, 2005; Office of the Gene Technology Regulator, 2008).

### *Purée*

Papaya purée is prepared from fully ripe peeled fruit with the seeds removed. Papaya flesh is pulped, passed through a sieve and thermally treated (Figure 6.2). Papaya purée is an important intermediate product in the manufacture of several products such as beverages, ice cream, jam and jelly (Brekke et al., 1972; Ahmed et al., 2002).

### *Nectar and beverages*

Papaya nectar is prepared from papaya puree and consumed either alone or with other fruit juices such as passion fruit juice and pineapple juice (Brekke et al., 1972). Canned papaya beverages should be stored at 24°C or below to maintain acceptable quality (Brekke et al., 1976).

### *Dehydrated products*

Drying and freeze drying are used to reduce the moisture content of papaya chunks and slices. Powdered or dried papaya can be used as a flavouring agent, meat tenderiser or as an ingredient in soup mixes (Singfield, 1998).

### *Seeds and leaves*

Papaya seeds are sometimes used to adulterate whole black pepper (Morton, 1987). Papaya leaves contain papain, a strong proteolytic enzyme. Crushed leaves may be used to tenderise meat; however, stomach trouble, purgative effects and abortion may result from consumption of the dried papaya leaves (Morton, 1987).

### *Papain*

Papaya latex is obtained by cutting the green fruit surface with glass, sharp bone or bamboo and collecting the exuding latex in porcelain or earthenware containers over a couple of days. The latex is then sun dried or oven dried, and ground into powder. A proteolytic enzyme, papain is purified from papaya latex and used in the food and feed industries, as well as the pharmaceutical and cosmetic industries (OGTR, 2008). Papain is used in food processing to tenderise meat, clarify beer and juice, produce chewing gum, coagulate milk, prepare cereals and produce pet food (Morton, 1987).

### *Papaya pomace, skins and leaves*

Papaya pomace, skins, leaves and other by-products of papaya processing may find use in animal feed applications (Babu et al., 2003; Fouzder et al., 1999; Munguti et al., 2006; Reyes and Fermin, 2003; Alobo, 2003; Ulloa et al., 2004).

Figure 6.2. **Papaya puree processing**

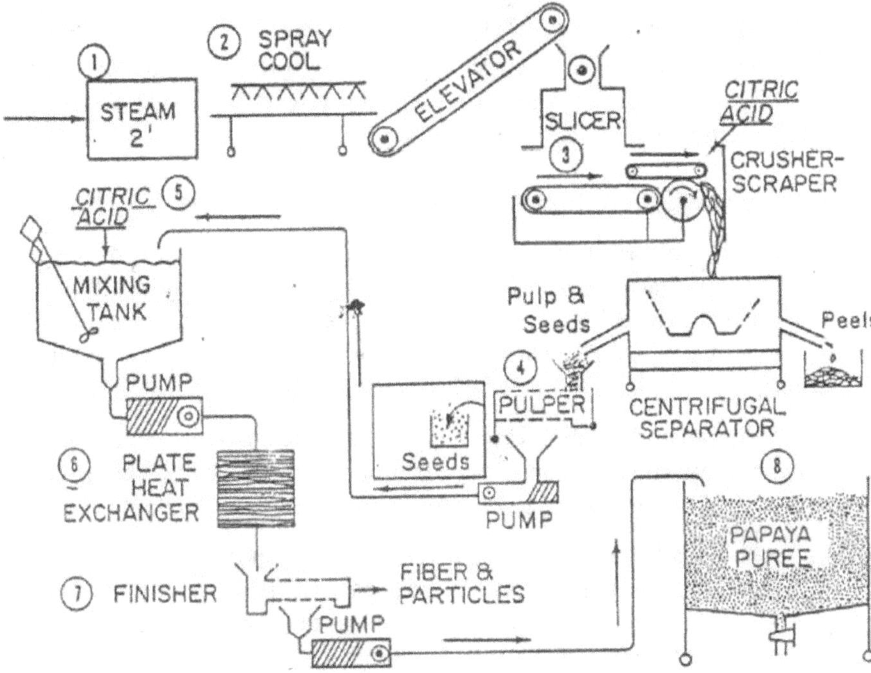


Notes:

1. Steaming whole ripe fruits for two minutes.
2. Spraying the fruits with cold water, slicing and rotating.
3. Separating pulp and seeds without breakage from the peel after acidification with citric acid.
4. Separating seeds the pulp.
5. Adjusted pH of the pulp in mixing tank to 3.4 to 3.5 by citric acid.
6. Heating the acidified pulp at 96°C for 2 minutes.
7. Removing fibre and seed specks from the purée.
8. Transferring papaya puree to containers for freezing.

*Source*: Brekke et al. (1972).

## *Appropriate comparators for testing new varieties*

This chapter suggests parameters that papaya breeders should measure when developing new modified varieties. The data obtained in the analysis of a new papaya variety should ideally be compared to those obtained from an appropriate near isogenic non-modified variety, grown and harvested under the same conditions.[1] The comparison can also be made between values obtained from new varieties and data available in the literature, or chemical analytical data generated from other commercial papaya varieties.

Components to be analysed include key nutrients, toxicants and allergens. Key nutrients are those which have a substantial impact in the overall diet of humans (food) and animals (feed). These may be major constituents (fats, proteins, and structural and non-structural carbohydrates) or minor compounds (vitamins and minerals). Similarly, the levels of known anti-nutrients and allergens should be considered.

Key toxicants are those toxicologically significant compounds known to be inherently present in the species, whose toxic potency and levels may impact human and animal health. Standardised analytical methods and appropriate types of material should be used, adequately adapted to the use of each product and by-product. The key components analysed are used as indicators of whether unintended effects of the genetic modification influencing plant metabolism has occurred or not.

### *Breeding characteristics screened by developers*

Papaya varieties (cultivars) have been developed by selection of desired fruit phenotypes (fruit shape, taste, size, flesh-colour, firmness and uniformity) as well as agronomic characteristics (disease resistance, fruit column compaction, yield) (Martin et al., 2006; Chan, 2007). Due to consumer preference and economic reasons, fruits from hermaphrodite plants are selected for consumption.

Recently, production of papain from papaya has been developed on an industrial scale. Therefore, high papain levels in papaya fruits could be a desired characteristic to be taken into consideration in future papaya breeding programmes (Magdalita et al., 2007).

Molecular techniques have been developed with the potential to aid papaya-breeding programmes. Generally, farmers grow excess papaya plants until flowering time when papaya sex can be determined. The hermaphrodite plants with desired fruit shape are preferable over male and female plants, which are consequently removed from the field. Sex-specific molecular markers have been developed that could potentially reduce the cost of growing and removing the unwanted plants (Parasnis et al., 1999; Deputy et al., 2002). Establishing genetic relationships among papaya varieties is important for the introduction of desired characteristics into papaya breeding programmes. Amplified fragment length polymorphism (AFLP) studies indicate limited genetic diversity among papaya cultivars (Kim et al., 2002). Recently, microsatellite markers that are capable of distinguishing DNA polymorphisms between close cultivars have been developed (Eustice et al., 2008) and the first draft of the papaya genome was published (Ming et al., 2008). Genes associated with fruit development and ripening may aid in the development of new varieties with desirable qualities.

## Nutrients

### *Constituents of papaya fruit*

At an unripe stage, papaya is consumed as a cooked vegetable while at a ripened stage it is consumed as a fruit.

Similar to other vegetables or fruits, the main constituent of papaya is water. The dry matter content increases during fruit development from unripe to ripe stages. Proximate nutrient content, fibre composition and total sugar composition of papaya fruit per 100 g of dry weight of edible portion are shown in Table 6.4.

Other components, including minerals, vitamins, fatty acids and amino acids, are presented in Tables 6.5-6.8.

Table 6.4. **Proximate, fibre and total sugar composition of papaya fruit**

per 100 g dry weight of edible portion

| Nutrient | Unit | Ripe | | | | Unripe | | Range of mean values **(Ripe fruits)** |
|---|---|---|---|---|---|---|---|---|
| | | USDA 2009[1] | Saxholt et al.[3] | Puwastien et al.[4] | Wills et al.[5] | USDA 2008[2] | Puwastien et al.[4] | |
| | | **Mean value, g per 100 g fresh weight** | | | | | | |
| Water | g | 88.83 | 86.5 | 89.1 | 89.3 | 92.16 | 92.6 | 86.5-89.3 |
| | | **Mean value, g per 100 g dry weight** | | | | | | |
| Protein | g | 5.46 | 4.4-5.2 | 8.26 | 3.74 | 5.48 | 10.8 | 3.74-8.26 |
| Total lipid (fat) | g | 1.25 | 1.5-2.2 | 0.92 | 0.93 | 1.3 | 1.35 | 0.92-2.2 |
| Ash | g | 5.46 | 3.7 | 4.59 | 2.80 | NR | 6.76 | 2.8-5.46 |
| Carbohydrate by difference | g | 87.8 | 73.3 | 86.2 | 64.5 | 87.5 | 81.1 | 64.5-87.8 |
| Total dietary fibre | g | 16.1 | 14.1-17.0 | 11.9 | 21.5 | 16.6 | 27.0 | 11.9-21.5 |
| Total sugars | g | 52.8 | 53.3 | NR | 64.5 | 52.7 | NR | 52.8-64.5 |

*Notes:* NR: not reported. Mean values reported on a dry weight basis were calculated from a fresh weight basis using the mean moisture level reported for each source. 1. Based on orange-fleshed papaya (possibly including genetically engineered varieties). 2. Based on papaya, green, cooked (possibly including genetically engineered varieties). 3. Refuse: 33% (seed and skin). 4. Percentage of refuse is not determined. 5. Based on orange-fleshed Australia type and refuse: 30% (seed and skin).

*Sources:* USDA Agricultural Research Service (2009, 2008); Saxholt et al. (2008); Puwastien et al. (2000); Wills et al. (1986).

Cultivation conditions vary depending on the climate, growing seasons, site of cultivation and papaya varieties. All these factors can influence the nutrient content of papaya (Hardisson et al., 2001; Chavasit et al., 2002; Wall, 2006; Marelli de Souza et al., 2008; Charoensiri et al., 2009). Differences in the nutrient content among cultivation sites and papaya varieties are shown in Table 6.9.

Stages of maturity affect the nutrient content of papaya fruits. For example, the vitamin C content of papaya increases with ripening (Table 6.6) (Bari et al., 2006; Hernandez et al., 2006). Consequently, when comparing the nutrient content of papaya fruits, it is important to compare fruits harvested and stored under similar conditions.

### *Proximate nutrient content, fibre and total sugars*

The major components of papaya dry matter are carbohydrates. The total dietary fibre content of ripe papaya fruit varies from 11.9-21.5 g/100 g dry matter (Puwastien et al., 2000; Wills et al., 1986; USDA Agricultural Research Service, 2009; Saxholt et al., 2008). The crude protein content ranges from 3.74-8.26 g/100 g dry matter and the total lipid content varies between 0.92 g and 2.2 g/100 g dry matter (Table 6.4).

### *Carbohydrates*

There are two main types of carbohydrates in papaya fruits, the cell wall polysaccharides and soluble sugars.

During an early stage of fruit development, glucose is the main sugar. The sucrose content increases during the ripening process and can reach levels up to 80% of total sugars (Paull, 1993). Among the major soluble sugars in ripe fruits (glucose, fructose and sucrose), sucrose is most prevalent. During fruit ripening, the sucrose content was shown to increase from 13.9 ± 5.0 mg/g fresh weight in green fruit to 29.8 ± 4.0 mg/g fresh weight in ripe fruits (Gomez et al., 2002).

### Minerals

The edible portion of the ripe papaya fruit contains both macrominerals and microminerals. The macrominerals include sodium, potassium, calcium, magnesium and phosphorus. The microminerals include iron, copper, zinc, manganese and selenium (Table 6.5).

Table 6.5. **Mineral content of papaya fruit**

per 100 g dry weight of edible portion

| Nutrient | Unit | Ripe | | | | | Unripe | | Range of mean values (Ripe fruits) |
|---|---|---|---|---|---|---|---|---|---|
| | | USDA 2009[1] | Saxholt et al. | Puwastien et al. | Sanchez Castillo et al. | Wills et al.[3] | USDA 2008[2] | Puwastien et al. | |
| Macrominerals | | | | | | | | | |
| Sodium (Na) | mg | 26.86 | 15.55-54.07 | 128.4 | 35.71 | 65.42 | 38 | 283.8 | 15.55-128.4 |
| Potassium (K) | mg | 2 300 | 1 370-1 622 | 1 238 | 2 309 | 1 308 | 2 066 | 2 743 | 1 238-2 309 |
| Calcium (Ca) | mg | 214.9 | 57.93-285.9 | 229.4 | 190.5 | 261.7 | 216 | 635.1 | 57.93-285.93 |
| Magnesium (Mg) | mg | 89.53 | 111.1-229.6 | NR | 95.24 | 130.8 | 89 | NR | 89.53-229.63 |
| Phosphorus (P) | mg | 44.76 | 63.41-92.59 | 146.8 | 95.24 | NR | 38 | 432.4 | 44.76-146.8 |
| Microminerals | | | | | | | | | |
| Iron (Fe) | mg | 0.90 | 1.93-14.81 | 12.84 | 3.57 | 4.67 | 0.90 | 8.11 | 0.9-14.81 |
| Copper (Cu) | mg | 0.14 | 0.12 | 0.18 | 0.83 | NR | 0.17 | 0.14 | 0.12-0.83 |
| Zinc (Zn) | mg | 0.63 | 0.39-0.62 | 0.92 | 0.60 | 2.80 | 0.64 | 0 | 0.39-2.80 |
| Manganese (Mn) | mg | 0.10 | 0.081 | NR | 0.24 | NR | NR | NR | 0.081-0.24 |
| Selenium (Se) | µg | 5.4 | NR | NR | NR | NR | 5.1 | NR | – |

*Notes:* NR: not reported. Means values reported on a dry weight basis were calculated from a fresh weight basis using the mean moisture level reported for each source, as shown in Table 6.4.

1. Based on orange-fleshed papaya (possibly including genetically engineered varieties). 2. Based on papaya, green, cooked (possibly including genetically engineered varieties). 3. Based on orange-fleshed Australia type and refuse: 30% (seed and skin).

*Sources:* USDA Agricultural Research Service (2009, 2008); Saxholt et al. (2008); Puwastien et al. (2000); Sanchez Castillo et al. (1998); Wills et al. (1986).

### Vitamins and precursors

Papaya is a source of carotenoids, vitamin C and folate (Table 6.6). A serving of 100 g of ripe papaya fruit contributes about 19% of the nutrient reference value (NRV) for folate (Codex Aimentarius Commission, 2006).

Papaya fruit also contains thiamin, riboflavin, niacin, pantothenic acid, vitamin B-6 and vitamin K (Bari et al., 2006; Adetuyi et al., 2008; Saxholt et al., 2008; USDA Agricultural Research Service, 2009).

Table 6.6. **Vitamin content of papaya fruit**

per 100 g dry weight of edible portion

| Nutrient | Unit | Ripe | | | | Unripe | | Range of mean values **(Ripe fruits)** |
|---|---|---|---|---|---|---|---|---|
| | | USDA 2009[1] | Saxholt et al. | Puwastien et al. | Wills et al.[3] | USDA 2008[2] | Puwastien et al. | |
| Vitamin C, total ascorbic acid | mg | 553.3 | 457.8 | 568.8 | 560.75 | 386.5 | 391.9 | 457.8-568.8 |
| Thiamin | mg | 0.242 | 0.200 | 0.275 | 0.28 | 0.191 | 0.54 | 0.200-0.28 |
| Riboflavin | mg | 0.286 | 0.237 | 0.459 | 0.28 | 0.26 | 0.41 | 0.237-0.459 |
| Niacin | mg | 3.03 | 2.504 | 2.75 | 2.80 | 2.717 | 4.05 | 2.504-3.03 |
| Pantothenic acid | mg | 1.95 | 1.615 | NR | NR | NR | NR | 1.615-1.95 |
| Vitamin B6 | mg | 0.17 | 0.141 | NR | NR | 0.153 | NR | 0.141-0.17 |
| Total folate | µg | 340.2 | 385.2-466.7 | NR | NR | 165 | NR | 340.2-466.7 |
| Folate, DFE | µg DFE | 340.2 | NR | NR | NR | 165 | NR | – |
| Vitamin B12 | µg | 0.00 | 0.00 | NR | NR | 0 | NR | – |
| Vitamin A, IU | IU | 9 794 | NR | NR | NR | NR | NR | – |
| Vitamin A, RAE | µg RAE | 492.4 | 145.9 | NR | NR | 369 | NR | 145.9-492.4 |
| Vitamin E (alpha-tocopherol) | mg | 6.54 | NR | NR | NR | 6.51 | NR | – |
| Vitamin K (phylloquinone) | µg | 23.28 | NR | NR | NR | 23.0 | NR | – |
| Carotene, beta | µg | 2 471 | 866-3 103 | 7 807 | 2 243 | 1 849 | 0 | 866-7 807 |
| Cryptoxanthin, beta | µg | 6 813 | NR | NR | NR | 5 089 | NR | – |
| Lutein + zeaxanthin | µg | 671.4 | NR | NR | NR | 497 | NR | – |

*Notes:* NR: not reported; DFE: dietary folate equivalent; RAE: retinol activity equivalent. Mean values reported on a dry weight basis were calculated from a fresh weight basis using the mean moisture level reported for each source, as shown in Table 6.4.

1. Based on orange-fleshed papaya (possibly including genetically engineered varieties). 2. Based on papaya, green, cooked (possibly including genetically engineered varieties). 3. Based on orange-fleshed Australia type and refuse: 30% (seed and skin).

*Sources:* USDA Agricultural Research Service (2009, 2008); Saxholt et al. (2008); Puwastien et al. (2000); Wills et al. (1986).

### *Carotenoids*

Carotenoids are responsible for the flesh colour of papaya fruit mesocarp.Red-fleshed papaya fruits contain five carotenoids: beta-carotene, beta-cryptoxanthin, beta-carotene-5-6-epoxide, lycopene and zeta-carotene. Yellow-fleshed papaya contains only three carotenoids: beta-carotene, beta-cryptoxanthin and zeta-carotene (Tables 6.9 and 6.10) (Chandrika et al., 2003).

As shown in Table 6.6, the content of beta-carotene varies from 866 µg/100 g dry matter to 7 807 µg/100g dry matter in ripe fruits (Puwastien et al., 2000; Saxholt et al., 2008). Differences in the methods of analysis have been shown to contribute to the variations in reported beta-carotene content (Rodriguez-Amaya et al., 2008).

### *Vitamin C*

Papaya is a source of vitamin C with amounts varying between the maturation stages (Table 6.6) (Bari et al., 2006; Hernandez et al., 2006). Variation in vitamin C content was also reported among papaya varieties (Table 6.9) (Franke et al., 2004; Wall, 2006).

### *Fatty acids*

Papaya contains a low level of fatty acids. Palmitic acid and linolenic acid are two major fatty acids in papaya (Table 6.7).

Chan and Taniguchi (1985) studied fatty acid composition changes in papaya pulp during fruit ripening and reported no significant difference in lipid composition with maturity of papaya fruits.

Table 6.7. **Fatty acid content of ripe papaya**

% of total fatty acids

| **Nutrient** | USDA[1] | Saxholt et al. | Range of values |
|---|---|---|---|
| Total saturated fatty acids | 38.4 | 38.9 | 38.4-38.9 |
| 12:0 lauric acid | 0.9 | 0.89 | 0.89-0.9 |
| 14:0 myristic acid | 6.3 | 6.2 | 6.2-6.3 |
| 16:0 palmitic acid | 28.5 | 28.3 | 28.3-28.5 |
| 18:0 stearic acid | 1.8 | 1.77 | 1.77-1.8 |
| Total monounsaturated fatty acids | 33.9 | 33.6 | 33.6-33.9 |
| 16:1 undifferentiated palmitoleic acid | 17.8 | 17.7 | 17.7-17.8 |
| 18:1 undifferentiated oleic acid | 16.1 | 15.9 | 15.9-16.1 |
| Total polyunsaturated fatty acids | 27.7 | 27.4 | 27.4-27.7 |
| 18:2 undifferentiated linoleic acid | 5.4 | 5.31 | 5.31-5.4 |
| 18:3 undifferentiated linolenic acid | 22.3 | 22.1 | 22.1-22.3 |

*Note:* 1. Based on orange-fleshed papaya (possibly including genetically engineered varieties).

*Sources:* USDA Agricultural Research Service (2009); Saxholt et al. (2008).

### *Proteins and amino acids*

Proteins constitute approximately 3.74-8.26 g/100 g of dry matter (Table 6.4). Aspartic acid is the most abundant amino acid in ripe fruits followed by glutamic acid (Table 6.8).

Table 6.8. **Amino acid content of ripe papaya**

% of total amino acids

| Nutrient | USDA[1] | Saxholt et al. | Blakesley | Range of mean values |
|---|---|---|---|---|
| Alanine | 5.6 | 5.7 | 5.7 | 5.6-5.7 |
| Arginine | 4.1 | 3.9 | 4.2 | 3.9-4.2 |
| Aspartic acid | 19.8 | 20.0 | 21.0 | 19.8-21.0 |
| Glutamic acid | 13.3 | 13.5 | 14.1 | 13.3-14.1 |
| Glycine | 7.2 | 7.1 | 7.7 | 7.1-7.7 |
| Histidine | 2.0 | 2.1 | 2.3 | 2.0-2.3 |
| Isoleucine | 3.2 | 3.2 | 3.4 | 3.2-3.4 |
| Leucine | 6.4 | 6.4 | 6.9 | 6.4-6.9 |
| Lysine | 10.1 | 10.3 | 8.4 | 8.4-10.3 |
| Methionine | 0.8 | 0.7 | 0.8 | 0.7-0.8 |
| Phenylalanine | 3.6 | 3.6 | 3.8 | 3.6-3.8 |
| Proline | 4.1 | 3.9 | 4.2 | 3.9-4.2 |
| Serine | 6.0 | 6.1 | 6.5 | 6.0-6.5 |
| Threonine | 4.4 | 4.3 | 4.6 | 4.3-4.6 |
| Tryptophan | 3.2 | 3.2 | NR | 3.2 |
| Tyrosine | 2.0 | 2.1 | 2.3 | 2.0-2.3 |
| Valine | 4.1 | 3.9 | 4.2 | 3.9-4.2 |

*Notes:* NR: not reported. 1. Based on orange-fleshed papaya (possibly including genetically engineered varieties).

*Sources:* USDA Agricultural Research Service (2009); Saxholt et al. (2008); Blakesley (1979).

## *Organic acids*

The major organic acids found in ripe papaya are (Hernandez et al., 2009):

- citric acid (335 ± 32 mg/100 g fresh weight),
- L-malic acid (209 ± 12 mg/100 g fresh weight),
- qiunic acid (52 ± 5 mg/100 g fresh weight),
- succinic acid (52 ± 3 mg/100 g fresh weight),
- tartaric acid (13 ± 2 mg/100 g fresh weight),
- oxalic acid (10 ± 1 mg/100 g fresh weight), and
- fumaric acid (1.1 ± 0.1 mg/100 g fresh weight).

Table 6.9. **Nutritive value of different varieties[1] of ripe papaya grown at different locations**

| Nutrient | Unit | Varieties | | |
|---|---|---|---|---|
| | | Sunrise[2] | Sunrise[3] | Kapoho[2] |
| | | Mean value, per 100 g fresh weight of edible portion | | |
| Water | g | 87.5 | 84.9 | 86.0 |
| | | Mean value, per 100 g dry weight of edible portion | | |
| Ascorbic acid | mg | 374.4 | 427.2 | 324.3 |
| β-carotene | μg | 644 | 2 717 | 1 042 |
| α-carotene | μg | ND | ND | ND |
| β-cryptoxanthin | μg | 2 307 | 6 092 | 3 045 |
| Lutein | μg | 878.4 | 857.6 | 1 701 |
| Lycopene | μg | 10 801 | 24 334 | ND |
| Vitamin A | μgRAE | 149.6 | 480.1 | 213.6 |
| Phosphorus (P) | mg | 40 | 52.98 | 57.14 |
| Potassium (K) | mg | 1 384 | 1 466 | 640.7 |
| Calcium (Ca) | mg | 99.2 | 131.8 | 70 |
| Magnesium (Mg) | mg | 199.2 | 216.6 | 137.1 |
| Sodium (Na) | mg | 51.2 | 92.7 | 40 |
| Iron (Fe) | mg | 3.36 | 3.05 | 2.07 |
| Manganese (Mn) | mg | 0.24 | 0.13 | 0.21 |
| Zinc (Zn) | mg | 0.56 | 0.60 | 0.64 |
| Copper (Cu) | mg | 0.56 | 0.53 | 0.79 |
| Boron (B) | mg | 1.12 | 1.32 | 0.93 |

*Notes:* ND: not detected; RAE: retinol activity equivalent. Mean values reported on a dry weight basis were calculated from a fresh weight basis using the mean moisture level as shown. Wall also reports data on other papaya varieties grown at various locations in Hawaii. 1. The Sunrise papaya variety is a red-fleshed variety while Kapoho is a yellow-fleshed variety. 2. Cultivated in the Kapoho area on the island of Hawaii, Hawaii. 3. Cultivated in the Moloaa area on the island of Kauai, Hawaii.

*Source*: Wall (2006).

Table 6.10. **Major provitamin A and non-provitamin A carotenoids in fruit pulp of yellow- and red-fleshed papaya**

μg/100 g dry matter

| Carotenoid | **Yellow flesh*** (n = 10) | **Red flesh*** (n = 10) |
|---|---|---|
| **Provitamin A carotenoids** | | |
| Beta-carotene | 140 ± 0.4 | 700 ± 0.7 |
| Beta cryptoxanthin | 1 540 ± 3.3 | 1 690 ± 2.9 |
| Beta-carotene-5-6-epoxide | ND | 290 ± 0.6 |
| Calculated retinol equivalent (μg/kg DW) | 1 516 ± 342 | 2 815 ± 305 |
| **Non-provitamin A carotenoids** | | |
| Lycopene | ND | 1 150 ± 1.8 |
| Zeta-carotene | 1 510 ± 3.4 | 990 ± 1.1 |

*Notes:* ND: not detected; DW: dry weight. * The varieties of papaya fruits were not specified by the authors.

*Source:* Chandrika et al. (2003).

### *Chemical composition of by-products from papaya processing*

Most papaya processing by-products are fed to buffalo, fish and poultry. The nutrients of major concern for buffalo are crude protein, crude fat (ether extractable), crude ash, carbohydrates, neutral detergent fibre (NDF), acid detergent fibre (ADF), calcium and phosphorus. The major nutrient considerations for fish feedstuff are apparent protein digestibility (APD) and amino acid levels in papaya leaf meal; however, APD is not expected to be routinely measured in feed stuff (OECD, 2008; Eusebio and Coloso, 2000). The composition of papaya processing by-products is shown in Table 6.11.

Table 6.11. **Chemical composition of papaya processing by-products**

% dry matter basis

| | Pomace | Dried skins | | Leaves | | Defatted papaya kernel flour[5] | Fresh papaya processing (pulp, peels and seeds)[6] |
|---|---|---|---|---|---|---|---|
| | 1 | 2 | 3 | 4 | 3 | | |
| Dry matter[1] | 92.2 | 87.41 | 83.9 ± 0.13 | 94.6 | 90.3 ± 0.29 | 92.5 ± 0.52 | > 80 |
| Crude protein | 18.44 | 22.9 | 17.9 ± 0.24 | 23.0 | 28.2 ± 0.5 | 32.4 ± 0.48 | 23.2 |
| Total fat (ether extract) | 4.73 | 3.68 | 1.8 ± 0.31 | 11.1 | 10.5 ± 0.25 | 0.7 ± 0.21 | NR |
| Crude fibre | 29.58 | 12.2 | 19.4 ± 0.22 | 11.4 | 13.0 ± 0.13 | 4.2 ± 0.06 | 18.2 |
| Nitrogen-free extract | 28.59 | 49.78 | 45.6 ± 0.40 | 38.5 | 32.9 ± 0.33 | NR | 29.5 |
| Crude ash | 18.66 | 11.44 | 15.4 ± 0.34 | 15.9 | 15.4 ± 0.12 | 5.3 | 8.6 |
| Acid insoluble ash | 4.04 | NR | NR | NR | NR | NR | NR |
| Calcium | 1.81 | NR | NR | NR | NR | NR | NR |
| Phosphorus | 0.61 | NR | NR | NR | NR | NR | NR |

*Notes:* NR: not reported. 1. Dry matter is reported as percentage of fresh weight.

*Sources:* 1. Babu et al. (2003); 2. Fouzder et al. (1999); 3. Munguti et al. (2006); 4. Reyes and Fermin (2003); 5. Alobo (2003); 6. Ulloa et al. (2004).

## Other constituents

### *Anti-nutrients*

Ripe papaya fruits (including peel and pulp) contain low amounts of anti-nutrients (tannin, phytate and oxalate). The mean levels of tannin, phytate and oxalate are 10.16 mg, 3.29 mg and 1.89 mg/100 g of dry matter, respectively (Onibon et al., 2007). A significant reduction in the levels of anti-nutrients was reported in papaya fruits stored at 27 ± 1°C and 10 ± 1°C. After eight days of storage at 27±1°C, the phytate content was reduced from 1.22% to 0.34% and the oxalate content from 0.45% to 0.13%. The content of tannin was reduced from 0.062% to 0.006% and 0.021% to an undetectable level, for condensed and hydrolysable tannin, respectively (Adetuyi et al., 2008).

### *Toxicants*

The major natural toxicants found in papaya are benzylglucosinolate (BG), benzyl isothiocyanate (BITC) and alkaloids. These substances are important for plant natural defence mechanisms (El Moussaoui et al., 2001). BITC is derived from BG by the action of the myrosinase enzyme. Although both BG and BITC are found in papaya peel, pulp and seed, the highest levels of BG and BITC are found in seeds,

1 269.3 ± 90.0 and 461.4 ± 14.2 µmol/100 g fresh weights respectively. The levels of BG and BITC in papaya pulp were < 3.0 and < 0.3 µmol/100 g fresh weight respectively (Nakamura et al., 2007). The concentration of BITC decreases in pulp and increases in seeds during fruit ripening (Tang, 1971). Wills and Widjanarko (1995) reported that BITC content decreased from 109 µg BITC/g when papaya is green to 10 µg BITC/g when papaya is fully ripe. Sheu and Shyu (1996) reported BITC content ranging from 5.4 µg to 33.6 µg/g fresh weight in pulp from four different papaya varieties. BITC content in papaya pulp is shown in Table 6.12.

Table 6.12. **BITC content of papaya pulp (µg/g fresh weight)**

| Developmental stage | Tang[1] | MacLeod and Pieris[1] | Wills and Widjanarko[1,2] | Sheu and Shyu[3,4] | Nagamura et al.[5] |
|---|---|---|---|---|---|
| Green | 74[6] | NR | 109 | NR | NR |
| Ripe | 4 | 0.0014 | 10 | 5.4–33.6 | < 0.447 |

*Notes:* NR: not reported.

1. BITC content was determined by gas chromatography. 2. Concentrations of BITC in green and ripe Australian papaya at 20°C. 3. Sheu and Shyu (1996) reported concentrations for Tainoung No. 2, Tainoung No. 5, Solo and Sunrise varieties. 4. BITC content was determined by solid phrase extraction and gas chromatography. 5. BITC content was determined by high-performance liquid chromatography. 6. Green immature papaya fruit weighing 187 g.

*Sources:* Tang (1971); MacLeod and Pieris (1983); Wills and Widjanarko (1995); Sheu and Shyu (1996); Nagamura et al. (2007).

Carpaine is a major alkaloid found in various parts of papaya, but is primarily found in leaves (Krishna et al., 2008). Papaya leaves contain the bitter alkaloids carpaine and pseudocarpaine, and must be boiled with several changes of water before consumption (Morton, 1987). Carpaine has been found in papaya leaves at concentrations between 1 000-1 500 mg/kg (Duke, 1992).

### *Allergens*

Papaya contains four cysteine endopeptidases including papain, chymopapain, glycyl endopeptidase and caricain. Papain is commonly found in papaya latex (Azarkan et al., 2003). The recorded level of papain in papaya latex is 51 000-135 000 mg/kg (Office of the Gene Technology Regulator, 2008).

Papain can induce IgE-mediated allergic reactions through oral, respiratory or contact routes of exposure. Occupational allergy to papain in exposed workers has been documented in a number of studies. The typical symptoms include bronchial asthma, rhinitis or both (Baur and Fruhmann, 1979; Baur et al., 1982; Niinimaki et al., 1993; Soto-Mera et al., 2000; Van Kampen et al., 2005). One case of a life-threatening anaphylaxis due to occupational exposure to papain was also reported (Freye, 1988).

Allergy to papaya-derived products unrelated to occupational exposure has also been described. Garcia-Ortega et al., (1991) showed that administration of chymopapain for chemonucleolysis (a medical procedure that involves the dissolving of the gelatinous cushioning material in an intervertebral disk by the injection of an enzyme such as chymopapain) resulted in sensitisation in some patients. In some sensitised chemonucleolysis patients, IgE specific to all four cysteine proteinases was detected (Dando et al., 1995). Mansfield and Bowers (1983) reported severe systemic allergic reactions mediated by papain-specific IgE in some individuals that ingested papain-containing meat tenderiser. In addition, two cases of allergy to papain in individuals using soft contact lens solution have been reported (Bernstein et al., 1984; Santucci et al., 1985).

Pollen from papaya flowers has been shown to induce respiratory allergy (Blanco et al., 1998; Singh and Kumar, 2002). Using RAST inhibition assay, Blanco et al. (1998) demonstrated that papaya pollen, papaya fruit and papain extracted from papaya contain common allergens. Papaya pollen in papaya planting areas can contribute to aeropollen and aeroallergen loads (Chakraborty et al., 2005).

Sensitisation to papaya does not typically occur from eating papaya fruit. However, once sensitised, individuals may suffer allergic reactions following any type of exposure to papaya or papaya-derived products (Morton, 1987).

## Suggested constituents to be analysed related to food use

Ripe papaya fruits and papaya products are consumed by humans for their flavour and nutritional value. Unripe papaya fruits are consumed both as a cooked vegetable and processed products.

The colour of the papaya fruit flesh is related to the carotenoids present in papaya. For example, three provitamin A caroteniods (beta-carotene, beta-carotene-5-6-epoxide, and beta-cryptoxanthin) and two non-provitamin A carotenoids (zeta-carotene and lycopene) are found in red ripe papaya fruits. Yellow ripe papaya fruits, however, only contain beta-carotene, beta-cryptoxanthin and zeta-carotene.

The constituents that should be analysed in ripe and unripe papaya fruits are shown in Table 6.13.

Table 6.13. **Suggested constituents to be analysed in the unripe and ripe papaya fruits**

| Parameter | Unripe/ Ripe papaya |
|---|---|
| Moisture[1] | X |
| Crude protein[1] | X |
| Total fat (ether extract)[1] | X |
| Ash[1] | X |
| Carbohydrate by difference[2] | X |
| Total dietary fibre | X |
| Total sugars | X |
| Total ascorbic acid | X |
| Beta-carotene | X |
| Beta-cryptoxanthin | X |
| Benzylisothiocyanate (BITC) | X |

*Notes:* 1. These components should be measured using a method suitable for the measurement of proximates. 2. Carbohydrates are calculated as follows: 100 – (water + crude protein + total fat + ash) g/100 g fresh weight.

## Suggested constituents to be analysed related to feed use

Papaya leaf, peel and pomace may be used as feedstuffs. Papaya peels from ripe and unripe fruits are by-products from processing papaya in kitchens, hotels or restaurants, while papaya pomace is discarded from fruit juice factories. The use of papaya by-products in animal feed is mainly limited to experimental studies and small-scale farms.

Hasan et al. (2007) reported the use of papaya leaves as a feed ingredient for Gouramy fish cultured in extensive aquaculture systems in Indonesia. Diets of green papaya leaves, an artificial diet containing 25% crude protein, or a 1:1 ratio mixture of green papaya leaves and artificial diet, were compared as potential diets for African giant land snail (*Archachatina marginata*). It was found that a 1:1 ratio mixture of green papaya leaves and artificial diet resulted in significantly higher body weight gain as well as other morphological parameters, including shell length, shell width and shell aperture of the animal (Ejidike, 2007). Papaya leaf meal in a diet formulated to contain 27% crude protein was evaluated for its potential as a feed ingredient for farmed abalone (*Haliotis asinine*) diets in the Philippines (Reyes and Fermin, 2003).

As a fish feedstuff, papaya leaf meal is comparable in amino acid and nutrient content to white cowpea and mungbean seed meals (Eusebio and Coloso, 2000). There was no difference in apparent protein digestibility value between papaya leaf meal and white cowpea and mungbean seed meals (Eusebio and Coloso, 2000).

Fouzder et al. (1999) reported that use of dried papaya skin in pullet diets at levels up to 90 g/kg diet showed no significant difference in growth including weight gain, feed intake, feed conversion ratio and protein efficiency ratio between test and control animals.

Papaya pomace was shown to provide rumen degradable dry matter and crude protein in buffaloes (Babu et al., 2003).

Papaya leaves and green fruits contain toxicants such as benzyl isothiocyanate (BITC) that can cause irritation of the mucus epithelial membrane. Munguti et al. (2006) reported that soaking in water and heat treatment destroys such toxic compounds in papaya and other plants. In the process of making papaya leaf meal, papaya leaves are soaked for 24 hours, drained, rinsed and air-dried prior to heat treatment and grinding (Eusebio and Coloso, 2000; Reyes and Fermin, 2003). Ulloa et al. (2004) reported that papaya meal (pulp, peel and seeds) has high gross energy content and high potential digestible energy levels and may be suitable for use as a fish feed.

The constituents suggested for analysis related to feed use are shown in Table 6.14.

Table 6.14. **Suggested constituents to be analysed in papaya for feed use**

| Parameter | Fruit | Leaves | Skins |
|---|---|---|---|
| Moisture[1] | X | X | X |
| Crude protein[1] | X | X | X |
| Crude fat[1] | X | X | X |
| Ash[1] | X | X | X |
| Carbohydrates[2] | X | X | X |
| Total dietary fibre[3] | X | X | X |
| Neutral detergent fibre[4] | X | X | X |
| Acid detergent fibre[4] | X | X | X |
| Amino acids | X | X | |
| Calcium | X | X | X |
| Phosphorus | X | X | X |
| Carpaine | | X | |
| Benzylisothiocynate (BITC) | X | X | X |

*Notes:* 1. These components should be measured using a method suitable for the measurement of proximates. 2. Carbohydrates are calculated as follows: 100 – (water + crude protein + total fat + ash) g/100g fresh weight. 3. Total dietary fibre analysis is more relevant for dietary considerations of non-ruminant animals. 4. Neutral detergent fibre and acid detergent fibre analyses are more relevant for dietary considerations of ruminant animals.

## Note

1. For additional discussion of appropriate comparators, see Codex Alimentarius Commission (2003; paragraphs 44 and 45).

## *References*

Adetuyi, F.O. et al. (2008), "Antinutrient and antioxidant quality of waxed and unwaxed pawpaw *Carica papaya* fruit stored at different temperatures", *African Journal of Biotechnology*, Vol. 7, No. 16, pp. 2 920-2 924, available at: www.ajol.info/index.php/ajb/article/download/59202/47504.

Ahmed, J. et al. (2002), "Thermal degradation kinetics of carotenoids and visual color of papaya puree", *Journal of Food Science*, Vol. 67, No. 7, pp. 2 692-2 695, http://dx.doi.org/10.1111/j.1365-2621.2002.tb08800.x.

Alobo, A.P. (2003), "Proximate composition and selected functional properties of defatted papaya (*Carica papaya* L.) kernel four", *Plant Food Human Nutrition*, Vol. 58, pp. 1-7.

Azarkan, M. et al. (2003), "Fractionation and purification of the enzymes stored in the latex of *Carica papaya*", *Journal of Chromatography B*, Vol. 790, Issues 1-2, pp. 229-238.

Babu, A.R. et al. (2003), "*In sacco* dry matter and protein degradability of papaya (*Carica papaya*) pomace in buffaloes", *Buffalo Bulletin*, Vol. 22, No. 1, pp.12-15.

Bari, L. et al. (2006), "Nutritional analysis of two local varieties of papaya (*Carica papaya*) at different maturation stages", *Pakistan Journal of Biological Sciences*, Vol. 9, Issue 1, pp. 137-140, http://dx.doi.org/10.3923/pjbs.2006.137.140.

Baur, X. and G. Fruhmann (1979), "Papain-induced asthma: Diagnosis by skin test, RAST, and bronchial provocation test", *Clinical Allergy*, Vol. 9, No. 1, pp. 75-81.

Baur, X. et al. (1982), "Clinical symptoms and results of skin test, RAST, and bronchial provocation test in thirty-three papain workers: Evidence for strong immunogenic potency and clinically relevant 'proteolytic effects of airborne papain'", *Clinical Allergy*, Vol. 12, No. 1, pp. 9-17.

Bernstein, D.I. et al. (1984), "Local ocular anaphylaxis to papain enzyme contained in a contact lens cleansing solution", *Journal of Allergy and Clinical Immunology*, Vol. 74, No. 3, Part 1, pp. 258-260.

Blakesley, C.N. et al. (1979), "Gamma irradiation of subtropical fruits 2 volatile components, lipids and amino acids of mango, papaya, and strawberry pulp", *Journal of Agricultural and Food Chemistry*, Vol. 27, No. 1, pp. 42-48.

Blanco, C. et al. (1998), "Carica papaya pollen allergy", *Annals of Allergy, Asthma and Immunology*, Vol. 81, No. 2, pp. 171-175.

Brekke, J.E. et al. (1976), "Effects of storage temperature and container lining on some quality attributes of papaya nectar", *Journal of Agricultural and Food Chemistry*, Vol. 24, No. 2, pp. 341-343, http://dx.doi.org/10.1021/jf60204a055.

Brekke, J.E. et al. (1972), "Papaya puree: A tropical flavor ingredient", *Journal of Food Products Development*, Vol. 6, pp. 36-37.

Campostrini, E. and O.K. Yamanishi (2001), "Estimation of papaya leaf area using the central vein length", *Scientia Agricola*, Vol. 58, No. 1, pp. 39-42, http://dx.doi.org/10.1590/S0103-90162001000100007.

Carter, S. (1997), *In Global Agricultural Marketing Management*, Food and Agriculture Organization, Regional Office for Africa.

Chakraborty, P. et al. (2005), "Aerobiological and immunochemical studies on *Carica papaya* L. pollen: An aeroallergen from India", *Clinical Allergy*, Vol. 60, No. 7, pp. 920-926.

Chan, Y.K (2007), "Stepwise priorities in papaya breeding", *ISHS Acta Horticulturae*, International Symposium on Papaya, Vol. 740, pp. 43-48.

Chan Jr., H.T. and M.H. Taniguchi (1985), "Changes in fatty acid composition of papaya lipids (*Carica papaya*) during ripening", *Journal of Food Science*, Vol. 50, Issue 4, pp. 1 092-1 094, http://dx.doi.org/10.1111/j.1365-2621.1985.tb13019.x.

Chandrika, U.G. et al. (2003), "Carotenoids in yellow- and red-fleshed papaya (*Carica papaya* L)", *Journal of the Science of Food and Agriculture*, Vol. 83, Issue 12, pp. 1 279-1 282, http://dx.doi.org/10.1002/jsfa.1533.

Charoensiri, R. et al. (2009), "Beta-carotene, lycopene, and alpha-tocopherol contents of selected Thai fruits", *Food Chemistry*, Vol. 113, pp. 202-207.

Chavasit, V. et al. (2002), "Change in β-carotene and vitamin A contents of vitamin A-rich foods in Thailand during preservation and storage", *Journal of Food Science*, Vol. 67, Issue 1, pp. 375-379, http://dx.doi.org/10.1111/j.1365-2621.2002.tb11413.x.

Codex (2006), *The Codex Guidelines on Nutrition Labelling*, CAC/GL 2-1985, Rev. 1-1993, Amd. 2003 and Amd. 2006.

Codex (2005), "Codex standard for papaya", *Codex Standard Series No. 183-1993, Rev. 1-2001, Amd. 1-2005.*

Codex (Codex Alimentarius Commission) (2003; with Annexes II and III adopted in 2008), *Guideline for the Conduct of Food Safety Assessment of Foods Derived from Recombinant DNA Plants*, CAC/GL 45/2003, available at: www.codexalimentarius.net/download/standards/10021/CXG_045e.pdf.

Dando, P.M. et al. (1995), "Immunoglobulin E antibodies to papaya proteinases and their relevance to chemonucleolysis", *Spine*, Vol. 20, No. 9, pp. 981-985.

Deputy, J.C. et al. (2002), "Molecular markers for sex determination in papaya (*Carica papaya* L.)", *Theoretical and Applied Genetics*, Vol. 106, No. 1, pp. 107-111.

Duke, J.A. (1992), *Handbook of Phytochemical Constituents of GRAS Herbs and Other Economic Plants*, CRC Press, Ann Arbor, Michigan, pp. 136-137.

Ejidike, B.N. (2007), "Influence of artificial diet on captive rearing of African giant land snail *Archachatina marginata Pulmonata*: Stylommatophora", *Journal of Animal and Veterinarian Advances*, Vol. 6, pp. 1 028-1 030.

El Moussaoui, A. et al. (2001), "Revisiting the enzymes stored in the laticifers of *Carica papaya* in the context of their possible participation in the plant defence mechanism", *Cellular and Molecular Life Sciences*, Vol. 58, No. 4, pp. 556-570.

Eusebio, P.S. and R.M. Coloso (2000), "Nutritional evaluation of various plant protein sources in diets for Asian sea bass *Lates calcarifer*", *Journal of Applied Ichthyology*, Vol. 16, Issue 2, pp. 56-60, http://dx.doi.org/10.1046/j.1439-0426.2000.00160.x.

Eustice, M. et al. (2008), "Development and application of microsatellite markers for genomic analysis of papaya", *Tree Genetics & Genomes*, Vol.4, Issue 2, pp.333-341, http://dx.doi.org/10.1007/s11295-007-0112-2.

FAOSTAT (2004, 2007, 2008), FAO Statistics online, http://faostat.fao.org (accessed 24 March 2010).

Fouzder, S.K. et al. (1999), "Use of dried papaya skin in the diet of growing pullets", *British Poultry Science*, Vol. 40, No. 1, pp. 88-90.

Franke, A.A. et al. (2004), "Vitamin C and flavonoid levels of fruits and vegetables consumed in Hawaii", *Journal of Food Composition and Analysis*, Vol. 17, Issue 1, pp. 1-35.

Freye, H.B. (1988), "Papain anaphylaxis: A case report", *Allergy Proceedings*, Vol. 9, No. 5, pp. 571-574.

Garcia-Ortega, P. et al. (1991) "Sensitization to chymopapain in patients treated with chemonucleolysis", *Med. Chin. (Barc)*, Vol. 96, No. 11, pp. 410-412.

Gomez, M. et al. (2002), "Evolution of soluble sugars during ripening of papaya fruit and its relation to sweet taste", *Journal of Food Science*, Vol. 67, Issue 1, pp. 442-447, http://dx.doi.org/10.1111/j.1365-2621.2002.tb11426.x.

Hardisson, A. et al. (2001), "Mineral composition of the papaya (*Carica papaya* variety sunrise) from Tenerife Island", *European Food Research and Technology*, Vol. 212, Issue 2, pp. 175-181.

Hasan, M.R. et al. (2007), "Study and analysis of feeds and fertilizers for sustainable aquaculture development", *FAO Fisheries Technical Papers*, No. 497, Food and Agriculture Organization, Rome, available at: www.fao.org/docrep/011/a1444e/a1444e00.htm.

Hernandez, Y. et al. (2009), "Factors affecting sample extraction in the liquid chromatographic determination of organic acids in papaya and pineapple", *Food Chemistry*, Vol. 114, Issue 2, pp. 734-741.

Hernandez, Y. et al. (2006), "Determination of vitamin C in tropical fruits: A comparative evaluation of methods", *Food Chemistry* Vol. 96, Issue 4, pp. 654-664.

Kim, M.S. et al. (2002), "Genetic diversity of *Carica papaya* as revealed by AFLP markers", *Genome*, Vol. 45, No. 3, pp. 503-512.

Krishna, K.L. et al. (2008), "Review on nutritional, medical and pharmacological properties of papaya (*Carica papaya* Linn.)", *Natural Product Radiance*, Vol. 7, No. 4, pp. 364-373, available at: http://nopr.niscair.res.in/bitstream/123456789/5695/1/NPR%207%284%29%20364-373.pdf.

MacLeod, A.J. and N.M. Pieris (1983), "Volatile components of papaya (*Carica papaya* L.) with particular reference to glucosinolate products", *Journal of Agricultural and Food Chemistry*, Vol. 31, No. 5, pp. 1 005-1 008, http://dx.doi.org/10.1021/jf00119a021.

Magdalita, P.M. et al. (2007), "Recent developments in papaya breeding in the Philippines", *ISHS Acta Horticulturae*, International Symposium on Papaya, Vol. 740, pp. 49-59.

Mano, R. et al. (2009), "Dietary intervention with Okinawan vegetables increased circulating endothelial progenitor cells in healthy young woman", *Atherosclerosis*, Vol. 204, No. 2, pp. 544-548.

Mansfield, L.E. and C.H. Bowers (1983), "Systemic reaction to papain in a non-occupational setting", *Journal of Allergy and Clinical Immunology*, Vol. 71, No. 4, pp. 371-374.

Marelli de Souza, L. et al. (2008), "L-ascorbic acid, β-carotene and lycopene content in papaya fruits (*Carica papaya*) with or without physiological skin freckles", *Scientia Agricola (Piracicaba, Brazil)*, Vol. 65, No. 3, pp. 246-250, http://dx.doi.org/10.1590/S0103-90162008000300004.

Martin, S.L.D. et al. (2006), "Partial diallel to evaluate the combining ability for economically important traits of papaya", *Scientia Agricola (Piracicaba, Brazil)*, Vol. 63, No. 6, pp. 540-546, http://dx.doi.org/10.1590/S0103-90162006000600005.

Matsuura, F.C.A.U. et al. (2004), "Sensory acceptance of mixed nectar of papaya, passion fruit and acerola", *Scientia Agricola (Piracicaba, Brazil)*, Vol. 61, No. 6, pp. 604-608, http://dx.doi.org/10.1590/S0103-90162004000600007.

Mendoza, E.M.T. (2007), "Development of functional foods in the Philippines", *Food Science and Technology Research*, Vol. 13, No. 3, pp. 179-186, http://dx.doi.org/10.3136/fstr.13.179.

Ming, R. et al. (2008), "The draft genome of the transgenic tropical fruit tree papaya (*Carica papaya* Linnaeus)", *Nature*, Vol. 452, No. 7190, pp. 991-997, http://dx.doi.org/10.1038/nature06856.

Morton, J.F. (1987), "Papaya", in: *Fruits of Warm Climates*, Julia F. Morton, Miami, Florida, pp. 336-346, available at: www.hort.purdue.edu/newcrop/morton/papaya_ars.html.

Mugula, J.K. et al. (1994) "Production and storage stability of non-alcoholic pawpaw beverage powder", *Plant Foods for Human Nutrition*, Vol. 46, No. 2, pp. 167-173.

Munguti, J.M. et al. (2006), "Proximate composition of selected potential feedstuffs for Nile tilapia (*Oreochromis niloticus* Linnaeus) production in Kenya", *Die Bodenkultur*, Vol. 57, No. 3, pp. 131-141, available at: https://diebodenkultur.boku.ac.at/volltexte/band-57/heft-3/munguti.pdf.

Nakamura, Y. et al. (2007), "Papaya seed represents a rich source of biologically active isothiocyanate", *Journal of Agricultural and Food Chemistry*, Vol. 55, No. 11, pp. 4 407-4 413.

Niinimaki, A. et al. (1993), "Papain-induced allergic rhinoconjunctivitis in a cosmetologist", *Journal of Allergy and Clinical Immunology*, Vol. 92, No. 3, pp. 492-493.

Odu, E.A. et al. (2006), "Occurrence of hermaphroditic plants of *Carica papaya* L. (Caricaceae) in southwestern Nigeria", *Journal of Plant Science*, Vol. 1, Issue 3, pp. 254-263, http://dx.doi.org/10.3923/jps.2006.254.263.

OECD (2008), "Consensus document on compositional considerations for new varieties of tomato: Key food and feed nutrients, toxicants and allergens", *Series on the Safety of Novel Foods and Feeds*, No. 17, OECD, Paris, available at: www.oecd.org/science/biotrack/46815296.pdf.

OECD (2005), "Consensus document on the biology of papaya (*Carica papaya*)", *Series on Harmonisation of Regulatory Oversight in Biotechnology*, No. 33, OECD, Paris, available at: www.oecd.org/science/biotrack/46815818.pdf.

Office of the Gene Technology Regulator (2008), "The biology of *Carica papaya* L. (papaya, papaw, paw paw)", Vers. 2: Feb. 2008, Australian Government, Dep. Health and Ageing, OGTR, available at: www.ogtr.gov.au/internet/ogtr/publishing.nsf/Content/papaya-3/$FILE/biologypapaya08.pdf.

Onibon, V.O. et al. (2007), "Nutritional and antinutritional composition of some Nigerian fruits", *Journal of Food Technology*, Vol. 5, Issue 2, pp. 120-122.

Parasnis, A.S. et al. (1999), "Microsatellite $(GATA)_n$ reveals sex-specific differences in papaya", *Theoretical and Applied Genetics*, Vol. 99, Issue 6, pp. 1 047-1 052.

Paull, R.E. (1993), *Pineapple and Papaya in Biochemistry of Fruit Ripening*, Chapman and Hall, Boundary Row, London.

Paull, R.E. et al. (2008), "Fruit development, ripening and quality related genes in the papaya genome", *Tropical Plant Biology*, Vol. 1, pp. 246-277, http://dx.doi.org/10.1007/s12042-008-9021-2.

Pesante, A. (2003), *Market Outlook Report: Fresh Papaya*, State of Hawaii, Department of Agriculture, Honolulu, Hawaii.

Picha, D. (2006), "Horticultural crop quality characteristics important in international trade", *ISHS Acta Horticulturae*, IV International Conference on Managing Quality in Chains – The Integrated View on Fruits and Vegetables Quality, Vol. 712, pp. 423-426.

Pollack, S. and A. Perez (2008), *Fruit and Tree Nuts Situation and Outlook Yearbook 2008*, Market and Trade Economics Division, Economic Research Service, United States Department of Agriculture, Washington, DC, FTS-2008.

Puwastien, P. et al. (2000), *ASEAN Food Composition Tables of Nutrition*, Institute of Nutrition, Mahidol University, Thailand, see: www.inmu.mahidol.ac.th/aseanfoods/publication.html.

Reyes, O.S. and A.C. Fermin (2003), "Terrestrial leaf meals or freshwater aquatic fern as potential feed ingredients for farmed abalone *Haliotis asinina* (Linnaeus 1758)", *Aquaculture Research*, Vol. 34, Issue 8, July, pp. 593-599, http://dx.doi.org/10.1046/j.1365-2109.2003.00846.x.

Rodriguez-Amaya, D.B. et al. (2008), "Updated Brazilian database on food carotenoids: Factors affecting carotenoid composition", *Journal of Food Composition and Analysis*, Vol. 21, Issue 6, pp. 445-463.

Sanchez-Castillo, C.P. et al. (1998), "The mineral content of Mexican fruits and vegetables", *Journal of Food Composition and Analysis*, Vol. 11, Issue 4, December, pp. 340-356.

Santucci, B. et al. (1985), "Contact urticaria from papain in soft lens solution", *Contact Dermatitis*, Vol. 12.

Saxholt, E. et al. (2008), *Danish Food Composition Database*, Version 7, Department of Nutrition, National Food Institute, Technical University of Denmark, www.foodcomp.dk/v7/fcdb_default.asp.

Sheu, F. and Y.T. Shyu (1996), "Determination of benzyl isothiocyanate in papaya fruit by solid phase extraction and gas chromatography", *Journal of Food and Drug Analysis*, Vol. 4, No. 4, pp. 327-334.

Singfield, P. (1998), "Report #19 1998: Papaya and Belize", Belize Development Trust, available at: www.belize1.com/BzLibrary/trust19.html.

Singh, A.B. and P. Kumar (2002), "Common environmental allergens causing respiratory allergy in India", *Indian Journal of Pediatrics*, Vol. 69, No. 3, pp. 245-250.

Sone, T. et al. (1998), "Comparison of diets among elderly female residents in 2 suburban districts in Chiang Mai Province/Thailand, in dry season –Survey on high and low-risk districts of lung cancer incidence", *Applied Human Science: J. Physio. Anthropo* Vol.17-2, pp. 49-56, http://dx.doi.org/10.2114/jpa.17.49.

Soto-Mera, M.T. et al. (2000), "Occupational allergy to papain", *Allergy*, Vol. 55, No. 10, pp. 983-984.

Stice, K.N. et al. (2010), "'Fiji red' papaya: Progress and prospects in developing a major agriculture diversification industry", *ISHS Acta Horticulturae*, Vol. 851, International Symposium on Papaya, pp. 423-426.

Tang, C.-S. (1971), "Benzyl isothiocyanate of papaya fruit", *Phytochemistry*, Vol. 10, Issue 1, pp. 117-121.

Ulloa, J.B. et al. (2004), "Tropical agricultural residues and their potential uses in fish feeds: The Costa Rican situation", *Waste Management*, Vol. 24, No. 1, pp. 87-97.

USDA Agricultural Research Service (2009), *USDA National Nutrient Database for Standard Reference*, Release 22, Nutrient Data Laboratory, available at: http://ndb.nal.usda.gov.

USDA, Agricultural Research Service (2008), *USDA Food and Nutrient Database for Dietary Studies*, FNDDS 3.0, available at: www.ars.usda.gov/News/docs.htm?docid=12068.

Van Kampen, V. et al. (2005), "Occupational allergies to papain", *Pneumologie*, Vol. 59, No. 6, pp. 405-410.

Wall, M.M. (2006), "Ascorbic acid, vitamin A and mineral composition of banana (*Musa* sp.) and papaya (*Carica papaya*) cultivars grown in Hawaii", *Journal of Food Composition and Analysis*, Vol. 19, Issue 5, pp. 434-445.

Wills, R.B.H. and S.B. Widjanarko (1995), "Changes in physiology, composition, and sensory characteristics of Australian papaya during ripening", *Australian Journal of Experimental Agriculture*, Vol. 35, No. 8, pp. 1 173-1 176, http://dx.doi.org/10.1071/EA9951173.

Wills, R.B.H. et al. (1986), "Composition of Australian foods -31. Tropical and sub-tropical fruit", *Food Technology in Australia*, Vol. 38, pp. 118-123.

Yon, R.M. (1994), *Papaya: Fruit Development, Postharvest Physiology, Handling and Marketing in ASEAN*, Food Technology Research Center, Malaysian Agricultural R & D. Kuala Lumpur, Malaysia.

# *Chapter 7*

# Sugarcane (*Saccharum* ssp. hybrids)

*This chapter, prepared by the OECD Task Force for the Safety of Novel Foods and Feeds with Australia as the lead country, deals with the composition of sugarcane (*Saccharum *ssp. hybrids). It contains elements that can be used in a comparative approach as part of a safety assessment of foods and feeds derived from new varieties. Background is given on sugarcane production, harvesting, processing and uses, followed by appropriate varietal comparators and characteristics screened by breeders. Nutrients in sugar, sugarcane juice, molasses, bagasse and whole cane, as well as other constituents (allergens, anti-nutrients and toxicants), are then detailed. The final sections suggest key products and constituents for analysis of new varieties for food use and for feed use.*

## Background

### *Introduction*

Sugarcane is one of the oldest cultivated plants (James, 2004) and has been described as one of the world's most efficient living collectors of solar energy, storing this energy in the form of fibre and fermentable sugars (FAO, 1988).

The sugarcane plant is a tall perennial tropical grass belonging to the genus *Saccharum*, and is closely related to other tropical grasses such as sorghum and maize. The plant forms a single unbranched stem that reaches an average height of three to four metres. The stem diameter ranges from three to five centimetres depending on the species and it is the stems (stalks or canes) from which sugar (sucrose) is extracted.

There are two confirmed wild species of *Saccharum*, and four domesticated ones (Bakker, 1999). The two wild species are *S. spontaneum* L. and *S. robustum* E.W. Brandes & Jeswiet ex Grassl. *S. spontaneum* can be found throughout the tropical areas of Africa as well as in Asia and Oceania, whereas *S. robustum* is restricted to Papua New Guinea and neighbouring islands.

The four domesticated species are *Saccharum officinarum* L. (the noble cane), *S. edule* Hassk., *S. barberi* Jeswiet and *S. sinense* Roxb. Noble canes are thought to be derived from *S. robustum* (Bakker, 1999). Noble canes have high sucrose content and a soft rind and were the original soft, sweet tasting chewing cane. Varieties of noble cane formed the basis of the earliest sugar production industries. Little, if any, noble cane is now grown for commercial sugar production. *S. edule* is restricted to Melanesia and Indonesia and is considered to be a mutant of *S. officinarum*. *S. barberi* has thin stalked hardy canes and is suited to semitropical and temperate climates. This species is believed to have arisen in India as a hybrid of *S. spontaneum* and *S. officinarum* (Bakker, 1999). Sugar was first manufactured from canes of this species. *S. sinense* has tall, vigorous, hardy canes and arose from hybridisation between *S. spontaneum* and *S. officinarum*.

Modern cultivated varieties of sugarcane are hybrids derived from breeding between the species of former commercial importance. The result of these breeding programmes is that modern hybrid sugarcane varieties incorporate the vigour and hardiness of *S. spontaneum* and *S. sinense* coupled with the high sugar content of *S. officinarum* and *S. barberi*.

### *Production*

Sugarcane, which is grown on approximately 24 million hectares in 102 countries in tropical and subtropical zones of both northern and southern hemisphere countries (FAOSTAT, 2009), is the world's leading sugar-producing crop, accounting for about 75% of world sugar supply (Dillon et al., 2007). The rest of the world's sugar supply is produced from sugar beet, which is grown in the temperate zones of the northern hemisphere (OECD, 2002).

Brazil is the world's largest sugarcane producer, having produced around 670 million tonnes in 2009 (FAOSTAT). Other major sugarcane producers are India, the People's Republic of China (hereafter "China"), Thailand, Mexico, Pakistan, Colombia, Australia, Argentina, the United States and other countries as listed in Table 7.1. Brazil, China, India and Thailand account for 50% of the world's sugar production and 59% of world sugar exports (USDA Foreign Agricultural Service, 2009).

While a large amount of sugarcane cultivation is directed towards sugar production, a number of countries also direct significant amounts into fuel ethanol production. In Brazil, for example, the recent trend has been to direct greater than 50% of the sugarcane crop into ethanol production (USDA, 2009).

Table 7.1. **Main sugarcane producing countries**

| Country | Production in 2009 (million metric tonnes, MMt) |
|---|---|
| Brazil | 671.4 |
| India | 285.0 |
| China (People's Republic of) | 116.2 |
| Thailand | 66.8 |
| Mexico | 49.5 |
| Pakistan | 50.0 |
| Colombia | 38.5[1] |
| Australia | 31.4 |
| Argentina | 29.9[1] |
| United States | 27.5 |
| Philippines | 22.9 |
| Indonesia | 26.5[1] |
| South Africa | 20.5[1] |
| Guatemala | 18.0 |
| Egypt | 17.0[1] |
| Viet Nam | 15.2 |
| Cuba | 14.9 |
| Peru | 10.1 |
| World | **1 661.3**[2] |

*Notes:* 1. FAO estimate. 2. May include official, semi-official or estimated data.

*Source:* FAOSTAT (2009).

## *Harvesting and processing*

### *Harvesting*

Sugarcane is harvested when its sucrose content is at its highest, and glucose and fructose content at their lowest. In Brazil, for example, industrial harvesting of sugarcane starts when the sucrose content is between 12.3% and 16% (Lavanholi, 2008). Traditionally the sugarcane is burnt before harvest to remove leaves, weeds and other trash that might interfere with milling; however, it is now relatively common for sugarcane to be harvested green. The leafy tops of the cane stalks are removed and the stalks are cut off at ground level and either transported whole or chopped into small lengths called billets before being delivered to the mill for processing. In some countries, sugarcane tops are a major harvesting by-product and are frequently used for livestock feed during the harvest season.

## *Processing*

The primary objective of sugarcane processing is to extract as much sucrose as possible from the plant stems. Processing, which is essentially a series of separations of non-sugars from sucrose, traditionally takes place in two stages: *i)* removal of juice from the cane stalks and extraction of cane or raw sugar; *ii)* refinement of raw sugar to white and brown refined products. In Brazil, most of the sugarcane mills integrate sugar and ethanol production, allowing some by-products of the sugar processing to be used as substrate for ethanol production.

Sugarcane juice is obtained by pressing sugarcane stalks; this is a part of both industrial and artisanal processing. The steps involved in industrial sugarcane processing are summarised below and in Figure 7.1 (for more detailed descriptions see also Clarke, 1988; Chen and Chou, 1993; Godshall, 2003). A number of foodstuffs are also derived from artisanal sugarcane processing, which is described below and in Figure 7.2. Extraction rates using artisanal processing tend not to be as efficient as industrial systems.

### Cane sugar production

Harvested sugarcane stems are chopped, shredded and then crushed using roller mills to extract the juice. Alternatively, the juice can be extracted using a diffuser (this is known as diffusion). Imbibition with water enhances the extraction of juice. Sucrose extraction using a diffusion system averages about 97-98%, compared to 90-91% using a traditional milling system (Godshall, 2003); extraction of non-sugars may, however, be higher with the diffusion system (Clarke, 1988). The fibrous material exiting the last mill or the drying mills after the diffuser and once all the cane juice has been extracted is called bagasse. Bagasse contains roughly 50% moisture, small amounts of residual sugar (1-3%) and the remainder being plant fibre (Paturau, 1989). Bagasse is primarily used as a fuel in the cane factory to generate power but when surplus exists it may also be directed to other uses, such as paper making and animal feed.

The collected juice is strained to remove large particles and then clarified using heat and lime – a process known as clarification. Lime is added to adjust the pH to prevent inversion of sucrose, and the temperature of the juice is raised to over 100°C. A heavy precipitate, called “mud”, forms which is separated from the juice in the clarifier, and then either returned to the diffuser or filtered to produce filtercake. Filtercake is the main processing waste from raw sugar production and contains about 15-30% fibre, 5-15% crude protein, 5-15% sugar, 5-15% crude wax and fats, and 10-20% ash (Paturau, 1989). Filtercake has minimal feed use and no food use and is mainly used as a soil conditioner/fertiliser. In some production systems, sulphur dioxide ($SO_2$) and small quantities of soluble phosphate may also be added. Sulphur dioxide is used to acidify the juice to coagulate the soluble solids and decrease the juice viscosity. These methods are often used in the production of direct consumption sugar.

Following clarification, the juice is concentrated using evaporation to produce syrup. The syrup is then further concentrated by boiling under vacuum until it becomes supersaturated, then seeded with crystalline sugar in a vacuum pan to initiate the crystallisation of sucrose from the mother liquor. The mixture of sugar and mother liquor is called massecuite.

Centrifugation is used to separate the sugar crystals from the massecuite. The resultant separated mother liquor is called molasses (called the “A molasses” or “first molasses”), which is typically subjected to further rounds of crystallisation to maximise the sugar yield. The molasses from the second round of crystallisation (called the "B molasses" or "second molasses") is of much lower purity than the first molasses. The final molasses, or "C molasses", is typically referred to as blackstrap molasses. Molasses is one of the main by-products of sugarcane processing. Molasses has a variety of food and feed uses, in addition to being a valuable raw material for the fermentation industry, where it is used principally to produce industrial ethanol, but also alcoholic beverages (rum), acetic acid, butanol, acetone, citric acid and glycerol (Paturau, 1989).

In places where the sugar and ethanol production are integrated, it is more common to direct the second or even the first molasses for fermentation to ethanol. Alternatively, some countries, especially in Latin America, use molasses to produce the distilled alcoholic beverage called rum.

The raw sugar is washed, dried and placed in large storage bins ready for refining. Raw sugar is typically about 98% pure. While the majority of raw sugar that is produced is destined for refining, in many countries a number of raw sugar products for direct consumption are also produced. These include the white raw sugar products known as plantation or mill white sugar and *blanco directo*, as well as the speciality brown sugar products known as *demerara* and *turbinado* sugar.

Typically, the processing of sugarcane yields about 70% water, 15% bagasse, 10% sugar, 3% molasses, and, if produced, 2% filtercake (Fuller, 2004).

## Refined sugar production

The aim of the refining process is to remove the colour and reduce the soluble ash concentration to acceptable levels. The process involved in refining raw sugar can vary from country to country but typically follows a series of basic steps. The first step in refining is to remove the surface layer of molasses from the sugar crystals (affination). This is achieved by washing the raw sugar with warm saturated syrup which softens the adhering molasses layer and then using centrifugation to separate the sugar crystals from the syrup (typically called the "affination syrup"). The affination syrup can be recycled, either by using it in a raw sugar washing step or by melting to recover additional sugar, leaving a final syrup known as refinery blackstrap molasses. Affination typically removes about 65-70% of the colour, ash and non-sucrose sugars present in the original raw sugar. The washed sugar crystals are dissolved in water to yield syrup often referred to as melt liquor. The melt liquor must then be decolourised before the refined sugar can be crystallised from the liquor.

Decolourisation is conducted in two stages: the primary stage involves a carbonation, sulphitation or phosphatation process. Carbonation consists of adding lime to the melt liquor and then bubbling carbon dioxide through the liquor to produce a calcium carbonate precipitate. Sulphitation consists of adding lime to the melt liquor and then bubbling sulphur dioxide through the liquor to produce a calcium sulphate precipitate. Phosphatation uses phosphoric acid, lime and a polyacrylamide flocculent to produce a calcium phosphate precipitate. The secondary stage involves the use of carbonaceous adsorbents (e.g. granular activated carbon) or ion exchange resins as decolourising agents. Crystallisation is the final step in the refining process, and typically follows the same sequence as used for the crystallisation of cane sugar, involving a series of crystallisation steps under vacuum.

The recovered sugar is dried and graded prior to packing, while the syrup is recycled for further recovery. The final syrup is used as the starting material for specialty products such as brown sugar and inverted syrups.

## Ethanol production

The juice used in ethanol production undergoes similar treatment as juice used in sugar production (Figure 7.1).

Fermentation is the most important phase in ethanol production. It starts with the preparation of the must, which is a sugar solution, whose concentration is adjusted so fermentation becomes more efficient. The must is prepared from molasses, juice and water, so that the mixture reaches a final concentration in the range of 16-23° Brix (% soluble solids). The must is then mixed with the yeast suspension and after 4-12 hours fermented wine is produced and sugars (sucrose, glucose and fructose) are converted into ethanol. The wine has an ethanol content of 4-7% and is centrifuged to recover the yeasts, which can be used again or incorporated into animal feed, after drying and deactivating. After the yeasts are separated, the wine undergoes a distillation process, producing a distillate, which is commonly designated as "phlegm" (at 40-50°GL),[1] and a residue designated as "vinasse", which goes to the fields and is used as fertiliser or as animal feed. The rectification phase is a dehydration process, involving fractional distillation in columns using multiple trays, which concentrates the ethanol in the phlegm from distillation so to obtain hydrated ethanol (96°GL) at the end and remove impurities, such as higher homologous alcohols, aldehydes, esters, amines, acids and bases. For the production of 99.7°GL anhydrous alcohol, cyclohexane is used as a dehydrating agent in an additional dehydration phase.

Figure 7.1. **Sugarcane industrial processing**

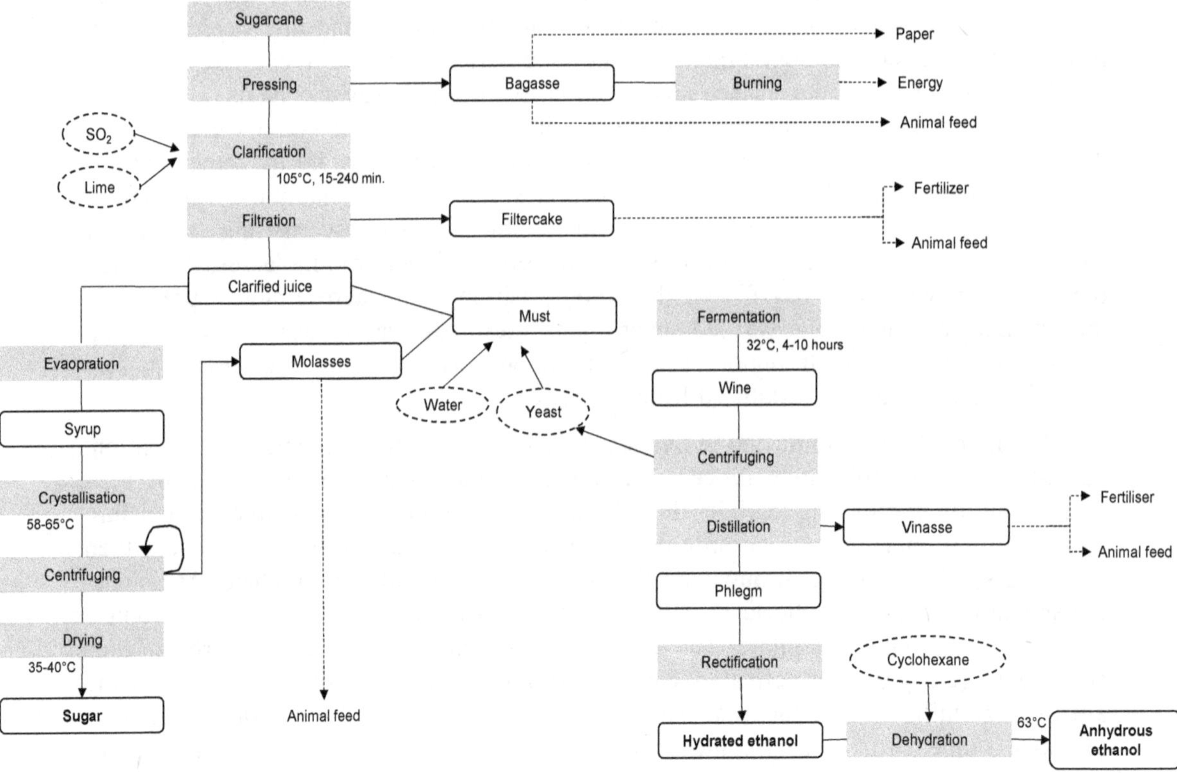


*Source*: Cheavegatti-Gianotto et al. (2011).

## Artisanal processing

Although most of the sugarcane production is devoted to sugar and ethanol production, there are some foodstuffs derived from artisanal sugarcane processing (Figure 7.2) which can be important regionally.

The most widespread of these is sugarcane candy commonly known as panela or rapadura.[2] India and Colombia are the major producers of panela (rapadura), accounting for 66% of world production, estimated at 13 million tonnes (FAO, 2007). Muscovado sugar, which is produced in a similar fashion to rapadura, differs from brown sugar because this last product is obtained by adding molasses to refined white sugar while the production of muscovado sugar does not include refining steps. Sugarcane syrup is produced through the concentration of the sugarcane juice, and is also called "liquid rapadura", due to its similarity with this product.

The production of artisanal sugarcane derivatives is more simplified compared to sugar and ethanol production as it entails very few refining steps (César et al., 2003).

Figure 7.2. **Sugarcane artisanal processing**

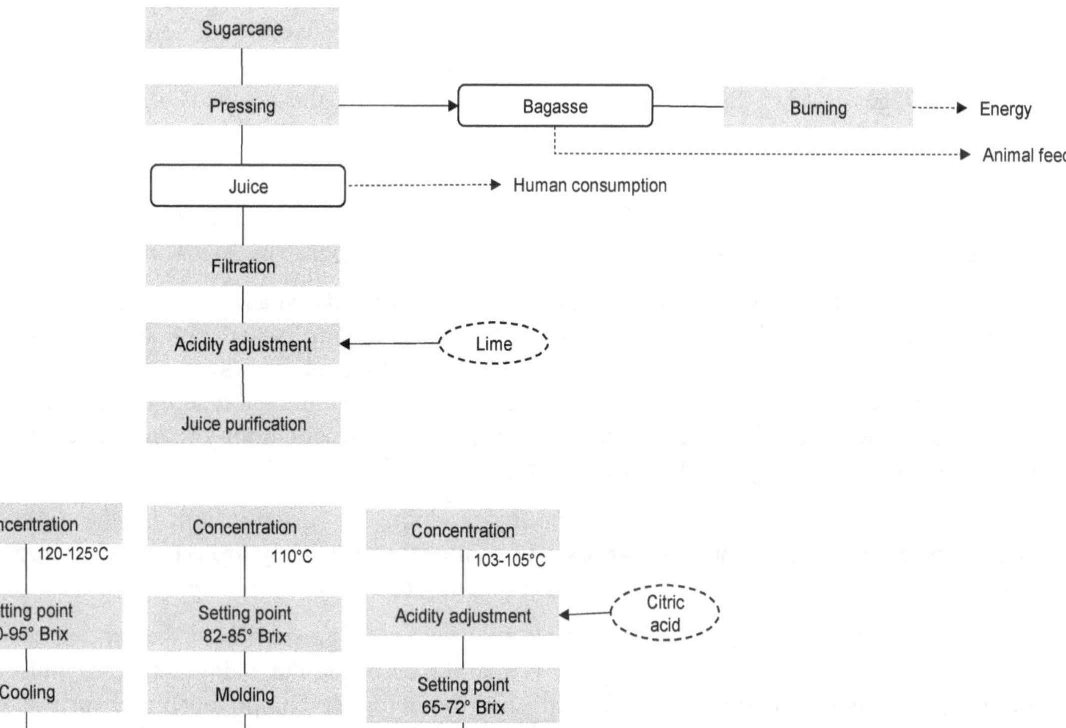


*Source*: Cheavegatti-Gianotto et al. (2011).

## *Uses*

The main products obtained from sugarcane processing are sugar (sucrose) and ethanol. Ethanol is used mainly as a biofuel.

Sugar, as the main food product obtained from sugarcane, is primarily used as a sweetener. Sugar is also used as a preservative, for example for jams and fruits.Beyond industrial sugar, sugarcane is also used to produce artisanal products such as sugarcane juice, muscovado sugar, sugarcane syrup, rapadura and other similar sugarcane candies. Sugarcane juice may also be fermented and then distilled to produce a type of rum called *cachaça*, which is Brazil's most popular distilled alcoholic beverage.

In some regions, sugarcane is grown specifically for fresh juice production. These varieties are distinct from the hybrid varieties grown for commercial sugar production. In Malaysia, for example, particular varieties of noble canes (*S. officinarum*), which have a softer and less fibrous stem, are grown specifically for fresh juice production (Yusof et al., 2000).

While most sugarcane production is intended for sugar and ethanol production, the crop is also cultivated in many countries to be fed to all classes of livestock (FAO, 1988). It is commonly used as feed when availability of conventional forage sources is scarce, for example, during drought conditions, or during winter when the productivity of other forages is low. Since sugarcane is available during the dry season, when it is needed most, it is commonly offered *in natura* to livestock during this period, but it is also possible to ensile it.

Sugarcane juice is also used as feed and is an excellent readily available carbohydrate source for all classes of livestock, but is mainly used for monogastrics, particularly pigs.

The sugarcane crop also produces a number of by-products (sugarcane tops, bagasse, filtercake, molasses and vinasse) after harvest and processing, which are increasingly being recognised as valuable feedstuffs. Feed products obtained from sugarcane are high in fibre and/or energy and therefore are primarily used in ruminant feeding, especially cattle. To meet nutritive requirements, feed rations containing sugarcane or its by-products are usually combined with other feed products.

Sugarcane tops, the major sugarcane by-product, are usually left in the field after harvest but are used for feed purposes in some countries. They are typically offered *in natura* and are highly palatable with good voluntary consumption indices.

Bagasse is primarily used as combustible fuel for power generation at the processing factory. When not used as fuel, bagasse is mainly used for the manufacture of pulp and paper products, building materials (utilising the cellulose component), and furfural and its derivatives (utilising the hemicellulose component) (Cheesman, 2005). Bagasse has also been recognised as a potential feedstuff for large ruminants where it has been used as a roughage ingredient in beef and dairy rations (Pate, 1979). Its use, however, is typically restricted to 15-30% of dry matter due to its low digestibility and palatability, high lignin and very low nitrogen contents. Digestibility can be improved through the use of various chemical or thermo-mechanical treatments, but hydrolysis by steam treatment is most commonly used. Bagasse palatability can also be improved through the addition of molasses.Until recently, bagasse did not have any food uses; however, new technology has enabled bagasse to be used as a source of dietary fibre in processed and baked foods (KFSU, 2009).

Molasses is primarily used to produce either alcohol (potable alcohol or industrial/fuel ethanol) or for animal feeding. In preparing silage for animal feed, the quality can be improved with silage additives such as fibre-degrading enzymes used alone, or in combination with a bacterial inoculant. For low sugar crops, such as grasses and legumes, the concentration of fermentable sugars can be raised with molasses, whey or cereal grains to facilitate the growth of lactic acid-producing bacteria. There are some minor food uses for molasses. Molasses is used as a sweetener and as syrup accompanying other foods, and also as the starting product for the preparation of other edible syrups such as treacle. Molasses is also fermented and then distilled to produce rum. Since the mid-1960s, bacterial fermentation of molasses is used in countries such as Brazil to produce monosodium glutamate, a flavour enhancer commonly referred to as MSG.

Filtercake is mainly used as fertiliser. As this by-product has moderate levels of protein, varying from 5.3-16%, its use for animal feed has been tried in many countries. However, filtercake also contains relatively high levels of wax (15%), which hampers its digestibility, limiting its use in animal feed to a minimum.

Vinasse, the residue produced from the ethanol distillation process, is almost completely used as fertiliser. It has only minimal use in animal feed because of its liquid and corrosive characteristics.

## *Appropriate comparators for testing new varieties*

This chapter suggests parameters that sugarcane breeders should measure when developing new modified varieties. The data obtained in the analysis of a new sugarcane variety should ideally be compared to those obtained from the original non-modified variety from which the new sugarcane variety was obtained,[3] grown and harvested under the same conditions.[4] The comparison can also be made between values obtained from new varieties and data available in the literature, or chemical analytical data generated from other commercial sugarcane varieties.

Components to be analysed include key nutrients and toxicants. Key nutrients are those which have a substantial impact in the overall diet of humans (food) and livestock (feed). These may be major constituents (fats, proteins, and structural and non-structural carbohydrates) or quantitatively more minor compounds (vitamins and minerals). Key toxicants are those toxicologically significant compounds known to be inherently present in the species, whose toxic potency and levels may have an impact on human and animal health. Standardised analytical methods and appropriate types of material should be used, adequately adapted to the use of each product and by-product. The key components analysed are used as indicators of whether unintended effects of the genetic modification influencing plant metabolism have occurred or not.

## *Breeding characteristics screened by developers*

The characteristics most commonly sought by sugarcane breeders are those that have the greatest economic importance and include productivity, disease resistance, as well as various quality parameters (Cox et al., 2000; Berding et al., 2004).

Productivity is measured as sucrose yield per hectare and is influenced by cane yield and sugar content. Sugar content is the most economically important of all the characteristics screened and is therefore an important objective of sugarcane breeding programmes; although the evidence indicates that most productivity gains to date have been delivered via improvements in cane yield (Berding et al., 2004; Jackson, 2005).

Disease resistance has historically been a major focus of sugarcane breeding programmes, with limited genetic variation for resistance or tolerance being available for most diseases of sugarcane (Berding et al., 2004). Most sources of resistance come from wild canes, specifically *S. spontaneum* (Walker, 1987). Major diseases of international distribution and importance include ratoon stunting disease (bacterial), leaf scald (bacterial), smut (fungal), red rot (fungal), rust (fungal)[5] and mosaic (viral)[6] (Rott and Girad, 2000).

Important sugarcane quality parameters include those used to determine millability and juice quality. For milling, the major influencing characteristic is cane fibre, where both fibre quantity and quality are of interest (Berding et al., 2004). Fibre quantity is routinely measured in selection trials, where varieties with excessively high or low fibre content are discarded (Cox et al., 2000). In addition, tests on milling performance are conducted on all varieties being propagated for potential release. These tests measure characteristics such as fibre length and shear strength.

The characteristics routinely measured to determine juice quality are Brix (% soluble solids) and Pol (apparent sucrose in juice) (Mackintosh, 2000). These measures, corrected for fibre content, allow determination of the levels of impurities in the juice (i.e. Brix minus Pol equals the total impurities in the juice), and also enables an estimation of the percentage of recoverable sucrose from the juice (referred to as commercial cane sugar [CCS]), or estimated recoverable crystal [ERC]). The CCS is calculated from measurements of Brix, pol and fibre (Bureau of Sugar Experiment Stations, 1984). While not a direct measurement of sucrose content, the CCS tends to be highly correlated with and similar to sucrose percentage on a fresh weight basis (Muchow et al., 1996). In Australia, the average CCS is about 13%, but values occasionally reach 17% or 18% (Jackson, 2005). Some countries use chemical "ripeners" (e.g. glyphosate) which can increase sucrose content by 0.5-2.0% in early harvested crops (Solomon and Li, 2004). Despite concerted efforts through conventional and molecular breeding, the stored sucrose content of elite sugarcane cultivars has remained static for several decades (Jackson, 2005).

## Nutrients

### *Sugar*

The Codex Standard for Sugars (Codex Alimentarius Commission, 2001) describes refined white sugar, intended for human consumption, as purified and crystallised sucrose (saccharose) with a polarisation not less than 99.7°Z.[7] Generally speaking, refined white sugar contains about 99.93% sucrose, with minor amounts of water, invert or reducing sugars (glucose and fructose), ash, colour components plus other organic non-sugar compounds (Clarke, 1988). Although these minor components typically make up less than 0.1% of sugar content, they may affect the quality of the sugar and its behaviour during storage (van der Poel et al., 1998). The sucrose content of raw sugar varies, but is mainly in the range of 97-99.5% sucrose.

### *Sugarcane juice*

Sugarcane juice is an opaque, viscous liquid of brownish to deep green colour, whose composition varies within limits according to the variety, age and health of the sugarcane, environment, agricultural planning (maturation, harvest period, handling, transport and storage), pests and diseases.

The chemical composition of sugarcane juice is given in Table 7.2. The extracted juice has high water content (about 85%) and contains mainly sucrose and reducing sugars like glucose and fructose.The sugar content is heavily influenced by the maturity of the cane at harvest, with sucrose content increasing with maturity and glucose and fructose content generally decreasing (Qudsieh et al., 2001). The protein content is negligible. In terms of the total amino acid content, the most abundant are aspartic acid, glutamic acid and alanine (van der Poel et al., 1998). The amino acid content of sugarcane juice is given in Table 7.3.

Table 7.2. **Composition of sugarcane juice**

| **Constituent** | **Unit** | **Crude, on-farm**[1] | | **Factory**[2] | *Range* |
|---|---|---|---|---|---|
| | | Perez | FAO and AFRIS[3] | Perez | |
| Moisture | % | 76-84 | 76.2 | 81-85 | 76-85 |
| Total sugars | % DM | 84-90 | NR | 77-85 | 77-90 |
| Ash | % DM | 2.5-2.8 | 0.93 | 3.3-4.8 | 0.9-4.8 |
| Crude protein | % DM | NR | 0.19 | NR | |
| Calcium (Ca) | % DM | NR | 0.06 | NR | |
| Phosphorus (P) | % DM | NR | 0.06 | NR | |

*Notes:* NR: not reported. DM: dry matter. 1. Juice is typically extracted using a simple motorised, draught powered or human operated roller mill. 2. Water is typically added. 3. Reported as single values.

*Sources:* Perez (1997); FAO and AFRIS (2009).

Table 7.3. **Amino acid composition of sugarcane juice**

| **Amino acid** | g/100 g dry matter |
|---|---|
| Aspartic acid | 0.08-0.13 |
| Glutamic acid | 0.03-0.06 |
| Alanine | 0.04-0.08 |
| Valine | 0.02-0.04 |
| Threonine | 0.01-0.05 |
| Isoleucine | 0.01-0.02 |
| Glycine | 0.01-0.02 |
| Leucine | Trace |
| Lysine | Trace |
| Serine | Trace |
| Arginine | Trace |
| Phenylalanine | Trace |
| Tyrosine | Trace |
| Histidine | Trace |
| Proline | Trace |
| Methionine | Trace |
| Tryptophan | Trace |

*Source*: Roberts and Martin (1959).

## *Molasses*

The composition of molasses tends to be highly variable. It is primarily influenced by the processing technology used rather than differences in plant composition.

All grades of molasses contain significant amounts of sugars. The chemical composition of final molasses is given in Table 7.4. In addition to high levels of sugars, molasses is also characterised by having no fat or fibre, and very little protein. Molasses products are low in phosphorus but are reasonably good sources of other minerals, such as calcium and potassium (Table 7.5), although the levels can be quite variable. The vitamin content of sugarcane is not considered to be of any nutritional significance due to the wide variation and low content of most of the important vitamins (Curtin, 1973).

Table 7.4. **Composition of final molasses**

| Constituent | Unit | FAO and AFRIS[1] | Curtin[1] | Wythes et al.[2] | NRC[1] | Chang-Yen et al.[2] | Figueroa and Ly[1] | Bortolussi and O'Neill[3] | Johnson and Miller[1] | Range |
|---|---|---|---|---|---|---|---|---|---|---|
| Moisture | % | 26 | 25 | 23.6 | 25 | 27.76 | 16.5 | 23.4 ± 0.09, 23.5 ± 0.1 | 31.1 | 16.5-31.1 |
| Crude protein | % DM | 4.2 | 3.0 | NR | 5.8 | NR | NR | NR | 4.86 | 3.0-5.8 |
| Ash | % DM | 8.6 | 8.1 | 13.6 | 13.1 | 11.28 | 9.8 | 17.5 ± 0.1, 17.6 ± 0.11 | 18.4 | 8.1-18.4 |
| Sucrose | % DM | NR | NR | 45.8 | NR | NR | 40.2 | 45.2 ± 0.12, 45.4 ± 0.13 | 34.8 | 34.8-45.8 |
| Total sugars | % DM | NR | 48 | 65.3 | NR | NR | 58.3 | 63.8 ± 0.13, 63.7 ± 0.14 | NR | 48-65.3 |

*Notes:* NR: not reported; DM: dry matter. 1. Reported as single values. 2. Values are means. 3. The values are means ± standard error for two sugarcane growing regions in Australia.

*Sources:* FAO and AFRIS (2009); Curtin (1973); Wythes et al. (1978); NRC (1982); Chang-Yen et al. (1983); Figueroa and Ly (1990); Bortolussi and O'Neill (2006); Johnson and Miller (2007).

Table 7.5. **Mineral composition of final molasses**

| Mineral | Unit | Curtin[1] | Wythes et al.[2] | NRC[1] | Johnson and Miller[1] | Range |
|---|---|---|---|---|---|---|
| Calcium (Ca) | % DM | 0.8 | 1.15 | 1.00 | 0.97 | 0.8-1.15 |
| Phosphorous (P) | % DM | 0.08 | 0.07 | 0.11 | 0.74 | 0.07-0.74 |
| Magnesium (Mg) | % DM | 0.35 | 0.61 | 0.43 | NR | 0.35-0.61 |
| Potassium (K) | % DM | 2.4 | 5.19 | 3.84 | 3.03 | 2.4-5.19 |
| Sodium (Na) | % DM | 0.2 | 0.1 | 0.22 | NR | 0.1-0.22 |
| Chloride | % DM | NR | 2.98 | 3.10 | NR | 2.98-3.10 |
| Sulphur (S) | % DM | 0.8 | 0.73 | 0.47 | NR | 0.47-0.8 |
| Copper (Cu) | mg/kg DM | NR | 10.7 | 79.0 | NR | 10.7-79.0 |
| Iron (Fe) | mg/kg DM | NR | 247 | 250.0 | NR | 247-250.0 |
| Manganese (Mn) | mg/kg DM | NR | 82 | 56.0 | NR | 56.0-82 |
| Zinc (Zn) | mg/kg DM | NR | 11.6 | 30.0 | NR | 11.6-30.0 |
| Cobalt (Co) | mg/kg DM | NR | 2.7 | 1.21 | NR | 1.21-2.7 |

*Notes:* NR: not reported; DM: dry matter. 1. Reported as single values. 2. Values are means.

*Sources:* Curtin (1973); Wythes et al. (1978); NRC (1982); Johnson and Miller (2007).

## *Bagasse*

Sugarcane bagasse typically contains approximately 40-50% moisture, and 1-3% sugar, with the remainder as fibre (Payne, 1991). The fibre fraction includes cellulose, hemicellulose and lignin.

The quantity and composition of bagasse varies with variety and maturity of the cane, harvesting practices (green or burnt cane, degree of removal of cane leaves and tops), and the milling process, particularly the amount and temperature of water used for imbibition (van der Poel et al., 1998).

The composition of bagasse is given in Tables 7.6 and 7.7.

Table 7.6. **Composition of bagasse**

| **Constituent** | **Unit** | Clarke | Pate[1] | Kaushal et al.[2] | de Carvalho[3] | dos Anjos et al. | Rabelo et al.[3] | Range |
|---|---|---|---|---|---|---|---|---|
| Moisture | % | NR | 49.0 | NR | 59.89 | 48.8[3] | 50, 60-65 | 48.8-65 |
| Crude protein | % DM | NR | 2.4 | 2.00, 1.54 | 2.32 | 0.8-2.32 | NR | 0.8-2.4 |
| Crude fibre | % DM | NR | 43.0 | NR | NR | 58.5[3] | NR | 43.0-58.5 |
| Cellulose | % DM | 45.3-58.4 | 41.6 | 44.1, 43.3 | NR | NR | 35.8 | 35.8-58.4 |
| Lignin | % DM | 14.3-22.3 | 12.6 | 12.2, 14.2 | NR | NR | 9.91, 20.2 | 9.9-22.3 |
| Hemicellulose | % DM | 22.3-31.8 | NR | 41.8, 42.4 | NR | NR | 16.4 | 16.4-42.4 |
| Acid detergent fibre | % DM | NR | 54.9 | 55.9, 59.8 | 38.34 | 54.4-64.89 | NR | 38.3-64.9 |
| Neutral detergent fibre | % DM | NR | 83.4 | 85.9, 85.7 | 59.02 | 88.3-93.72 | NR | 59.0-93.7 |
| Ether extract | % DM | NR | 0.86 | 0.72, 0.86 | 0.07 | 0.6-1.68 | NR | 0.07-1.7 |
| Ash | % DM | 1.0-3.9 | 1.70 | 3.05, 2.10 | 1.22 | NR | 1.6, 2.2 | 1.0-3.9 |

*Notes:* NR: not reported; DM: dry matter. 1. Values are means of two samples. 2. Reported as single values from two different sugar mills in India. 3. Reported as single values.

*Sources:* Clarke (1978); Pate (1979); Kaushal et al. (1980); de Carvalho (2006); dos Anjos et al. (2008); Rabelo et al. (2010).

Table 7.7. **Mineral composition of bagasse**

| **Mineral** | **Unit** | Pate[1] | Kaushal et al.[2] | Range of values |
|---|---|---|---|---|
| Calcium (Ca) | % DM | 0.15 | 0.274, 0.161 | 0.15-0.274 |
| Phosphorous (P) | % DM | 0.09 | 0.0032, 0.0018 | 0.0018-0.09 |
| Sulphur (S) | mg/kg DM | NR | 1 375, 925 | 925-1 375 |
| Sodium (Na) | mg/kg DM | NR | 29, 56 | 29-56 |
| Potassium (K) | mg/kg DM | NR | 108, 78 | 78-108 |
| Magnesium (Mg) | mg/kg DM | NR | 535, 375 | 375-535 |
| Zinc (Zn) | mg/kg DM | NR | 31, 22 | 22-31 |
| Iron (Fe) | mg/kg DM | NR | 345, 220 | 220-345 |
| Copper (Cu) | mg/kg DM | NR | 52, 8 | 8-52 |
| Manganese (Mn) | mg/kg DM | NR | 30, 18 | 18-30 |

*Notes:* NR: not reported; DM: dry matter. 1. Values are means of two samples. 2. Reported as single values.

*Sources:* Pate (1979); Kaushal et al. (1980).

## *Whole cane*

Sugarcane is considered a semi-perennial since it has to be replanted, on average, every four years. This means that a plant is in the field all year round and, therefore, subject to seasonal variation in its nutrient composition.

The most important constituent in sugarcane is sucrose, which is typically measured in the plant stalk. Sucrose content can be quite variable, typically ranging from 9-20% (fresh weight basis) (Berding, 1997). On a dry weight basis, sucrose content in the stalk can reach as high as 60%. Reported ranges for dry matter sucrose content of varieties grown in Australia include 39.2-59.7% (Berding, 1997) and 30-55% (Inman-Bamber et al., 2009).

Sugarcane is typically harvested when the maturation index (MI), which is the ratio between the Brix of the stalks tip and base, ranges between 0.85 and 1.0. Maturation indexes over 1.0 indicate that the sugarcane is losing its energetic potential due to the sucrose inversion process (dos Anjos et al., 2008).

In certain countries, such as Australia, the main feed product derived from sugarcane production is sugarcane tops, which are left in the field after harvest. In other countries, such as Brazil, it is the whole plant (tops and stalks) that is used as a feed product. In terms of their use as feed, there is no agreed stage of maturity or age when whole cane or tops are harvested, which again can lead to wide variation in reported composition.

In the case of sugarcane tops, composition will also depend on the point at which the top is cut from the cane (Fuller, 2004). Typically, sugarcane tops consist of three distinct parts – the leaves, the bundle leaf sheath and variable amounts of immature cane (Naseeven, 1988). As sugarcane tops include the green leaves and the upper young portion of the stalk, they contain a reasonable amount of protein compared to other types of sugarcane forages, e.g. chopped whole sugarcane (Dixon, 1977) (Table 7.8).

Table 7.8. **Composition of sugarcane tops**

| Constituent | Unit | Dixon[1] | Preston[2] | Mahatab et al.[2] | Naseeven[3] | Rangnekar[4] | Range |
|---|---|---|---|---|---|---|---|
| Moisture | (%) | 68.7 | 73.1 | NR | 71.0 ± 2.3 | NR | 68.7-73.1 |
| Crude protein | (% DM) | 4.0 | NR | 5.60 | 5.9 ± 0.7 | 6.2 | 4.0-6.2 |
| Crude fibre | (% DM) | 36.3 | NR | 33.31 | 33.5 ± 2.1 | 30.9 | 30.9-36.3 |
| Ether extract | (% DM) | 1.5 | 0.84 | 1.70 | 1.7 ± 0.3 | 1.5 | 0.8-1.7 |
| Ash | (% DM) | 9.2 | 7.87 | 5.93 | 8.5 ± 2.1 | 8.5 | 5.9-9.2 |
| Nitrogen-free extract | (% DM) | 49.0 | NR | 53.46 | 50.3 ± 3.9 | 52.9 | 49.0-53.5 |

*Notes:* NR: not reported. 1. Values obtained from pooled samples. 2. Reported as single values. 3. The values are means ± standard deviation. 4. Values are means.

*Sources:* Dixon (1977); Preston (1977); Mahatab et al. (1981); Naseeven (1988); Rangnekar (1988).

Likewise, a moderate level of crude protein exists in whole sugarcane, but only if harvested at a very young age (Pate et al., 1984). This, however, is counteracted by the lower digestibility of young sugarcane compared to mature sugarcane, which has a lower fibre and increased sucrose content (Pate, 1979). The composition of mature whole sugarcane is given in Table 7.9.

Table 7.9. **Composition of mature whole sugarcane**

| Constituent | Unit | Banda and Valdez[1] | Kung Jr. and Stanley[2] | Pate et al.[3] | Rangnekar[3] | Oliveira et al. | Pereira et al.[4] | Fernandes et al. | Azêvedo et al. | de Souza França | Santos et al. | dos Anjos et al. | Range |
|---|---|---|---|---|---|---|---|---|---|---|---|---|---|
| Moisture | % | 77.8 ± 1.76 | 68.50 | 74.2 | 70.0 | 72.46-74.06 | 72.20 | 70.5-80.9 | 69.3-76.9 | 73.10-77.55 | 68.53-70.71 | 67.46-75.37 | 67.5-80.9 |
| Crude protein | % DM | 2.89 ± 0.35 | 1.79 | 2.3 | 2.3 | 2.17-2.62 | 2.50 | 2.2-3.2 | NR | 1.89-3.34 | 3.51-4.08 | NR | 1.8-4.1 |
| Crude fibre | % DM | 25.0 ± 1.66 | 27.7 | 22.7-35.9 | 30.1 | NR | NR | NR | NR | NR | NR | NR | 22.7-35.9 |
| Ether extract | % DM | 0.81 ± 0.22 | 1.13 | NR | 1.3 | NR | NR | NR | NR | NR | NR | NR | 0.8-1.3 |
| NDF | % DM | NR | NR | 52.7 | NR | 40.86-42.31 | 57.83 | 44.8-51.2 | 43.8-52.6 | 46.90-54.33 | 48.60-56.88 | 39.4-77.6 | 39.4-77.6 |
| ADF | % DM | 33.4 ± 1.48 | 34.2 | 35.4 | NR | 25.51-27.17 | NR | NR | 24.3-31.6 | NR | 26.24-36.88 | 24.95-54.37 | 24.3-54.4 |
| Ash | % DM | NR | 3.94 | 4.3 | 6.2 | 1.29-1.43 | NR | 1.2-1.8 | NR | NR | NR | NR | 1.2-6.2 |

*Notes:* NR: not reported; DM: dry matter; NDF: neutral detergent fibre; ADF: acid detergent fibre.

1. Values are means ± standard error. 2. Values obtained from pooled samples. 3. Values are means. 4. Reported as single values.

*Sources:* Banda and Valdez (1976); Kung Jr. and Stanley (1982); Pate et al. (1984); Rangnekar (1988); Oliveira et al. (2007); Pereira et al. (2000); Fernandes et al. (2001); Azêvedo et al. (2003); de Souza França (2005); Santos et al. (2006); dos Anjos et al. (2008).

## Other constituents

### *Sugarcane allergens*

There are no reports in the literature of food-related allergic reactions to sugarcane. There are also no known or putative food, respiratory or contact allergens listed for sugarcane in the Food Allergy Research and Resource Program (FARRP) *Protein AllergenOnline Database* (Version 10).[1]

A small number of literature reports exist of sugarcane pollen acting as an airborne allergen (e.g. Agata et al., 1994; Chakraborty et al., 2001). In countries such as Australia, however, it is reported that commercial sugarcane cultivars rarely flower or produce seed in the field, therefore exposure to sugarcane has not been associated with any reports of allergic responses (Office of the Gene Technology Regulator, 2008).

### *Anti-nutrients and toxicants*

There are virtually no reports in the literature relating to the presence of anti-nutrients in sugarcane.

In terms of anti-nutritional properties, sugarcane generally has low digestibility due to its high fibre content (dos Anjos et al., 2008). This is the case for both monogastrics and ruminants. Bagasse, in particular, has very poor digestibility and may also have a depressing effect on feed intake. Lignin is the key element that limits the digestibility of fibre. In ruminants, lignin is thought to interfere with microbial degradation of fibre polysaccharides by acting as a physical barrier (Buxton and Redfearn, 1997).

According to one unconfirmed report, sugarcane contains the cyanogenic glycoside, dhurrin (β-D-glucopyranosyloxy-(S)-p-hydroxymandelonitrile), which is the same cyanogenic glycoside found in *Sorghum* spp. (De Rosa et al., 2007). The concentration of cyanogenic glycosides in plants varies with the variety, stage of growth, season, time of day and certain environmental as well as agronomic factors (e.g. application of fertiliser). Generally, however, young plants, new shoots and regrowth often contain the highest concentrations of cyanogenic glycosides (Knight and Walter, 2001). Extensive processing of sugarcane will naturally reduce levels of any dhurrin and therefore of exposure to hydrogen cyanide through consumption of sugarcane by animals or humans.

Cyanogenic glycosides themselves are relatively non-toxic (EFSA, 2004; 2007). However, when plant tissues are damaged or stressed, this can result in the hydrolysis of the cyanogenic glycosides by the bacterial enzyme β-glucosidase, leading to the release of free hydrogen cyanide (HCN), which is potentially toxic to both animals – especially ruminants – and humans. Enzymatic conversion of cyanogenic glycosides is enhanced when the plant is chewed, crushed, frozen, wilted or subjected to drought (Knight and Walter, 2001).

Little data are available on the dhurrin content of sugarcane. Foliar extracts from young sugarcane seedlings have been reported to contain dhurrin at the level of 4.3 mg/g fresh weight (range 3.4-5.6 mg/g fresh weight) after wounding (De Rosa et al., 2007). Theoretically, this amount of dhurrin may yield a level of HCN which is potentially harmful to livestock. However, it is not known how representative this reported level is of sugarcane varieties in general, nor are data available on the HCN potential of mature leaves, which are more likely to be fed in whole or in part to livestock. Only a single inconclusive report could be found of cyanide poisoning of livestock (cattle) attributed to feeding of sugarcane under extreme conditions of prolonged drought (Seifert and Beller, 1969). Due to a lack of detail in this report, the information cannot be confirmed. Moreover, the absence of other substantiated reports in the literature suggests that, in practice, the feeding of sugarcane to livestock does not represent a risk in terms of cyanide toxicity. The presence of dhurrin in sugarcane is also most unlikely to represent a risk to humans because extensive processing will reduce or remove both dhurrin and hydrogen cyanide prior to consumption.

## Suggested constituents to be analysed related to food use

### *Key products consumed by humans*

The main food product derived from sugarcane is sugar, which is almost pure sucrose with low traces of reducing sugar. Other food products are molasses, sugarcane juice and various candies.

Although unprocessed sugarcane as a whole is not very often used for human consumption, in some producer countries it is common for sugarcane to be consumed *in natura*, where the harvested stalk is sucked to extract the juice; however, there is almost no intake of its indigestible fibre content. Fresh sugarcane juice is also sold by many street vendors in Southeast Asia, South Asia and Latin America and in some countries may also be bottled for local distribution. It is also gaining popularity in countries such as Australia, where it can be purchased fresh from juice bars, cafes and restaurants. The juice must be consumed soon after extraction as it is rapidly oxidised. The oxidation, which is caused by the activity of polyphenol oxidase, can be reduced using thermal and chemical pre-treatments of stalks prior to juice extraction, significantly prolonging the shelf life of the juice (Eissa et al., 2010).

Few other food uses currently exist for sugarcane, primarily because of the fibrous nature of the stalk. However, recently sugarcane bagasse has been used as a source of dietary fibre for human consumption (KFSU, 2009). Steam, heat and pressure treatment is used to break down the cellulose and hemicellulose in the bagasse, which is then dried and milled as edible plant fibre.

### *Suggested analyses for food use*

Sugarcane's main contribution to the human diet is sugar, mainly in the form of sucrose, and this is primarily obtained through the consumption of refined sugar, with lesser contributions from products such as molasses, candies and sugarcane juice, depending on the country. Sugarcane is not a significant source of other nutrients, although developments in processing and biotechnology may see this change in the future.

While sugar is the main food product derived from sugarcane, analyses of the composition of sugar would be of little value for comparative assessment as sugar is composed almost entirely of sucrose, with only trace amounts of other substances. Analyses of other food products such as molasses would be equally uninformative as molasses composition, in particular, is highly dependent on the refining process used and therefore may be highly variable. These processed products should therefore not be used as the basis for the comparison of different varieties of sugarcane.

As sugarcane is not a significant source of other nutrients, it is recommended that only major constituents be measured for the purpose of comparison, and that these be measured in whole cane (comprising stalks and leaves). The exception to this is sucrose content, which is traditionally measured in the stalk only. Little data are available for sucrose content of whole sugarcane. Since the level of key constituents may vary with the maturity of sugarcane, it is recommended that the analytes to be compared are measured in plants harvested at a similar stage of maturity.

The key constituents suggested to be analysed in sugarcane intended for human consumption are shown in Table 7.10.

Table 7.10. **Suggested constituents to be analysed for food use**

| Constituent | Whole sugarcane |
|---|---|
| Moisture | X |
| Crude protein | X |
| Fat (ether extract) | X |
| Crude fibre | X |
| Ash | X |
| Sucrose | X (stalk) |

## Suggested constituents to be analysed related to feed use

### *Key products consumed by animals*

To compensate for its low mineral and protein levels and low dry matter digestibility, sugarcane is commonly used in combination with other, richer nutritional feeds, or has its composition improved by addition of nitrogen and sulphur salts during feed formulation. It is also possible to improve the digestibility of sugarcane, using sodium hydroxide treatments for example, which break down the fibre content.

Sugarcane tops are used for feed purposes in some countries and are highly palatable. They are mostly fed to large ruminants, but because of their low nutritional quality, they are typically only offered to animals following physical, chemical or biological pre-treatment to increase their nutritional quality. In Australia, sugarcane tops are often conserved as hay during the harvest season (June to December) and fed to cattle during drought conditions (McKenzie and Griffiths, 2007). Sugarcane tops can also be ensiled, and generally are comparable to fresh tops in terms of their feeding value (Deville et al., 1979).

Fresh chopped whole sugarcane is often fed to cattle in sugarcane growing regions. In these situations, the crop must be harvested daily as sugarcane "sours" rapidly and becomes unpalatable if left for any length of time after chopping (Kung and Stanley, 1982). In studies undertaken in Florida, where fresh-chopped whole sugarcane was fed at levels from 20% to 77% of the diet dry matter (with the remainder supplied by corn, citrus pulp and cottonseed meal), the rate of gain, feed utilisation and carcass quality decreased as the percentage of sugarcane in the diet increased (Pate et al., 1984). Fresh-chopped sugarcane has been found to have only 70% the value of corn silage when used as a major diet ingredient (Creek and Squire 1976). The best results are achieved when sugarcane is fed at moderate levels (30-40%).

Whole sugarcane can be ensiled like other forage crops, but its nutritive value is significantly reduced. This is largely because of the sugar content, which is fermented readily to ethanol, and the high moisture content, which produces excessive seepage losses (Pate et al., 1984). In order to avoid alcoholic fermentation, which decreases nutritional content, palatability and animal consumption, it is necessary to add preservatives such as quick-lime, urea, sodium hydroxide, potassium sorbate or *Lactobacillus buchneri* to the material to be ensiled.

Sugarcane tends not to be used for grazing as the sugarcane stool[9] can be destroyed by overgrazing or grazing for extended periods.

The main sugarcane derivatives that are fed to animals are sugarcane juice and molasses. The fermentable carbohydrates in sugarcane juice (sucrose, glucose and fructose) are completely digestible by both ruminant and non-ruminant livestock and are increasingly being used in tropical countries as a viable alternative to starch in cereal grains (Preston, 1988). As sugarcane juice contains virtually no protein, such diets are supplemented with protein extracted from soybean meal or fishmeal (Speedy et al., 1991) or other sources such as cassava (Preston, 1988). When fed at 40% of the dry matter intake, gains of up to 800 g/day in pigs have been achieved (Speedy et al., 1991).

Molasses is often used to supplement cattle grazing poor-quality roughages when energy intake is a limiting factor. However, molasses is a poor source of protein and needs to be supplemented with urea as a non-protein source of nitrogen for sustaining high levels of production. Molasses is also extremely palatable to livestock and therefore is often used to mask unpalatable feed ingredients. Its physical properties also enable it to improve ration composition by minimising fines, dustiness and ingredient separation. For these latter two uses, only low concentrations (5-10%) are required (Preston, 1983). Production responses in cattle to molasses fed at 25-30% of total dry matter intake are about 70% that of grain; molasses efficiency drops off at levels greater than 25-30% of the diet and in rations where there are inadequate levels of roughage and protein (Ashwood, 2008).

### *Suggested analyses for feed use*

The composition of sugarcane by-products such as molasses tends to be highly variable and influenced heavily by the processing technology used. It is therefore recommended that these processing by-products not be used as the basis for the comparison of sugarcane varieties.

Sugarcane is generally fed to livestock as either sugarcane tops or as whole sugarcane, stalks and leaves together. Therefore, analyses should be done either of sugarcane tops or of whole sugarcane, depending on the prevailing feeding practice. Since the level of key constituents may vary with the maturity of sugarcane, it is recommended that the analytes to be compared are measured in plants harvested at a similar stage of maturity.

The key constituents suggested to be analysed in sugarcane intended for animal consumption are shown in Table 7.11. Acid detergent fibre and neutral detergent fibre are relevant analytes particularly for ruminant feed.

Table 7.11. **Suggested constituents to be analysed for animal feed**

| Constituent | Sugarcane tops | Whole sugarcane |
|---|---|---|
| Moisture | X | X |
| Crude protein | X | X |
| Fat (ether extract) | X | X |
| Ash | X | X |
| Crude fibre[1] | X | |
| Acid detergent fibre | | X |
| Neutral detergent fibre | | X |
| Sucrose | | X (stalk) |

*Note:* 1. Crude fibre is typically a component of proximates analysis.

## Notes

1. Alcohol by volume, referred to as degrees Gay-Lussac, or °GL.
2. Also called *papelón*, *raspadura*, *chancaca*, *atado dulce*, *piloncillo*, *empanizao*, *panocha*, *gur* and *jaggery*, depending on the production country.
3. Typically this would be a near isogenic line, however, this term is not appropriate in the case of sugarcane breeding, because no backcrossing is done.
4. For additional discussion of appropriate comparators, see Codex A.C. (2003: par 44 and 45).
5. Common rust and orange rust.
6. Sugarcane mosaic virus (SCMV) and sorghum mosaic virus (SrMV).
7. oZ (sugar degrees) is the unit of the International Sugar Scale.

1. See: www.allergenonline.org.

9. The cluster of cane stalks arising from germination of sugarcane setts, or the regrowth which comes from the buds remaining in the stubble after fully grown stalks are harvested.

## *References*

Agata, H. et al. (1994), "Sensitization to sugar cane pollen in Okinawan allergic children", *Asian Pacific Journal of Allergy Immunology*, Vol. 12, No. 2, pp. 151-154.

Ashwood, A. (2008), "By-product feeds for beef cattle", *Brahman News* Dec.2008 No.161, Austral. Brahman Breeders' Association Ltd,: www.brahman.com.au/technical_information/nutrition/ByProductfeeds.html.

Azêvedo, J.A.G. et al. (2003), "Avaliação da divergência nutricional de variedades de cana-de-açúcar (*Saccharum* spp.)" ["Evaluation of the nutritional divergence of the sugarcane (*Saccharum* spp.) varieties"], *Revista Brasileira de Zootecnia*, Vol. 32, No. 6, pp. 1 431-1 442, http://dx.doi.org/10.1590/S1516-35982003000600018.

Bakker, H. (1999), *Sugar Cane Cultivation and Management*, Kluwer Acad./Plenum Publishers, New York.

Banda, M. and R.E. Valdez (1976), "Effect of stage of maturity on nutritive value of sugar cane", *Tropical Animal Production*, Vol. 1, pp. 94-97.

Berding, N. (1997), "Clonal improvement of sugarcane based on selection for moisture content: Fact or fiction", in: Egan, B.T. (ed.), *Proceedings of the 1997 Conference of the Australian Society of Sugar Cane Technologists*, Watson Ferguson and Company, Brisbane, Australia, pp. 245-253.

Berding, N. et al. (2004), "Plant improvement of sugarcane", in: James, G. (ed.), *Sugarcane*, Second Edition, Blackwell Science, Victoria, pp. 20-53.

Bortolussi, G. and C.J. O'Neill (2006), "Variation in molasses composition from eastern Australian sugar mills", *Australian Journal of Experimental Agriculture*, Vol. 46, No. 11, pp. 1 455-1 463, http://dx.doi.org/10.1071/EA04124.

Bureau of Sugar Experiment Stations (1984), *The Standard Laboratory Manual for Australian Sugar Mills, Vol. 1. Principles and Practices*, Bureau of Sugar Experiment Stations, Indooroopilly, Queensland, AUS.

Buxton, D.R. and D.D. Redfearn (1997), "Plant limitations to fiber digestion and utilization", *Journal of Nutrition*, Vol. 127, No. 5, pp. 814S-818S.

César, M.A.A. et al. (2003), "Pequenas indústrias rurais de cana-de-açúcar: Melado, rapadura e açúcar mascavo", in: da Silva, F.C. et al. (eds.), *Brasília Embrapa Informação Tecnológica*, Vol. 1, pp. 53-82.

Chakraborty, P. et al. (2001), "Differences in concentrations of allergenic pollens and spores at different heights on an agricultural farm in West Bengal, India", *Annals of Agricultural and Environmental Medicine*, Vol. 8, No. 2, pp. 123-130.

Chang-Yen, I. et al. (1983), "Chemical analysis of seven nutrient elements in some sugar-cane products and by-products", *Tropical Agriculture (Trinidad)*, Vol. 60, No. 1, pp. 41-43.

Cheavegatti-Gianotto, A. et al. (2011), "Sugarcane (*Saccharum X officinarum*): A reference study for the regulation of genetically modified cultivars in Brazil", *Tropical Plant Biology*, Vol. 4, No. 1, pp. 62-89, http://dx.doi.org/10.1007/s12042-011-9068-3.

Cheesman, O.D. (2005), *Environmental Impacts of Sugar Production: The Cultivation and Processing of Sugarcane and Sugar Beet*, CABI Publishing, United Kingdom.

Chen, J.C.P and C.C. Chou (1993), *Cane Sugar Handbook: A Manual for Cane Sugar Manufacturers and their Chemists*, 12th Edition, John Wiley & Sons, New York.

Clarke, M.A. (1988), "Sugarcane processing: Raw and refined sugar manufacture", in: Clarke, M.A. and M.A. Godshall (eds.), *Chemistry and Processing of Sugar Beet and Sugarcane*, Elsevier Science Publishers B.V., Amsterdam, pp. 162-175.

Clarke, M.A. (1978), "Cane sugar", in: Kirk, R.E. and D.F. Othmer (eds.), *Encyclopaedia of Chemical Technology*, 3rd Edition, Vol. 3, John Wiley & Sons, New York, pp. 434-438.

Codex Alimentarius Commission (2003, Annexes II and III adopted in 2008), *Guideline for the Conduct of Food Safety Assessment of Foods Derived from Recombinant DNA Plants*, CAC/GL 45/2003, available at: www.codexalimentarius.net/download/standards/10021/CXG_045e.pdf (accessed March 2011).

Codex Alimentarius Commission (2001) "Codex Standard for Sugars", *Codex Standard Series No. 212-1999*, www.codexalimentarius.net/download/standards/338/CXS_212e_u.pdf (accessed March 2011).

Cox, M. et al. (2000), "Cane breeding and improvement", in: Hogard, M. and P. Allsopp (eds.), *Manual of Cane Growing*, Bureau of Sugar Experimental Stations, Indooroopilly, Australia, pp. 91-108.

Creek, M.J. and H.A. Squire (1976), "Use of a slaughter technique for technical and economic evaluation of sugarcane and maize silage based rations", *Tropical Animal Production*, Vol. 1, pp. 56-65.

Curtin, L.V. (1973), "Effect of processing on the nutritional value of molasses", in: *Effect of Processing on the Nutritional Value of Feeds, Proceedings of a Symposium*, Gainsville, Florida, 11-13 January 1972, National Academy of Science, pp. 286-296.

de Carvalho, G.G.P. et al. (2006), "Valor nutritivo do bagaço de cana-de-açúcar amonizado com quatro doses de uréia", *Pesuqisa Agropecuária Brasileira*, Vol. 41, No. 1, pp. 125-132, available at: http://seer.sct.embrapa.br/index.php/pab/article/view/7114/4159.

De Rosa Jr., V.E. et al. (2007), "Sugarcane dhurrin: Biosynthetic pathway regulation and evolution", *Proceedings of the International Society of Sugar Cane Technologists*, ICC, Vol. 26, pp. 958-962.

de Souza França, A.F. et al. (2005), "Avaliação do potencial produtivo e das características químico-bromatológicas de nove variedades de cana-de-açúcar irrigada", *Livestock Research for Rural Development*, Vol. 17, No. 1, available at: www.lrrd.org/lrrd17/1/souz17007.htm (accessed March 2011).

Deville, J. et al. (1979), "The production of silage from sugar cane tops and its use as fodder for cattle", *Tropical Animal Production*, Vol. 4, No. 2, pp. 134-137.

Dillon, S.L. et al. (2007), "Domestication to crop improvement: Genetic resources for *sorghum* and *saccharum* (andropogoneae)", *Annals of Botany*, Vol. 100, No. 5, pp. 975-989, http://dx.doi.org/10.1093/aob/mcm192.

Dixon, F.M. (1977), "Sugarcane for beef production: Derinded sugarcane and chopped cane compared with hay and citrus pulp", *Tropical Animal Production*, Vol. 3, No. 2, pp. 104-108.

dos Anjos, I.A. et al. (2008), "Cana-de-açúcar como forrageira", in: Dinardo-Miranda, L.L. et al. (eds.), *Cana-de-açúcar*, Instituto Agronômico, Campinas.

EFSA (European Food Safety Authority) (2007), "Opinion of the Scientific Panel on Contaminants in the Food Chain (CONTAM) related to cyanogenic compounds as undesirable substances in animal feed", *The EFSA Journal*, Vol. 434, pp. 1-67, http://dx.doi.org/10.2903/j.efsa.2007.434.

EFSA (2004), "Opinion of the Scientific Panel on Food Additives, Flavourings, Processing Aids and Materials in Contact with Food (AFC) on hydrocyanic acid in flavourings and other food ingredients with flavouring properties", *The EFSA Journal*, Vol. 105, pp. 1-28.

Eissa, H.A. et al. (2010), "Preservation of sugarcane juice by canning 1. Effect of thermal and chemical pre-treatments on the enzymatic browning of sugarcane juice", *Journal of American Science*, Vol. 6, No. 9, pp. 883-888, at: www.jofamericanscience.org/journals/am-sci/am0609/100_3647am0609_883_888.pdf.

FAO (2007), "Panela production as a strategy for diversifying incomes in rural areas of Latin America", *AGSF Working Document*, No. 6, Food and Agriculture Organization, Rome, available at: www.fao.org/docrep/016/ap307e/ap307e.pdf.

FAO (1988), "Sugarcane as feed", in: Sansoucy, R. et al. (eds.), *FAO Animal Production and Health Papers*, No. 72, Food and Agriculture Organization, Rome.

FAO and AFRIS (Animal Feed Resources Information System), "Sugarcane and by-products", www.fao.org/ag/aga/agap/frg/AFRIS/index_en.htm (accessed May 2009).

FAOSTAT (2009), FAO Statistics online database, http://faostat.fao.org (accessed March 2011).

FARRP (Food Allergy Research and Resource Program), *Protein AllergenOnline Database*, Version 10, University of Nebraska-Lincoln, available at: www.allergenonline.org (accessed March 2011).

Fernandes, A.M. et al. (2001), "Estimativas da produção de leite por vacas holandesas mestiças, segundo o Sistema CNCPS, em dietas contendo cana-de-açúcar com diferentes valores nutritivos", *Revista Brasileira de Zootecnia*, Vol.30 No.4, pp.1 350-1 357, http://dx.doi.org/10.1590/S1516-35982001000500031.

Figueroa, V. and J. Ly (1990), *Alimentación Porcina No Convencional*, Colección GEPLACEA, Serie DIVERSIFICACION, Mexico, pp. 215.

Fuller, M.F. (ed.) (2004), *Encyclopedia of Farm Animal Nutrition*, CABI Publishing.

Godshall, M.A. (2003), "Sugar and other sweeteners", in: Kent, J.A. (ed.), *Riegel's Handbook of Industrial Chemistry*, 10th Edition, Kluwer Academic Press, New York, pp. 329-361.

Inman-Bamber, N.G. et al. (2009), "Source-sink differences in genotypes and water regimes influencing sucrose accumulation in sugarcane stalks", *Crop & Pasture Science*, Vol. 60, No. 4, pp. 316-327, http://dx.doi.org/10.1071/CP08272.

Jackson, P.A. (2005), "Breeding for improved sugar content in sugarcane", *Field Crops Research*, Vol. 92, Issues 2-3, pp. 277-290.

James, G.L. (2004), "An introduction to sugarcane", in: James, G. (ed.), *Sugarcane*, Second Edition, Blackwell Science, Victoria, pp. 1-19.

Johnson, B.J. and W.F. Miller (2007), *Effect of Cane Molasses on Absorptive Capacity of Rumen Papillae in Dairy Cows During the Dry Period and Early Lactation.*

Kaushal, J.R. et al. (1980), "Chemical composition of sugarcane bagasse and filter-mud and the effect of source and alkali treatment on the *in vitro* dry matter digestibility of bagasse", *Indian Journal of Dairy Science*, Vol. 34, No. 4, pp. 458-462.

KFSU (2009), "Dietary fibre from sugarcane", website at: http://www.ats.business.gov.au/companies-and-technologies/health-and-biotechnology/medical-devices/kfsu-pty-ltd (accessed March 2011.

Knight, A.P. and R.G. Walter (2001), *A Guide to Plant Poisoning of Animals in North America*, Teton NewMedia.

Kung Jr., L. and R.W. Stanley (1982), "Effect of stage of maturity on the nutritive value of whole-plant sugarcane preserved as silage", *Journal of Animal Science*, Vol. 54, No. 4, pp. 689-696, http://dx.doi.org/10.2134/jas1982.544689x.

Lavanholi, M.G.D.P. (2008), "Qualidade da cana-de-açúcar como matéria-prima para produção de açúcar e álcool", in: Dinardo-Miranda, L.L. et al. (eds.), *Cana-de-açúcar*, Instituto Agronomico de Campinas, Campinas.

Mackintosh, D. (2000), "Sugar milling", in: Hogard, M. and P. Allsopp (eds.), *Manual of Cane Growing*, Bureau of Sugar Experimental Stations, Indooroopilly, Australia, pp. 369-377.

Mahatab, S.N. et al. (1981), "The chemical composition and nutritive quality of sugarcane tops", *Livestock Adviser*, Vol. 6, No. 6, June, pp. 5-8.

McKenzie, J. and C. Griffiths (2007), "Cane tops as cattle fodder", *Primefacts*, No. 314, New South Wales Department of Primary Industry, New South Wales, Australia, available at: www.dpi.nsw.gov.au/__data/assets/pdf_file/0009/110160/cane-tops-as-cattle-fodder.pdf.

Muchow, R.C. et al. (1996), "Effect of nitrogen on the time-course of sucrose accumulation in sugarcane", *Field Crops Research*, Vol. 47, Issues 2-3, pp. 143-153.

Naseeven, R. (1988), "Sugarcane tops as animal feed", in: Sansoucy, R. et al. (eds.), *Sugarcane as Feed, FAO Animal Production and Health Papers*, No. 72, Food and Agriculture Organization, Rome, available at: www.fao.org/docrep/003/s8850e/S8850E10.htm.

NRC (1982), *United States – Canadian Tables of Feed Composition: Nutritional Data for United States and Canadian Feeds*, Third Revision, National Academies Press, Washington, DC, available at: www.nap.edu/catalog/1713/united-states-canadian-tables-of-feed-composition-nutritional-data-for.

OECD (2002), "Consensus document on compositional considerations for new varieties of sugar beet: Key food and feed nutrients and anti-nutrients", *Series on the Safety of Novel Foods and Feeds*, No. 3, OECD, Paris, available at: www.oecd.org/env/ehs/biotrack/46815157.pdf.

Office of the Gene Technology Regulator (2008), "The biology of the *Saccharum* spp. (sugarcane)", Version 2: February 2008, Australian Government, Department of Health and Ageing, Office of the Gene Technology Regulator.

Oliveira, J.C. et al. (2007), *Demonstração dos Custos da Cadeia Produtiva da Rapadura: Estudo realizado no Vale do São Francisco*, Custos e @gronegócio on line 3 (Edição Especial) pp. 79-99, available at: www.custoseagronegocioonline.com.br/edicaoespecial2007.html (accessed March 2011).

Pate, F.M. (1979), "Improving the nutritive value of sugarcane bagasse with steam-pressure treatment", *1979 Florida Beef Report*, University of Florida, pp. 46-50, available at: http://animal.ifas.ufl.edu/beef_extension/reports/1979/docs/pate_treatment.pdf.

Pate, F.M. et al. (1984), "Sugarcane as a cattle feed: Production and utilisation", in: Gilbert, R.A. (ed.), *Florida Sugarcane Handbook*, University of Florida, Gainesville, Florida.

Paturau, J.M. (1989), *By-Products of the Cane Sugar Industry: An Introduction to their Industrial Utilization*, 3rd Edition, Elsevier, Amsterdam.

Payne, J.H. (1991), *Cogeneration in the Sugar Industry*, Elsevier, Amsterdam.

Pereira, E.S. et al. (2000) "Determinação das frações protéicas e de carboidratos e taxas de degradação *in vitro* da cana-de-açúcar, da cama de frango e do farelo de algodão", *Revista Brasileira de Zootecnia*, Vol. 29, No. 6, pp. 1 887-1 893, http://dx.doi.org/10.1590/S1516-35982000000600039.

Perez, R. (1997), "Feeding pigs in the tropics", *FAO Animal Production and Health Papers*, No. 132, Food and Agriculture Organization, Rome, at: www.fao.org/docrep/003/w3647e/w3647e00.HTM.

Preston, T.R. (1988), "Fractionation of sugarcane for feed and fuel", in: Sansoucy, S. et al. (eds.), *Sugarcane as Feed, FAO Animal Production and Health Papers*, No. 72, Food and Agriculture Organization, Rome, available at: www.fao.org/docrep/003/s8850e/s8850e27.htm.

Preston, T.R. (1983), "The use of sugarcane and by-products for livestock", in: Shemilt, L.W. (ed.), *Chemistry and World Food Supplies: The New Frontiers CHEMRAWN II*, Presentations at the Int. Conf. on Chemistry and World Food Supplies, Manila, Philippines, 6-10 Dec. 1982, Permagon Press, Oxford.

Preston, T.R. (1977), "Nutritive value of sugar cane for ruminants", *Tropical Animal Production*, Vol. 2, No. 2, pp. 125-142.

Qudsieh, H.Y.M. et al. (2001), "Physico-chemical changes in sugarcane (*Saccharum officinarum* var. yellow cane) and the extracted juice at different portions of the stem during development and maturation", *Food Chemistry*, Vol. 75, Issue 2, November, pp. 131-137.

Rabelo, S.C. et al. (2010), "Aproveitamento de resíduos industriais", in: Santos, F. et al. (eds.), *Cana-de-açúcar: Bioenergia, Açúcar e Álcool – Tecnologia e Perspectivas*, Viçosa, MG, pp. 465-486.

Rangnekar, D.V. (1988), "Integration of sugarcane and milk production in Western India", in: Sansoucy, R. et al. (eds.), *Sugarcane as Feed, FAO Animal Production and Health Papers*, No. 72, Food and Agriculture Organization, Rome, available at: www.fao.org/docrep/003/s8850e/s8850e17.htm.

Roberts, E.J. and L.F. Martin (1959), "Progress in determining organic non-sugars of sugarcane juice that affect sugar refining", in: Deitz, V.R. (ed.), *Proceedings of the 6th Technical Session on Bone Char (1959)*, Bone Char Research Project, Inc., pp. 67-99.

Rott, P. and J.C. Girad (2000), "Sugarcane producing countries, locations and their diseases", in: Rott, P. et al. (eds.), *A Guide to Sugarcane Diseases*, CIRAD-ISSCT, Montpellier France, pp. 323-336.

Santos, R.V. et al. (2006), "Composição química de cana-de-açúcar (*Saccharum* spp.) e das silagens com diferentes aditivos em duas idades de corte", *Ciencia e Agrotecnologia*, Vol. 30, pp. 1 184-1 189.

Seifert, H.S. and K.A. Beller (1969), "Hydrocyanic acid poisoning in cattle caused by grazing on sugarcane (*Saccharum officinarum*) and supplemental feeding of fruits of the algarrobo tree (*Prosopis juliflora*), *Berliner und Münchener tierärztliche Wochenschrift*, Vol. 82, No. 82, pp. 88-91.

Solomon, S. and Y. Li (2004), "Chemical ripening of sugarcane: Global progress and recent developments in China", *Sugarcane Agriculture*, Vol. 6, No. 4, pp. 241-249.

Speedy, A.W. et al., (1991), "A comparison of sugarcane juice and maize as energy sources in diets for growing pigs with equal supply of essential amino acids", *Livestock Research for Rural Development*, Vol. 3, No. 1, pp. 65-74, available at: www.lrrd.org/lrrd3/1/speedy.htm.

USDA Foreign Agricultural Service (2009), *Sugar: World Production Supply and Distribution*, May.

van der Poel, P.W. et al. (1998), *Sugar Technology: Beet and Cane Sugar Manufacture*, Verlag Dr. Albert Bartens KG, Berlin.

Walker, D.I.E.T. (1987), "Breeding for disease resistance", in: Heinz, D.J. (ed.), *Sugarcane Improvement through Breeding*, Elsevier, Amsterdam, pp. 455-502.

Wythes, J.R. et al. (1978), "Nutrient composition of Queensland molasses", *Australian Journal of Experimental Agriculture and Animal Husbandry*, Vol. 18, No. 94, pp. 629-634.

Yusof, S. et al. (2000), "Changes in quality of sugarcane juice upon delayed extraction and storage", *Food Chemistry*, Vol. 68, Issue 4, pp. 395-401.

# Chapter 8

# Low erucic acid rapeseed (canola)

*This chapter, prepared by the OECD Task Force for the Safety of Novel Foods and Feeds with Canada as the lead country, deals with the composition of low erucic acid rapeseed (canola). It updates and revises the original publication on canola composition issued in 2001. It contains elements that can be used in a comparative approach as part of a safety assessment of foods and feeds derived from new varieties. Background is given on low erucic acid rapeseed history, production, processing and use, followed by appropriate varietal comparators and characteristics screened by breeders. Nutrients in low erucic acid rapeseed seed and meal, as well as other constituents (anti-nutrients and toxicants, allergens), are then detailed. The final sections suggest key products and constituents for analysis of new varieties for food use and for feed use.*

## Background

### *History of low erucic acid rapeseed*

Oilseed rape species used to produce low erucic acid rapeseed oil and meal are derived from the *Brassica* genus of the Cruciferae (Brassicaceae) family, also known as the mustard or cabbage family. Oilseed rape was first cultivated in India about 4 000 years ago, and large-scale production was first reported in Europe in the 13th century. The world's supply of low erucic acid rapeseed is largely derived from two species, *B. napus* L.[1] and *B. rapa* L., and to a lesser extent from the mustard *B. juncea* (L.) Czern. Oil from low erucic acid oilseed rape (*B. napus* or *B. rapa* and now *B. juncea*) is also referred to in some countries as canola oil, canola quality mustard oil (*B. juncea*), zero erucic mustard (ZEM) oil, 0-rapeseed oil, low erucic acid oilseed rape (LEAR) oil, double-zero rapeseed oil, 00-Raps oil (in German), 00-colza oil or "colza simple 0" (in French), and non-specifically as: rapeseed oil, *huile de colza*/colza oil (European French/English), turnip rape oil (oil from *B. rapa*) and mustard oil. The non-specific terms apply to rapeseed oil but are sometimes incorrectly used to describe low erucic acid oils (canola oils) from *Brassica* species.

Interest in rapeseed breeding intensified in Canada soon after the crop was introduced from Europe in the 1940s. The initial efforts were directed towards improving agronomic characteristics and oil content. Nutritional experiments conducted as early as 1949 indicated that consumption of large amounts of rapeseed oil with high levels of erucic acid (C22:1) could be detrimental to animals (Boulter, 1983). Concerns about the nutritional safety of rapeseed oil and its potential impact on human health stimulated plant breeders to search for "genetically controlled" low levels of erucic acid in rapeseed. After ten years of backcrossing and selection to transfer the low erucic acid trait into agronomically adapted cultivars, the first low erucic acid varieties of *B. napus* and *B. campestris* were released in 1968 and 1971 respectively (Przybylski et al., 2005). *B. campestris* was later changed by taxonomists to *B. rapa* to reflect its original designation (Bell, 1995). In the late 1970s, the name "canola" was adopted in North America to distinguish the new plant, low erucic acid, from other types of rapeseed. In regions of the world other than Europe, the terms "canola" and "low erucic acid rapeseed" are used interchangeably.

In the 1990s, low glucosinolate *B. juncea* was developed at Agriculture and Agri-Food Canada through an interspecific cross between an Indian *B. juncea* line containing only 3-butenyl-type glucosinolate, and a low-glucosinolate, zero erucic acid *B. rapa* line. The original interspecific F1 generation was then backcrossed to Indian *B. juncea* (Love et al., 1990). Further breeding programmes were then initiated to combine the low glucosinolate characteristics with zero erucic acid and increased oil content of *B. juncea*. In 2001, Health Canada approved the food use of low erucic acid rapeseed oil derived from three "canola-quality" *B. juncea* varieties.

The term "canola" has therefore been registered and adopted by many countries to describe the oil (and seeds[2] and plants) obtained from the species *B. napus*, *B. rapa* and *B. juncea*. Canola must contain less than 2% erucic acid in the oil and less than 30 µmol/g glucosinolates (any one or any mixture of 3-butenyl glucosinolate, 4-pentenyl glucosinolate, 2-hydroxy-3-butenyl glucosinolate and 2-hydroxy-4-pentenyl glucosinolate) in the air-dried, oil-free meal. Throughout this chapter, the term "low erucic acid rapeseed" refers to low erucic acid-low glucosinolate rapeseed, or canola.

## *Production*

Low erucic acid rapeseed is the oilseed with the second-highest commodity production globally (after soybean), with a volume of 60.62 million metric tonnes (MMt), and the third-largest source of plant-based oil (after palm and soybean), with a volume of 22.35 MMt in 2009-10 (see Table 8.1). During the past 30 years, this crop has passed peanut, sunflower and cottonseed in worldwide plant-based oil production.

Canola is produced extensively in Europe, Canada, Asia and Australia, and to a more limited extent in the United States. By region in 2009, the European Union was the world's largest producer of low erucic acid rapeseed with a production of 21.4 MMt, followed by the People's Republic of China (hereafter "China") at 13.5 MMt, Canada at 11.8 MMt, India at 7.2 MMt and Australia at 1.9 MMt (see Table 8.2).

By country, Canada is the largest exporter of low erucic acid rapeseed seed and oil, accounting for 41.8% and 29.8% respectively of world exports. The United States is the largest single importing country of low erucic acid rapeseed oil, estimated at 1.0 MMt for 2008. The United States is Canada's largest export market for low erucic acid rapeseed oil; however, its market share is still only about 5%, or 500 000 tonnes of the over 10 million tonnes of all oil sources consumed annually (Agriculture and Agri-Food Canada, 2006). By country, Japan is the world's largest importer of rapeseed seed, estimated at 2.3 MMt for 2008 (Table 8.2).

The majority of low erucic acid rapeseed production in China is crushed for domestic oil and meal use, although small amounts of exports do occur. Low erucic acid rapeseed oil is second to soybean oil in China and represents approximately 30% of the domestic market (Agriculture and Agri-Food Canada, 2006).

Globally, transgenic low erucic acid rapeseed varieties were grown on 5.9 million hectares in 2008 compared to 5.5 million hectares in 2007. Cultivation areas are found predominantly in Canada and the United States. In Canada, transgenic varieties represented 87% of its total low erucic acid rapeseed crop in 2007. Australia cultivated transgenic rapeseed for the first time in 2008 (GMO Compass, n.d.). Transgenic varieties are also cultivated in Chile (James, 2011).

Table 8.1. **Commodity view of major oilseed and plant-based oil production, 2009-10**

Millions metric tonnes (MMt)

| Crop | Oilseed production, 2009-10 | Plant-based oil production, 2009-10 |
|---|---|---|
| Copra | 5.88 | |
| Coconut | | 3.62 |
| Cottonseed | 39.22 | 4.66 |
| Olive | | 2.91 |
| Palm | | 45.86 |
| Palm kernel | 12.22 | 5.50 |
| Peanut | 32.98 | 4.67 |
| Rapeseed | 60.62 | 22.35 |
| Soybean | 211.96 | 38.76 |
| Sunflower | 30.39 | 11.66 |

*Source*: Adapted from USDA, Foreign Agricultural Service (2011).

Table 8.2. **World production, imports and exports, 2008**

Millions metric tonnes (MMt)

| | Rapeseed production[1] | Exports rapeseed | Exports rapeseed oil | Imports rapeseed | Imports rapeseed oil |
|---|---|---|---|---|---|
| Australia | 1.9 | 0.5 | 0.1 | | |
| Canada | 11.8 | 6.7 | 1.3 | 0.1 | |
| China (People's Republic of) | 13.5 | | | 1.3 | 0.3 |
| European Union | 21.4 | 8.2 | 2.7 | 8.4 | 2.7 |
| India | 7.2 | | | | |
| Japan | | | | 2.3 | |
| United States | 0.7 | 0.5 | 0.2 | 1.0 | 1.0 |
| World | 61.6 | 15.9 | 4.3 | 16.0 | 4.4 |

*Note:* 1. Data for rapeseed production are for 2009.

*Source*: FAOSTAT (2011).

The *B. napus* varieties are produced in areas with longer growing seasons, while *B. rapa* is grown in short season areas. The *B. juncea* varieties have been shown to mature early, and to be more heat and drought tolerant, as well as higher yielding and more resistant to blackleg (a fungal disease), than *B. napus* and *B. rapa.* These characteristics make *B. juncea* well adapted to the semi-arid growing conditions of the Canadian prairies (Potts et al., 1999).

## *Processing*

Canola seed is traditionally crushed and solvent extracted in order to separate the oil from the meal. The process usually includes seed cleaning, seed pre-conditioning and flaking, seed cooking/conditioning, pressing the flake to mechanically remove a portion of the oil, solvent extraction of the presscake to remove the remainder of the oil, oil and meal desolventizing, degumming and refining of the oil, and toasting of the meal. Canola seed can also be subject to cold-press extraction (i.e. no heat or solvent). The main steps of the solvent extraction process are schematised in Figure 8.1.

### *Seed cleaning*

The seed is cleaned to remove plant stalks, grains from other plant species and other materials from the bulk of the seed. Aspiration, indent cleaning, sieving or some combination of these is used in the cleaning process. Dehulling of the seed is, at present, not a commercial process.

### *Seed pre-conditioning and flaking*

Many crushing plants in colder climates preheat the seed to approximately 35°C through grain dryers in order to prevent shattering which may occur when cold seed from storage enters the flaking unit (Unger, 1990). The cleaned seed is first flaked by roller mills set for a narrow clearance to physically rupture the seed coat. The objective here is to rupture as many cell walls as possible without damaging the quality of the oil. The thickness of the flake is important, with an optimum of 0.3-0.4 mm. Flakes thinner than 0.2 mm are very fragile while flakes thicker than 0.4 mm result in lower oil yield.

Figure 8.1. **Prepress solvent extraction process**

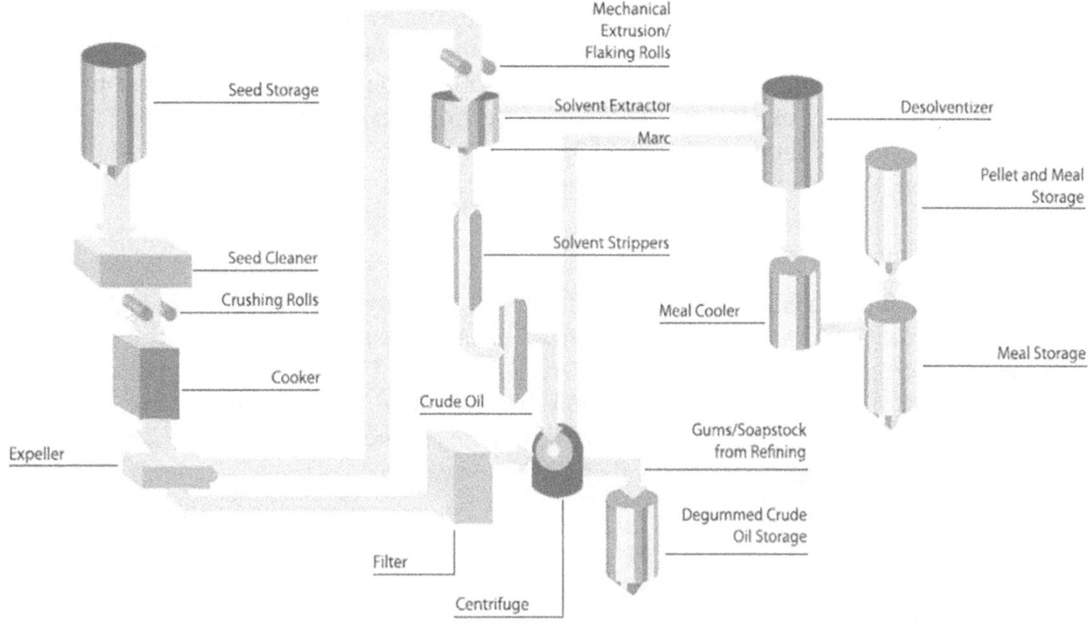


*Source*: Canola Council of Canada (2011).

## *Seed cooking/conditioning*

Flakes are cooked/conditioned by passing them through a series of steam-heated drum or stack-type cookers. Cooking serves to thermally rupture oil cells which have survived flaking, reduce oil viscosity and thereby promote coalescing of oil droplets, increase the diffusion rate of prepared oil cake and denature hydrolytic enzymes. Cooking also adjusts the moisture of the flakes, which is important in the success of subsequent pre-pressing operations. At the start of cooking, the temperature is rapidly increased to 80-90°C. The rapid heating serves to inactivate the myrosinase enzyme present in canola. This enzyme can hydrolyze the small amounts of glucosinolates present in canola and will produce undesirable breakdown products, which affect both oil and meal quality.

The cooking cycle usually lasts 15-20 minutes and the temperatures usually range between 80°C and 105°C, with an optimum of about 88°C. In some countries, especially China, cooking temperatures of up to 120°C have been traditionally used when processing high glucosinolate rapeseed to volatize some of the sulphur compounds which can cause odours in the oil. However, these high temperatures can negatively affect meal protein quality.

## *Pressing*

The cooked canola seed flakes are then pressed in a series of low pressure continuous screw presses or expellers. This action removes most of the oil while avoiding excessive pressure and temperature. The objective of pressing is to reduce the oil content of the seed from about 42% to 14-20%, making the solvent extraction process more economical and efficient, while producing acceptable quality presscake.

### *Solvent extraction*

Since the pressing is not able to remove all of the oil from the canola seed, the presscake is solvent extracted to remove the remaining oil. The cake from the expellers, containing between 14% and 20% oil, is sometimes broken into uniform pieces prior to solvent extraction. In solvent extraction, hexane specially refined for use in the vegetable oil industry is used. After a series of extractions, the marc (hexane saturated meal) that leaves the solvent extractor contains less than 1% oil.

### *Desolventizing of oil and meal*

The micella and meal are "stripped" of solvent, to recover solvent-free oil and meal. The micella containing the oil is desolventized using evaporator equipment. The solvent is removed from the marc in a desolventizer-toaster. This is done in a series of compartments or kettles within the desolventizer, often by injection of live steam, followed by final stripping and drying at a temperature of 103-107°C. The final, solvent-free meal contains about 1% oil and 8-10% moisture.

### *Degumming of oil*

The "crude" oil from the two extraction stages (physical and chemical) is usually blended and then degummed before being stored for sale or further processing. Degumming removes phosphatides co-extracted with the oil, which tend to separate from the oil as sludge during storage. The phosphatide content of crude oil varies, but is usually in the order of 1.25% (or 500 ppm if measured as phosphorus). Two degumming methods are in use: *i)* using water to precipitate phosphatides; *ii)* using an acid such as citric, malic, or phosphoric and water (super-degumming).

### *Alkali and physical refining of oil*

Degummed oil is further purified in a process of refining. One of two methods are used, namely, alkali refining, especially with water degummed oil, and physical refining with acid-water degummed oil. Alkali refining is the most common process used, even with acid-water degummed oil. Physical refining is a relatively new development. While it is very economical, physical refining requires well-degummed oil of moderate chlorophyll and free fatty acid content. Alkali refining reduces soap, free fatty acid and phosphorus levels. The further removal of free fatty acids is done by steam distillation in a deodoriser. This simultaneously deodorises the oil. Because deodorisation is the last process normally carried out on edible oils, this step may be delayed until other processes, such as hydrogenation of the oil, have been done. Alkali-refined oil contains chlorophylloid compounds which give the oil a green colour, and catalyze oil oxidation. These compounds are removed by adsorptive bleaching with acid-activated clays.

### *Effects of processing on meal quality*

The quality of the meal can be enhanced or diminished by altering the processing conditions in the crushing plant. Minimum processing temperatures (see in following section) are needed in order to deactivate myrosinase enzyme, which, if not destroyed, will break down glucosinolates into their toxic metabolites in the animal's digestive tract. The canola crushing process can also cause thermal degradation of 30-70% of glucosinolates in the meal (Daun and Adolphe, 1997). However, if temperatures are too high for too long a period, then the protein quality of the meal can decrease. There can be considerable variation in temperatures used during canola processing.

In these cases, it is important for canola meal users to consider the protein quality of the meal used for animal feed.

Some of the by-products of canola processing are sometimes added back into the canola meal. In the case of added gums and soap stocks, these oil-rich components will increase the energy content of the meal. In the case of added screenings and foreign material, the meal quality will decrease as the fibre content increases. These differences in processing practices may be identified as part of quality control programmes.

*Temperature*

Deactivation of myrosinase enzyme is best accomplished during the canola seed cooking stage. The early research of Youngs and Wetter (1969) regarding steps to minimise glucosinolate hydrolysis by myrosinase has become the operating practice for processors around the world. Moisture content of the seed during processing should be between 6% and 10%. Above 10% moisture, glucosinolate hydrolysis will proceed rapidly, and below 6% moisture the myrosinase enzyme is only slowly inactivated by heat. In addition, the temperature must be raised to 80-90°C as rapidly as possible during seed cooking. Myrosinase catalyzed hydrolysis of glucosinolates will proceed with increasing temperature until the enzyme is deactivated so that a slow rate of heating favours glucosinolate hydrolysis.

Excessive heating during processing can result in reduced animal digestibility of some amino acids, particularly lysine. Processors must exercise strict process control to ensure amino acid damage is minimised by not overheating the meal in the desolventizer-toaster. Examination of meal quality at various processing stages in several Canadian crushing plants revealed that canola meal is a uniform and high-quality product until it enters the desolventizer-toaster phase (Newkirk and Classen, 2000). During this stage, crude protein and lysine digestibility and lysine content were significantly reduced and the apparent metabolisable energy was numerically lower. This research by Newkirk and Classen suggests that the commonly used temperatures in the desolventizer-toaster stage of 105°C cause some protein damage. They found that processing with a maximum temperature of 95°C in the desolventizer-toaster significantly increases lysine digestibility, to similar levels found in soybean meal. Also, traditional toasting causes the meal to become much darker in colour. This is a quality concern for some feed manufacturers, whose customers prefer using light-coloured ingredients.

***Use***

Low erucic acid rapeseed seeds are processed into two major products: oil and meal. The oil and meal are then further manufactured into a wide variety of products for human and agricultural use as well as industrial use. Human food use of whole seeds and flour of low erucic acid rapeseed have been reported anecdotally, and a sensory evaluation of canola greens has been reported (Miller-Cebert et al., 2009).

The oil is used in food processing as well as for home cooking and baking. Refined low erucic acid rapeseed oil is widely used in both salad and cooking oil products, and is also acceptable in hydrogenated products such as margarine and shortenings (Przybylski et al., 2005; Malcolmson and Vaisey-Genser, 2001). In Canada, low erucic acid rapeseed oil represents about 68% of the edible plant-based oil consumed. It is widely used in both salad and cooking oil products (representing nearly 90% of these products), as well as in hydrogenated products such as margarine (representing 45% of these products) and shortenings (representing 50% of these products) (Malcolmson and

Vaisey-Genser, 2001). In the United States, low erucic acid rapeseed oil represents 7-8% of total oil consumption, and is used in all food products requiring an oil source. The oil is also used in a wide variety of non-food products such as dust de-pressants, de-icer for airplanes, suntan oils, biodiesel and bioplastics (Manitoba Canola Growers Association, 2008). By-products such as soap stock are also manufactured from the oil.

Food use of protein fractions from low erucic acid rapeseed meal has not been reported to any great extent (Tan et al., 2011). However, patents have recently been granted in Canada (e.g. Canadian patent CA 2553640) (Canadian Patent Database, 2011), and a firm has notified the US Food and Drug Administration (FDA) of certain uses of particular canola protein isolates that the firm has concluded are generally recognised as safe (GRN No. 327) (FDA, 2010).

The meal left after extraction of oil from the seed is used as a high (36-44%) protein feed source for all classes of livestock, poultry and fish. Prior to the late 1970s, the use of this oilseed processing by-product as an animal feed was limited by the presence of glucosinolates in the seed. Glucosinolates themselves are generally considered to be innocuous; however, the hydrolysis products have negative effects on animal production. The low palatability and the adverse effects of glucosinolates metabolites due to their antithyroid activity led to the development of varieties of rapeseed which have combined low levels of both glucosinolates and erucic acid (also known as "double zero" varieties). On a unit weight basis, canola meal has 55-65% of the value of 47% protein soybean meal for feeding broiler growers, 65-75% for feeding growing swine, and 75-85% for dairy cattle (Canola Council of Canada, 2009).

Low erucic acid rapeseed meal is typically balanced with other protein ingredients (e.g. soybean meal, field peas). Because low erucic acid rapeseed meal contains 30% hulls, it has a high fibre content, which limits its use in monogastric diets (to approximately 15% of the total diet). Higher inclusion rates are practical in ruminant rations, especially for dairy cows. Low erucic acid rapeseed meal can be used as the sole protein supplement for ruminants. De-hulled low erucic acid rapeseed meal has the potential to compete with soybean meal in swine and poultry diets. Meals derived from *B. juncea* have been shown to contain more crude protein and less total dietary fibre on a dry basis than either *B. napus* or *B. rapa* (Simbaya et al., 1995; Newkirk et al., 1997).

Because the oil is highly unsaturated, the amount that can be added to a ration may limit the use of meal from low erucic acid rapeseed meal high in residual oil (i.e. that has been cold-pressed) (Downey, 2007). Excessive levels of supplementation may also be undesirable as the protein requirements of the animal would be exceeded and nitrogen excretion would be increased. Typical rates of inclusion of seed, oil and meal from low erucic acid rapeseed into feed (for different animals) are shown in Table 8.3.

With the increase in market demand for low erucic acid rapeseed oil for the biodiesel market, a significant increase in the supply of low erucic acid rapeseed meal is expected. Properties of the meal arising from biodiesel production are also likely to be different if the oil is derived using cold-press extraction procedures.

## *Appropriate comparators for testing new varieties*

This chapter suggests parameters that breeders of low erucic acid rapeseed should measure when developing new modified varieties. The data obtained in the analysis of a new variety of low erucic acid rapeseed should ideally be compared to those obtained from an appropriate near isogenic non-modified variety, grown and harvested under the same conditions.[3] The comparison can also be made between values obtained from new varieties and data available in the literature, or chemical analytical data generated from commercial varieties of low erucic acid rapeseed.Components to be analysed include key nutrients, anti-nutrients and toxicants. Key nutrients are those which have a substantial impact in the overall diet of humans (food) and animals (feed). These may be major constituents (fats, proteins, and structural and non-structural carbohydrates) or minor compounds (vitamins and minerals). Similarly, the levels of known anti-nutrients and allergens should be considered. Key toxicants are those toxicologically significant compounds known to be inherently present in the species, whose toxic potency and levels may impact human and animal health. Standardised analytical methods and appropriate types of material should be used, adequately adapted to the use of each product and by-product. The key components analysed are used as indicators of whether unintended effects of the genetic modification influencing plant metabolism have occurred or not.

Table 8.3. **Recommended maximum rates of inclusion of low erucic acid rapeseed in feeds**

| Animal | Ingredient | | | |
|---|---|---|---|---|
| | Low erucic acid rapeseed seed | Low erucic acid rapeseed meal | Low erucic acid rapeseed cold-pressed meal | Low erucic acid rapeseed oil |
| Beef[1] | | | | |
| – Cow | 6-10 | 30 | 15 | 3 |
| – Feedlot | 6 | 30 | 15 | 3 |
| Dairy[1] | | | | |
| – Lactating | 3 | 25 | 10 | 3 |
| – Dry | 3 | 25 | 10 | 3 |
| – Calves | ND | 20 | 15 | 3 |
| Swine[2] | | | | |
| – Nursery | ND | 5 | | 3 |
| – Grower | 12-14 | 15 | 15 | 3 |
| – Finisher | 1214 | 15 | 15 | 3 |
| – Sow | 12 | 15 | | 3 |
| Poultry[2] | | | | |
| – Starter | ND | 5 | | 4 |
| – Grower | 10 | 15 | | 4 |
| – Finisher | 10 | 20 | | 4 |
| – Layers | 10 | 10 | | 3 |
| Fish[2] | | | | |
| – Trout/salmon | 20 | 20 | | 10 |
| – Catfish | 30 | 30 | | 10 |
| – Tilapia | 15 | 15 | | 10 |

*Notes:* ND: not determined.

1. Percentage of concentrates on dry matter basis. 2. Percentage of complete feed on dry matter basis.

*Sources*: Hickling (2005); McAllister et al. (1999); Racz and Christensen (2004); Van Barneveld and Ed-King (2002).

## *Breeding characteristics screened by developers*

Phenotypic characteristics provide important information related to the suitability of new varieties for commercial distribution. Selecting new varieties is initially based on parental data. Plant breeders developing new varieties of low erucic acid rapeseed evaluate many parameters at different stages in the developmental process. Typical goals include increasing agronomic flexibility and productivity, capturing niche markets and/or offering end-users more options. Included in this list might be features such as improved yields and yield stability, maturity, winter-hardiness, disease and pest resistance, lodging resistance and specific product attributes. New varieties must meet minimum criteria for yield, oil content, protein content, fatty acid profile, glucosinolate content and disease resistance. In response to concerns about trans fat in partially hydrogenated vegetable oils, low erucic acid rapeseed breeders continue work to develop lines that produce oils with a high oleic and low linolenic acid content.

Herbicide-resistant transgenic low erucic acid rapeseed was first introduced in Canada in 1995. In 2006, over 80% of the acreage of low erucic acid rapeseed in Canada was sown with transgenic varieties. The early stages of transgenic development in low erucic acid rapeseed in Canada focused mainly on herbicide tolerance and the evaluation of transgenic pollination control. The focus of development has shifted to hybrids over the past few years and now the major traits of interest include stress tolerance, metabolic pathway enhancement, biotic stress resistance as well as fatty acid composition modifications.

## Nutrients

### *Composition of low erucic acid rapeseed*

Low erucic acid rapeseed consists mainly of lipids, proteins and fibre. Lipids and protein are quantitatively the most important fractions and account for more than 60% of the seed weight. The average composition of low erucic acid rapeseed is presented in Table 8.4. The data are taken from 2006 to 2009 quality reports from Canada and Australia.

Table 8.4. **Canadian and Australian average composition of low erucic rapeseed seed, oil and meal, 2006-09**

| | Unit | 2006 | | 2007 | | 2008 | | 2009 | |
|---|---|---|---|---|---|---|---|---|---|
| | | CA[4] | AU[7] | CA[4] | AU[8] | CA[5] | AU[9] | CA[6] | AU |
| Oil content in seed | % | 44.6[1] | 42.2[2] | 43.4[1] | 44.0[2] | 44.3[1] | 41.8[2] | 44.5[1] | .. |
| Protein content in oil free meal[1] | % | 41.0[1] | 40.1[3] | 41.2[1] | 40.0[3] | 40.3[1] | 41.0[3] | 38.7[1] | .. |
| Total glucosinolates in seed[1] | µmol/g | 10.0[1] | 4.0[2] | 10.0[1] | 8.0[2] | 10.6[1] | 10.0[2] | 9.6[1] | .. |
| Erucic acid in oil | % | 0.05 | 0.1 | 0.04 | 0 | 0.01 | < 0.1 | 0.01 | .. |
| Linoleic acid in oil | % | .. | 20.2 | 19.3 | 20.4 | 18.4 | 20.3 | 18.8 | .. |
| Linolenic acid in oil | % | 9.9 | 11.1 | 9.8 | 11.0 | 9.1 | 10.7 | 10 | .. |
| Oleic acid in oil | % | 62.0 | 60.0 | 61.5 | 59.7 | 63.2 | 60.0 | 62.2 | .. |
| Total saturated fatty acids in oil | % | 7.0 | 7.2 | 7.0 | 7.4 | 7.1 | 7.6 | 6.8 | .. |
| Iodine value (calculated) | | 113.0 | 116.8 | 113.0 | 116.6 | 111.0 | 115.7 | 114 | .. |

*Notes:* CA: Canada, mean values from samples taken from three Canadian provinces; AU: Australia, mean values from samples taken from four Australian states; ..: not available.

1. 8.5% moisture basis. 2. 6% moisture basis. 3. 10% moisture basis.

*Sources*: Agriculture and Agri-Food Canada [4] (2008); [5] (2009); [6] (2010); Seberry [7] (2007); [8] (2008); [9] (2009).

## *Fatty acids*

Dietary fat serves several important nutritional functions. It is an important source of energy as well as the source of essential fatty acids that are important constituents of cell membranes. Fat serves as a precursor for many biologically active compounds and as a carrier for the fat-soluble vitamins (Przybylski et al., 2005).

Low erucic acid rapeseed oil consists of 91.8-99.0% triglycerides, up to 3.5% phospholipids, 0.5-1.8% free fatty acids, 0.5-1.2% non-saponifiable matter including 700-1000 mg/kg total tocopherols and 5-35 mg/kg pigments and 5-25 mg/kg sulphur (Przybylski et al., 2005).Low erucic acid rapeseed oil has the lowest content of saturated fatty acids (*ca.* 7%) of the vegetable oils (Gunstone, 2005) and it is also characterised by a relatively high level of monounsaturated fatty acids and an appreciable amount of alpha linolenic acid (alpha C18:3) (Przybylski et al., 2005). Fatty acid profiles and levels for low erucic acid rapeseed oil have been defined in the Codex Standard for Named Vegetable Oils (Codex Alimentarius Commission, 2005). Samples falling within the appropriate ranges specified in Table 8.5 are in compliance with this standard.

Fatty acid profiles for rapeseed oil and low erucic acid rapeseed oil from the Codex Standard are presented in Table 8.5.

Table 8.5. **Codex Standard for fatty acid composition of rapeseed oil and low erucic acid rapeseed oil**

% of total fatty acids

| **Fatty acid** | Common name | **Rapeseed** | **Low erucic acid rapeseed** |
|---|---|---|---|
| C6:0 | Caproic | ND | ND |
| C8:0 | Caprylic | ND | ND |
| C10:0 | Capric | ND | ND |
| C12:0 | Lauric | ND | ND |
| C14:0 | Myristic | ND-0.2 | ND-0.2 |
| C16:0 | Palmitic | 1.5-6.0 | 2.5-7.0 |
| C16:1 | Palmitoleic | ND-3.0 | ND-0.6 |
| C17:0 | Heptadecanoic | ND-0.1 | ND-0.3 |
| C17:1 | Heptadecenoic | ND-0.1 | ND-0.3 |
| C18:0 | Stearic | 0.5-3.1 | 0.8-3.0 |
| C18:1 | Octadecenoic (oleic) | 8.0-60.0 | 51.0-70.0 |
| C18:2 | Linoleic | 11.0-23.0 | 15.0-30.0 |
| C18:3 | Linolenic | 5.0-13.0 | 5.0-14.0 |
| C20:0 | Arachidic | ND-3.0 | 0.2-1.2 |
| C20:1 | Gadoleic (eicosenoic) | 3.0-15.0 | 0.1-4.3 |
| C20:2 | Ecosadienoic | ND-1.0 | ND-0.1 |
| C22:0 | Behenic | ND-2.0 | ND-0.6 |
| C22:1 | Erucic | > 2.0-60.0 | ND-2.0 |
| C22:2 | Docosadienoic | ND-2.0 | ND-0.1 |
| C24:0 | Lignoceric | ND-2.0 | ND-0.3 |
| C24:1 | Nervonic (tetracosenoic) | ND-3.0 | ND-0.4 |

*Note:* ND: non-detectable, defined as ≤ 0.05%.

*Source*: Adapted from Codex Alimentarius Commission (2005).

Minor fatty acids occur in low erucic acid rapeseed oil at a range of about 0.01-0.1%, except for palmitoleic acid (C16:1) which is around 0.6%. Conjugated linoleic acid (C18:2) may also be found in the oil often as artefacts of refining and deodorisation. The refining process is also a source of *trans*-isomers of fatty acids that occur as artefacts caused by the isomerization of one or more of the double bonds of *cis* linolenic acid (*cis* C18:3). Such *trans*-isomers can be found in any oil containing linolenic acid (C18:3) and may account for 1% or more of the parent fatty acid.

### *Vitamin K*

Low erucic acid rapeseed oil is a source of Vitamin K1 (phylloquinone) and the vitamin K1 content of the oil has been described in several publications (Table 8.6).

Rapeseed, soybean and olive oils are good sources of phylloquinone, and contain 50-200 µg vitamin K1/100 g oil. These vegetable oils are categorised as the second-most substantial contributors of vitamin K1 to the human diet after green leafy vegetables (FAO/WHO, 2002). The vitamin K1 content of low erucic acid rapeseed oil has been shown to be significantly affected by processing and storage conditions (temperature, exposure to light, etc.) (Ferland and Sadowski, 1992; Gao and Ackman, 1995). Therefore when considering the vitamin K1 content of low erucic acid rapeseed oil, it may be useful to take into account the state of processing and the storage conditions.

Table 8.6. **Vitamin K1 levels in low erucic acid rapeseed oil**

µg per 100 g of oil

| Reference | **Vitamin K1 (Phylloquinone)** |
|---|---|
| Ferland and Sadowski | 141 |
| Gao and Ackman | 125 |
| Shearer et al. | 123 |
| Piironen et al. | 150 |
| | 130[1] |
| Cook et al. | 108[2] |
| | 97[3] |
| Bolton-Smith et al. | 112.5 |
| Kamao et al. | 92 |
| USDA-ARS | 71.4 |

*Notes:* These measurements were obtained by various types of HPLC-based analytical methodologies. These data were obtained from analysis of oil available for retail sale.

1. Cold-pressed oil. 2. Sample prepared by enzymatic digestion and extraction. 3. Sample prepared by direct extraction.

*Sources:* Ferland and Sadowski (1992); Gao and Ackman (1995); Shearer et al. (1996); Piironen et al. (1997); Cook et al. (1999); Bolton-Smith et al. (2000); Kamao et al. (2007); USDA Agricultural Research Service (2011).

### *Tocopherols and sterols*

The main non-saponifiable components of vegetable oils are tocopherols and sterols. Tocopherols, which include Vitamin E, are natural antioxidants and their level in plants is governed by the level of unsaturated fatty acids. A simple increase in unsaturation will result in the formation of higher levels of antioxidants to protect the oil (Przybylski et al., 2005). The distribution of natural tocopherols varies with the different vegetable oils both quantitatively and in the amount of different isomers (Table 8.7). Low erucic acid rapeseed contains mostly alpha- and gamma-tocopherols usually at a 1:2 ratio.

Table 8.7. **Codex Standard for levels of tocopherols in low erucic acid rapeseed oil**

mg/kg

| **Tocopherol** | Low erucic acid rapeseed oil |
|---|---|
| Alpha-tocopherol | 100-386 |
| Beta-tocopherol | ND-140 |
| Gamma-tocopherol | 189-753 |
| Delta-tocopherol | ND-22 |
| Total | 430-2 680 |

*Note:* ND: non-detectable, defined as ≤ 0.05%.

*Source*: Adapted from Codex Alimentarius Commission (2005).

Besides the tocopherols, the sterols are the other non-saponifiable components of vegetable oils. Sterols are found in low erucic acid rapeseed in two forms in equal amounts, free and esterified sterols. The amount of total sterols present in the oil is approximately twice that found in soybean oil and slightly lower than the amount found in corn oil. Total sterols range from 450 mg to 1 130 mg/100 g of oil. The proportions of major sterols are presented in Table 8.8. Although the refining, bleaching and deodorisation of the oil reduces the levels of both tocopherols and sterols (Przybylski et al., 2005), low erucic acid rapeseed oil is still a source of these compounds.

Table 8.8. **Codex Standard of major sterols in low erucic acid rapeseed oil**

% of total sterols

| **Sterol** | Low erucic acid rapeseed oil |
|---|---|
| Cholesterol | ND-1.3 |
| Brassicasterol | 5.0-13.0 |
| Campesterol | 24.7-38.6 |
| Stigmasterol | 0.2-1.0 |
| Beta-sitosterol | 45.1-57.9 |
| Delta-5-avenasterol | 2.5-6.6 |
| Delta-7-stigmastenol | ND-1.3 |
| Delta-7-avenasterol | ND-0.8 |
| Others | ND-4.2 |

*Note:* ND: non-detectable, defined as ≤ 0.05%.

*Source*: Adapted from Codex Alimentarius Commission (2005).

### *Pigments*

Pigments in oilseeds impart undesirable colour to the oil and can promote oxidation in the presence of light as well as inhibit catalysts used for hydrogenation (Przybylski et al., 2005). Chlorophylls without phytol such as chlorophyllides and pheophorbides may present a nutritional effect because of their phototoxicity, which may be followed by photosensitive dermatitis (Endo et al., 1992). A bleaching step in the processing of low erucic acid rapeseed oil removes chlorophyll-related pigments and other colour bodies. In order to mitigate the "poisoning" effect of catalysts during hydrogenation, grading standards for low erucic acid rapeseed seed specify tolerance levels for the number of "green seeds" permitted. Lots which exceed the maximum tolerance level are rejected.

### *Trace elements*

Maximum permitted levels for iron, copper, lead and arsenic for low erucic acid rapeseed oil are provided in the Codex Standard for Named Vegetable Oils (Codex Alimentarius Commission, 2005). These are generally removed to trace levels during processing. Divalent sulphur components, which are decomposition products of glucosinolates, are found in crude low erucic acid rapeseed oil in ranges of 15-35 mg/kg. Refining, bleaching and deodorising steps reduce these levels to 9 mg/kg or lower (Przybylski et al., 2005).

### *Other identity characteristics of oil*

Non-specific measurements such as saponification values, unsaponifiable matter, iodine values and Crismer values are not considered to be necessary in the context of a comparative safety assessment. These measurements are required to compare with the Codex Standard for Named Vegetable Oils (Codex Alimentarius Commission, 2005).

## *Composition of low erucic acid rapeseed seed and meal*

Low erucic acid rapeseed meal is the by-product that remains after lipid extraction. Unlike other oilseeds, the hull is usually not separated from the seed. Table 8.9 provides typical nutritional profiles for low erucic acid rapeseed seed and meal.

Table 8.9. **Range in proximate and fibre composition of low erucic acid rapeseed seed and meal**[1]

| Component | Low erucic acid rapeseed seed[2] | | | Low erucic acid rapeseed meal[3] | | |
|---|---|---|---|---|---|---|
| | Samples | Mean | Range | Samples | Mean | Range |
| Moisture % fresh weight | 91 | 5.6 | 3.2-8.1 | 1 584 | 9.3 | 7.1-11.5 |
| Crude protein % | 91 | 24.7 | 21.3-28.1 | 1 560 | 39.9 | 35.6-44.3 |
| Fat % | 77 | 40.3 | 35.6-44.9 | 644 | 7.4 | 0.3-14.5 |
| Ash % | 10 | 5.0 | 4.1-5.9 | 285 | 7.4 | 6.1-8.7 |
| Crude fibre % | 1 | 9.1 | | 89 | 9.5 | 7.7-11.2 |
| Acid detergent fibre % | 15 | 19.4 | 11.9-26.8 | 890 | 20.8 | 17.6-23.9 |
| Neutral detergent fibre % | 15 | 26.7 | 18.7-34.7 | 949 | 30.1 | 25.6-34.6 |

*Notes:* 1. Dry matter basis, unless otherwise noted. 2. CanolasSeed accumulated crop years: 05/01/2000 through 04/30/2010. 3 Canola meal, dry accumulated crop years: 05/01/2000 through 04/30/2010.

*Source*: Adapted from Dairy One Cooperative Inc. (2010).

As can be seen from Table 8.9, there is a considerable range in the proximate composition of the seed and meal, some of which can be traced to the regional variability in the seed (Racz and Christensen, 2004) as well as to the method used to extract oil (Bonnardeaux, 2007). Regional and environmental variability in the composition of the seed is demonstrated in data presented by Pritchard et al. (2000), where a substantially lower range (17.4-23.0% dry matter) of crude protein content is reported.

Levels of vitamins and minerals are given in Tables 8.10 and 8.11.

Table 8.10. **Vitamin composition of low erucic acid rapeseed meal**

mg/kg, dry matter basis

| Vitamin | Low erucic acid rapeseed meal |
|---|---|
| Biotin | 0.98-1.1 |
| Choline | 6 700.0 |
| Folic acid | 0.8-2.3 |
| Niacin | 160.0 |
| Pantothenic acid | 9.5 |
| Pyridoxine | 7.2 |
| Riboflavin | 5.8 |
| Thiamin | 5.2 |
| Vitamin E | 13.0-14.0 |

*Sources*: Adapted from Hickling (2001) and Bell (1995).

Table 8.11. **Range in mineral composition of low erucic acid rapeseed meal**

Dry matter basis

| Mineral | Unit | Low erucic acid rapeseed meal [1] | | |
|---|---|---|---|---|
| | | Samples | Mean | Range |
| Calcium (Ca) | % | 589 | 0.74 | 0.49-0.99 |
| Phosphorus (P) | % | 597 | 1.12 | 0.94-1.29 |
| Magnesium (Mg) | % | 556 | 0.53 | 0.39-0.68 |
| Potassium (K) | % | 557 | 1.28 | 1.11-1.46 |
| Sodium (Na) | % | 557 | 0.06 | 0.00-0.31 |
| Sulfur (S) | % | 379 | 0.71 | 0.54-0.89 |
| Chloride | % | 137 | 0.12 | 0-0.27 |
| Iron (Fe) | ppm | 553 | 243.02 | 56.85-429.19 |
| Zinc (Zn) | ppm | 553 | 61.25 | 10.53-111.96 |
| Copper (Cu) | ppm | 553 | 5.92 | 0-24.24 |
| Manganese (Mn) | ppm | 553 | 64.06 | 15.25-112.86 |
| Molybdenum (Mo) | ppm | 553 | 0.93 | 0.31-1.55 |

*Note:* 1. Canola meal, dry accumulated crop years: 05/01/2000 through 04/30/2010.

*Source*: Adapted from Dairy One Cooperative Inc. (2010).

The amino acid composition and ranges over all geographic locations of low erucic acid rapeseed seed and meal are given in Table 8.12. The amino acid composition of low erucic acid rapeseed meal compares generally very well with that of soybean meal.

Soybean meal has higher lysine content and low erucic acid rapeseed meal contains more of the sulphur-containing amino acids, methionine and cystine.

Table 8.12. **Mean and/or range of amino acid composition of low erucic acid rapeseed seed and meal**

% of dry matter basis

| Amino acid | Fickler | | | | Bell et al. | | | Newkirk et al. | | Canola Council of Canada |
|---|---|---|---|---|---|---|---|---|---|---|
| | Low erucic acid rapeseed seed | | Low erucic acid rapeseed meal | | *B. napus* meal | *B. rapa* meal | *B. juncea* meal | NTCM[1] | TCM[2] | Low erucic acid rapeseed meal |
| | Mean | Range | Mean | Range | Mean | Mean | Mean | Mean | Mean | Mean |
| Alanine | 0.86 | 0.71-1.09 | 1.54 | 1.19-1.81 | 1.70 | 1.75 | 1.88 | 1.74 | 1.71 | 1.57 |
| Arginine | 1.19 | 0.93-1.55 | 2.07 | 1.37-2.65 | 2.15 | 2.13 | 2.53 | 2.34 | 2.59 | 2.08 |
| Aspartate + asparagine | | | | | | | | 2.90 | 2.83 | 2.61 |
| Aspartic acid | 1.48 | 1.20-2.03 | 2.50 | 1.96-3.47 | 2.68 | 2.73 | 3.02 | | | |
| Cystine | 0.46 | 0.32-0.52 | 0.85 | 0.58-1.13 | 0.97 | 0.83 | 0.90 | 0.92 | 0.93 | 0.86 |
| Glutamate + glutamine | | | | | | | | 6.45 | 7.13 | 6.53 |
| Glutamic acid | 3.23 | 3.23-4.35 | 6.11 | 4.22-7.60 | 5.92 | 5.60 | 6.02 | | | |
| Glycine | 0.99 | 0.82-1.29 | 1.78 | 1.36-2.07 | 1.92 | 1.87 | 2.00 | 1.95 | 1.92 | 1.77 |
| Histidine | 0.53 | 0.41-0.68 | 0.96 | 0.65-1.25 | 1.03 | 1.01 | 1.12 | 1.24 | 1.21 | 1.12 |
| Isoleucine | 0.76 | 0.62-1.02 | 1.38 | 1.02-1.62 | 1.03 | 1.18 | 1.28 | 1.73 | 1.69 | 1.56 |
| Leucine | 1.34 | 1.07-1.77 | 2.46 | 1.80-2.84 | 2.47 | 2.50 | 2.69 | 2.80 | 2.76 | 2.54 |
| Lysine | 1.14 | 0.96-1.50 | 1.76 | 1.13-2.36 | 2.03 | 2.05 | 2.08 | 2.35 | 2.16 | 2.00 |
| Methionine | 0.38 | 0.27-0.52 | 0.69 | 0.50-0.84 | 0.79 | 0.76 | 0.75 | 0.77 | 0.81 | 0.74 |
| Methionine + cystine | 0.84 | 0.64-1.19 | 1.56 | 1.11-1.97 | | | | | | 1.60 |
| Phenylalanine | 0.79 | 0.64-1.07 | 1.42 | 1.06-1.70 | 1.72 | 1.66 | 1.77 | 1.53 | 1.50 | 1.38 |
| Proline | 1.13 | 0.85-1.53 | 2.16 | 1.43-3.19 | 2.59 | 2.43 | 2.66 | 2.39 | 2.34 | 2.15 |
| Serine | 0.83 | 0.69-1.12 | 1.49 | 1.16-1.87 | 1.99 | 1.95 | 2.05 | 1.59 | 1.57 | 1.44 |
| Threonine | 0.86 | 0.74-1.17 | 1.51 | 1.12-1.67 | 1.40 | 1.49 | 1.54 | 1.74 | 1.71 | 1.58 |
| Tryptophan | 0.27 | 0.20-0.37 | 0.48 | 0.35-0.58 | 0.29 | 0.41 | 0.23 | | | 0.48 |
| Tyrosine | | | | | 1.14 | 1.07 | 1.14 | | | 1.16 |
| Valine | 0.99 | 0.80-1.33 | 1.77 | 1.33-2.09 | 1.33 | 1.49 | 1.57 | 2.18 | 2.1 | 1.97 |

*Notes:* 1. NTCM: non-toasted canola meal. 2. TCM: toasted canola meal.

*Sources:* Fickler (2005); Bell et al. (1998); Newkirk et al. (2003); Canola Council of Canada (2009).

## Other constituents

### *Anti-nutrients and toxicants*

Glucosinolates are considered anti-nutritional factors in low erucic acid rapeseed meal. On their own they are innocuous, but when cells of the seed are ruptured, glucosinolates come in contact with myrosinase. The myrosinase enzyme hydrolyzes the glucosinolates releasing sulphur, glucose and isothiocyanates. The isothiocyanates are goitrogenic, reducing the ability of the thyroid to absorb iodine (Downey, 2007). These metabolites of glucosinolates can affect animal performance and can be toxic to the liver and kidneys (Tripathi and Mishra, 2007). Heating during processing of the meal eliminates most of the myrosinase, but is not completely effective in eliminating the effects of glucosinolates because some intestinal microflora also produces myrosinase

(Tripathi and Mishra, 2007). Isothiocyanates are bitter compounds, and can also reduce palatability. Mean levels of glucosinolates in seed and meal are presented in Table 8.13.

Table 8.13. **Mean levels of glucosinolates of low erucic acid rapeseed seed and meal**

µmol/g

| Toxicant | Newkirk et al. | | Bell | | Bell et al. | | |
|---|---|---|---|---|---|---|---|
| | NTCM[1] | TCM[2] | Seed | Meal | *B. napus* meal | *B. rapa* meal | *B. juncea* meal |
| *Total glucosinolates* | 26.0 | 31.0 | 38.42 | 21.06 | | | |
| 3-Butenyl | 3.40 | 1.94 | 7.44 | 4.97 | 3.2 | 3.4 | 22.6 |
| 4-Pentenyl | 0.67 | 0.38 | 2.55 | 1.67 | 0.4 | 2.6 | 1.7 |
| 2- Hydroxy-3-butenyl | 6.28 | 3.64 | 13.44 | 8.82 | 7.4 | 6.7 | 3.5 |
| 2-Hydroxy-4-pentenyl | 0.2 | 0.2 | 0.99 | 0.74 | 0.1 | 1.0 | 0.1 |
| 3-Indolylmethyl | 0.58 | 0.22 | 0.63 | 0.38 | 1.1 | 0.2 | 0.1 |
| 4-Hydroxy-3-indoylmethyl | 4.20 | 0.78 | 13.37 | 4.48 | 9.2 | 4.2 | 4.0 |
| *Contaminant glucosinolates* | | | | | | | |
| 2-propenyl (allyl) | 0.52 | 0.37 | 1.41 | 1.05 | | 0.2 | 0.3 |
| 4-Hydroybenzyl | | | 2.31 | 2.25 | | | |

*Notes:* 1. NTCM: non-toasted canola meal. 2. TCM: toasted canola meal.

*Sources:* Newkirk et al. (2003); Bell (1995); Bell et al. (1998).

Low erucic acid rapeseed contains several phenolic compounds. Sinapine is the choline ester of sinapic acid and is the principle phenolic compound found in low erucic acid rapeseed. Levels in the meal have been reported to be in the range of 0.7-1.1% for North American and European plant varieties (Kowslowska et al., 1990), and 1.5% in Australian varieties (Bonnardeaux, 2007). Sinapine is converted into trimethylamine by intestinal microflora that is then absorbed. Most animals have the ability to convert the trimethylamine to trimethylamine oxide, a compound easily excreted. However, some animals, in particular laying hens, cannot readily catabolise trimethylamine, resulting in higher than normal levels in tissues and eggs, imparting a fishy odour and flavour.

Tannins are more complex phenolic compounds that can bind proteins and some complex carbohydrates and can reduce digestibility. Levels in low erucic acid rapeseed are typically 1-3% (Kozlowska et al., 1990). Some analytical methods include the simpler phenols, such as sinapine, and may therefore overestimate the amounts of tannins (Kozlowska et al., 1990).

Phytic acid (known as inositol hexakisphosphate [IP6], or phytate when in salt form) is the principal storage form of phosphorus in many plant tissues. Because of phytic acid binding capabilities, bio-availability of phosphorus from low erucic acid rapeseed is less available for monogastric animals because they lack the digestive enzyme, phytase, required to separate phosphorus from the phytate molecule. Phytic acid has also strong binding affinity to important minerals such as calcium, magnesium, iron and zinc, thus reducing the absorption of these minerals.

Anti-nutrient levels in low erucic acid rapeseed meal as a percent of oil-free meal are shown in Table 8.14.

Table 8.14. **Anti-nutrients of low erucic acid rapeseed meal**

% of oil-free meal

| Anti-nutrient | Bell | Canola Council of Canada | Kozlowska et al. | Bonnadeaux |
|---|---|---|---|---|
| Tannins | 1.5 | 1.5 | 1-3 | |
| Sinapine | 0.7-3.0 | 1.0 | 0.7-1.1 | 1.5 |
| Phytic acid | 2.0-5.0 | 3.3 | | |

*Sources:* Bell (1995); Canola Council of Canada (2009); Kozlowska et al. (1990); Bonnadeaux (2007).

### *Allergens*

There are several published studies reporting sensitivity and allergenicity of adults to *Brassica* species; however, most describe rare cases of respiratory symptoms due to occupational exposure (Suh et al., 1998; Alvarez et al., 2001), or residence in proximity to areas of intense canola cultivation (Trinidade et al., 2010). Discussion of occupational exposure is outside the scope of this chapter.

There are also published studies investigating the potential for *B. rapa* and *B. napus* to be food allergens in children. In one report, 1 887 children presenting primarily with atopic dermatitis (a symptom frequently associated with food allergy) were screened for *Brassica* sensitivity in a skin prick test, of which 206 (10.9%) tested positive (Poikonen et al., 2006). Allergic reaction was confirmed in 89% of these cases by oral challenge (lip swab and ingestion) with crushed seeds of *B. rapa* (*ibid.*). It was also observed that sensitisation to canola in children is associated with multiple allergies to other foods and pollens (Poikonen et al., 2008), and monosensitive patients are very rare.

Parallel studies identified the likely major IgE-reactive antigens in seeds (Puumalainen et al., 2006) and characterised potential cross-reactivity with related mustard plants, which are known food allergens (Poikonen et al., 2009). Because protein is either at very low levels or absent in low erucic acid rapeseed oil, the significance of the results of these allergenicity studies in determining the safety of consumption of low erucic acid rapeseed oil by the general population is likely low (Gylling, 2006).

Food allergy to low erucic acid rapeseed oil has not been reported in the scientific literature.

## Suggested constituents to be analysed related to food use

### *Low erucic acid rapeseed oil*

Globally, low erucic acid rapeseed oil has the potential to help consumers achieve dietary goals because it has the lowest concentration of saturated fatty acids (7% of total fatty acids) of all oils commonly consumed globally.

The successful reduction in erucic acid content has led to continued interest in compositional modifications to low erucic acid rapeseed oil. Subsequent mutagenesis of low erucic acid rapeseed led to the development of low erucic acid rapeseed oil with the linolenic acid content reduced from approximately 10% to less than 3%. Although high levels of linolenic acid are desirable from a nutritional point of view, they are undesirable in terms of chemical stability. High levels of polyunsaturated fatty acids lead to oxidative rancidity, a reduction in shelf life of the oil, and the development of off-flavours and odours after prolonged storage or repeated frying use (Przybylski et al., 2005). Reducing the level of linolenic acid also reduces the need for partial hydrogenation of edible oils used in the liquid form.

Other recent developments in low erucic acid rapeseed oil include the application of mutagenesis to produce high levels of oleic acid (i.e. from 60% to 75% total fatty acid content). The resulting high oleic acid producing cultivar was then crossed to low-linolenic cultivars to create high oleic/low linolenic lines. High oleic oils resemble the fatty acid composition of olive oil more closely than that of traditional low erucic acid rapeseed. Recombinant DNA technology has been applied to increase the levels of lauric (39%) and myristic acids (14%) in low erucic acid rapeseed oil. These oils have been developed for use in confectionery coatings, coffee whiteners, whipped toppings and centre filling fats. Low erucic acid rapeseed oil with stearic acid levels as high as 40% are being developed as replacements for hydrogenated fats in baked products. Oils with approximately 10% palmitic acid levels that result in improved crystallization in margarine products have also been developed and are being marketed in North America, Europe and Asia. These oils have also been developed through the use of recombinant DNA technology (Przybylski et al., 2005).

## *Recommendation of key components to be analysed*

For human nutrition, it is important to assess the fatty acid composition, vitamin E and vitamin K1 content of the oil. Constituents to be analysed are suggested in Table 8.15.

Because low erucic acid rapeseed meal may be used in the production of protein isolates, key nutrients in the protein fraction would include protein and amino acid composition, both of which could be analysed in either seed or meal. Because there are several different processes that may be used to produce canola protein isolate (Tan et al., 2011), compositional analysis of the seed or meal may be of greater utility than compositional analysis of specific individual protein isolates.

Table 8.15. **Suggested constituents to be analysed in low erucic acid rapeseed for human food**

| Constituent | Seed or meal | Oil |
|---|---|---|
| Crude protein[1] | X | |
| Crude fat[1] | X | |
| Ash[1] | X | |
| Amino acids | X | |
| Fatty acids[2] | X | X |
| Vitamin K1[2] | X | X |
| Vitamin E[2] | X | X |
| Glucosinolates | X | |
| Tannins | X | |
| Sinapine | X | |
| Phytic acid | X | |

*Notes:* 1. These components should be measured using a method suitable for the measurement of proximates. 2. Measurement of this component can be conducted in seed and/or oil.

The complete fatty acid profile (including C6:0 to C24:0) should be quantified in low erucic acid rapeseed oil for the purpose of compositional comparison between a modified low erucic acid rapeseed and appropriate comparators (e.g. commercial low erucic acid rapeseed varieties).

## Suggested constituents to be analysed related to feed use

### *Low erucic acid rapeseed for feed*

Low erucic acid rapeseed is used as a protein source for all classes of livestock, poultry and fish. The protein content of the meal is lower than that found in the meal from other oilseeds such as sunflower or soybean, because the hull of the low erucic acid rapeseed is typically not removed. Consequently, the fibre content is higher than in other oilseed meals. Low erucic acid rapeseed oil is frequently used to increase the energy density of diets and to improve palatability by reducing dust. Low erucic acid rapeseed oil would be used at 3-10% of the total ration, depending on the animal species.

Low erucic acid rapeseed meal is often blended with other sources of protein in feed ration balancing schemes. The meal is recognised as an excellent source of methionine and cystine, but contains less lysine than soybean meal. The digestibility of amino acids from low erucic acid rapeseed meal by pigs and poultry tends to be in the 75-85% range, about 10% lower than soybean meal (Hickling, 2001).

Processing methods in countries like Canada are reasonably standard (Hickling, 2005), and there is little variation in the amount of oil in low erucic acid rapeseed meal. However, this can be more variable in some parts of the world (Van Barneveld and Ed-King, 2002) and higher oil levels dilute the amounts of other nutrients in the final product. There may also be varietal and environment-influenced differences in the protein content of seeds. It is therefore advisable to routinely analyse low erucic acid rapeseed meal for fat and crude protein.

In most countries, a maximum fibre level in the form of acid detergent fibre and neutral detergent fibre is stated for finished feed products. Low erucic acid rapeseed meal can make a significant contribution to the fibrousness of feeds, particularly for non-ruminants, and can be the limiting factor regarding rate of inclusion in diets. Fibre analyses may be required if levels must meet a guarantee.

The mineral and vitamin composition of low erucic acid rapeseed meal is comparable to the mineral composition of other oilseeds. Minerals and vitamins are often added to livestock diets in stock quantities as premixes or base mixes, which de-emphasises the minerals and vitamins in the meal. One exception is phosphorus. The phosphorus in low erucic acid rapeseed meal is only about 30-50% available, due to the presence of phytic acid.

### *Recommendation of key nutrients and anti-nutrients to be analysed*

Proximate and fibre (acid detergent fibre and neutral detergent fibre) analyses are generally used by animal nutritionists to evaluate feed ingredients and to formulate least-cost rations for livestock, poultry and fish. Protein, fat and fibre are the key indicators of livestock feed quality. Amino acids and digestibility must also be considered when formulating rations based on low erucic acid rapeseed meal. The amino acid profile is a key indicator of protein quality. It is additionally advisable to provide analytical results for calcium and phosphorus, as shown in Table 8.16.

Table 8.16. **Suggested constituents to be analysed in low erucic acid rapeseed for feed use**

| Constituent | Seed or meal | Oil |
|---|---|---|
| Crude protein[1] | X | |
| Crude fat[1] | X | |
| Ash[1] | X | |
| Amino acids | X | |
| Fatty acids[2] | X | X |
| Acid detergent fibre | X | |
| Neutral detergent fibre | X | |
| Calcium | X | |
| Phosphorus | X | |
| Tannins | X | |
| Glucosinolates | X | |
| Sinapine | X | |
| Phytic acid | X | |

*Notes:* 1. These components should be measured using a method suitable for the measurement of proximates. 2. Measurement of this component can be conducted in seed and/or oil.

# Notes

1. For information on the environmental considerations for the safety assessment of oilseed rape, see OECD (2012).
2. In this document, seed refers to seed for human and animal consumption as opposed to seed for sowing.
3. For additional discussion of appropriate comparators, see Codex Alimentarius Commission (2003; paragraphs 44 and 45).

# *References*


Agriculture and Agri-Food Canada (2010), *Quality of Western Canadian Canola: 2009*, Canadian Grain Commission, available at: www.grainscanada.gc.ca/canola/harvest-recolte/2009/hqc09-qrc09-eng.pdf (accessed 1 April 1 2010).

Agriculture and Agri-Food Canada (2009), *Quality of Western Canadian Canola: 2008*, Canadian Grain Commission, available at: www.grainscanada.gc.ca/canola/harvest-recolte/2008/canola-2008-eng.pdf (accessed 1 April 2010).

Agriculture and Agri-Food Canada (2008), *Quality of Western Canadian Canola: 2007*, Canadian Grain Commission, available at: http://publications.gc.ca/collections/collection_2012/ccg-cgc/A92-14-2007-eng.pdf (accessed 27 March 2015).

Agriculture and Agri-Food Canada (2006), *Canola Oils: Situation and Outlook*, Vol. 19, No. 17, online bi-weekly bulletin.

Alvarez, M.J. et al. (2001), "Oilseed rape flour: Another allergen causing occupational asthma among farmers", *Allergy*, Vol. 56, No. 2, pp. 185-188.

Bell, J.M. (1995), "Meal and by-product utilization in animal nutrition", in: Kimber, D.S. and D.I. McGregor, *Brassica Oilseeds*, CAB International, United Kingdom.

Bell, J.M. et al. (1998), "Nutritional composition and digestibility by 80 kg to 100 kg pig of prepress solvent-extracted meals from low glucosinolate *B. juncea*, *B napus* and *B rapa* seed and of solvent-extracted soybean meal", *Canadian Journal of Animal Science*, Vol. 78, No. 2, pp. 199-203, http://dx.doi.org/10.4141/A97-094.

Bolton-Smith, C. et al. (2000), "Compilation of a provisional UK Database for the phylloquinone (Vitamin K1) content of food", *British Journal of Nutrition*, Vol. 83, No. 4, pp.389-399.

Bonnardeaux, J. (2007), *Uses for Canola Meal*, Department of Agriculture and Food, Government of Western Australia, available at: http://archive.agric.wa.gov.au/objtwr/imported_assets/content/sust/biofuel/usesfor canolameal_report.pdf.

Boulter, G.S. (1983), "The history and marketing of rapeseed oil in Canada", Chapter 3 in: Kramer, J.K.G. et al. (eds.), *High and Low Erucic Acid Rapeseed Oils, Production, Usage, Chemistry, and Toxicological Evaluation*, Academic Press.

Canadian Patent Database (2011), "Patent summary: CA 2553640 – Novel canola protein isolate" Canadian Intellectual Property Office, http://brevets-patents.ic.gc.ca/opic-cipo/cpd/eng/patent/2553640/summary.html?type=number_search (accessed 20 February 2011).

Canola Council of Canada (2011), schematic of prepress solvent extraction process (accessed October 2011).

Canola Council of Canada (2009), *Canola Meal: Feed Industry Guide, 4th Edition, 2009*, Canadian International Grains Institute, Canola Council of Canada, Winnipeg, Manitoba, Canada, available at: www.canolacouncil.org/media/516716/canola_meal_feed_guide_english.pdf.

Codex Alimentarius Commission (2005), "Codex Standard for Named Vegetable Oils, Vol. 8", *Codex Standard Series No. 210-2005*, Rome.

Codex Alimentarius Commission (2003, Annexes II and III adopted in 2008), *Guideline for the Conduct of Food Safety Assessment of Foods Derived from Recombinant DNA Plants*, CAC/GL 45/2003, available at: www.codexalimentarius.net/download/standards/10021/CXG_045e.pdf (accessed Oct 2011).

Cook, K.K. et al. (1999), "HPLC analysis for trans-vitamin K1 and dihydro-vitamin K1 in margarines and margarine-like products using the C30 stationary phase", *Food Chemistry*, Vol. 67, Issue 1, pp. 79-88.

Dairy One Cooperative Inc., Feed Composition Library (2010), *Accumulated crop years: 05/01/2000 through 04/30/2010.*

Daun, J.K. and D. Adolphe (1997), "A revision to the canola definition", *GCIRC Bulletin*, July 1997, pp. 134-141.

Downey, R.K. (2007), "Rapeseed to canola: Rags to riches", in: Eaglehan, A. and R.W.F. Hardy (eds.), *Economic Growth through New Products, Partnerships and Workforce Development*, National Agricultural Biotechnology Council, Ithaca, New York.

Endo, Y. et al. (1992), "Characterization of chlorophyll pigments present in canola seed, meal and oil", *Journal of the American Oil Chemists Society*, Vol. 69, Issue 6, pp. 564-568.

FAO/WHO (2002), *Human Vitamin and Mineral Requirements-Chapter 10 Vitamin K*, Report of a Joint FAO/WHO Expert Consultation, Bangkok, Thailand, FAO and WHO, Rome, available at: www.fao.org/docrep/004/y2809e/y2809e00.htm.

FAOSTAT (2011), FAO Statistics online website, http://faostat.fao.org, (accessed February 2011).

FDA (2010), GRAS Notice Inventory, *GRN No. 327 Cruciferin-rich canola/ rapeseed protein isolate and napin-rich canola/rapeseed protein* (Date of filing: 09 March 2010), available at: www.accessdata.fda.gov/scripts/fcn/fcnDetailNavigation.cfm?rpt=grasListing&id=327.

Ferland, G. and J.A. Sadowski (1992), "Vitamin K1 (phylloquinone) content of edible oils: Effects of heating and light exposure", *Journal of Agricultural and Food Chemistry*, Vol. 40, No. 10, pp. 1 869-1 873, http://dx.doi.org/10.1021/jf00022a028.

Fickler, J. (2005), *Amino Dat 3.0 Platinum*, Degussa AG Feed Additives.

Gao, Z.H. and R.G. Ackman (1995), "Determination of vitamin K1 in canola oils by high performance liquid chromatography with menaquinone-4 as an internal standard", *Food Research International*, Vol. 28, Issue 1, pp. 61-69.

GMO Compass (n.d.), www.gmo-compass.org/eng/agri_biotechnology/gmo_planting/257 global_gm_planting_2008.html (accessed October 2011).

Gunstone, F.D. (2005), "Vegetable oils", in: Shahidi, F. (ed.), *Bailey's Industrial Oil & Fat Products, Vol. 1: Edible Oil and Fat Products: General Applications, 6th Edition*, John Wiley & Sons, Inc. New York.

Gylling, H. (2006), "Rapeseed oil does not cause allergic reactions", *Allergy*, Vol. 61, Issue 7, p.895, http://dx.doi.org/10.1111/j.1398-9995.2006.01119.x.

Hickling, D. (2005), "Canola quality review", *Proceedings of the 2005 Annual Convention Canola Council of Canada*, Winnipeg, Manitob, Canada.

Hickling, D. (2001), *Canola Meal Feed Industry Guide, 3rd Edition*, Canadian International Grain Institute.

James, C. (2011), "Global status of commercialized biotech/GM crops: 2010", *ISAAA Brief 42-2010: Executive Summary*, International Service for the Acquisition of Agri-Biotech Applications, available at: www.isaaa.org/resources/publications/briefs/42/executivesummary/default.asp (accessed June 2011).

Kamao, M. et al. (2007), "Vitamin K content of foods and dietary vitamin K intake in Japanese young women", *Journal of Nutritional Science and Vitaminology (Tokyo)*, Vol. 53, No. 6, pp. 464-470.

Kozlowska, H. et al. (1990), "Phenolic acids and tannins in rapeseed and canola", in: Shahidi, F. (ed.), *Canola and Rapeseed: Production, Chemistry, Nutrition and Processing Technology*, Springer, New York.

Love, H.K. et al. (1990), "Development of low glucosinolate mustard", *Canadian Journal of Plant Science*, Vol. 70, No. 2, pp. 419-424, http://dx.doi.org/10.4141/cjps90-049.

Malcolmson, L. and M. Vaisey-Genser (2001), *Canola Oil, Properties and Performance*, Canola Council of Canada, Winnipeg, Manitoba, Canada.

McAllister, T.A. et al. (1999), "Feeding value for lambs of rapeseed meal arising from biodiesel production", *Animal Science*, Vol. 68, No. 1, pp. 183-194.

Manitoba Canola Growers Association (2008), *Canola: Canada's Oil*, available at: http://canolagrowers.com/resource/canola-canadas-oil (accessed 16 April 2010).

Miller-Cebert, R.L. et al. (2009), "Panelists liking of canola (*Brassica napus*) greens", *Nutrition and Food Science*, Vol. 39, No. 6, pp. 627-635.

Newkirk, R.W. and H.L. Classen (2000), "The effects of standard oil extraction and processing on the nutritional value of canola meal for broiler chickens", *Poultry Science*, Vol. 79, Supplement 1, p. 10.

Newkirk, R.W. et al. (2003), "The digestibility and content of amino acids in toasted and non-toasted canola meals", *Canadian Journal of Animal Science*, Vol. 83, pp. 131-139.

Newkirk, R.W. et al. (1997), "Nutritional evaluation of low glucosinolate mustard meals (*Brassica juncea*) in broiler diets", *Poultry Science*, Vol. 76, No. 9, pp. 1 272-1 277.

OECD (2012), "Consensus document on the biology of the *Brassica* crops (*Brassica* spp.)", *Series on Harmonisation of Regulatory Oversight in Biotechnology*, No.54, OECD, Paris, available at www.oecd.org/officialdocuments/publicdisplaydocumentpdf/?cote=env/jm/mono(2012)41&doclanguage=en.

Piironen, V. et al. (1997), "Determination of phylloquinone in oils, margarines and butter by high-performance liquid chromatography with electrochemical detection", *Food Chemistry*, Vol. 59, Issue 3, July, pp. 473-480.

Poikonen, S. et al. (2009), "Sensitization and allergy to turnip rape: A comparison between the Finnish and French children with atopic dermatitis", *Acta Paediatrica*, Vol. 98, No. 2, pp. 310-315, http://dx.doi.org/10.1111/j.1651-2227.2008.01020.x.

Poikonen, S. et al. (2008), "Sensitization to turnip rape and oilseed rape in children with atopic dermatitis: A case-control study", *Pediatric Allergy and Immunology*, Vol. 19, No. 5, pp. 408-411, http://dx.doi.org/10.1111/j.1399-3038.2007.00666.x.

Poikonen, S. et al. (2006), "Turnip rape and oilseed rape are new potential food allergens in children with atopic dermatitis", *Allergy*, Vol. 61, No. 1, pp. 124-127.

Potts, D.A. et al. (1999), "Canola-quality *Brassica juncea*, a new oilseed crop for the Canadian prairies", in: Wratten, N. and P.A. Salisbury (eds.), *Proceedings from the 10th International Rapeseed Congress*, Canberra, Australia, available at: www.regional.org.au/au/gcirc/4/70.htm (accessed April 2010).

Pritchard, F.M. et al. (2000), "Environmental effects on seed composition of Victorian canola", *Australian Journal of Experimental Agriculture*, Vol. 40, No. 5, pp. 679-685.

Przybylski, R. et al. (2005), "Canola oil", in: Shahidi, F. (ed.), *Bailey's Industrial Oil & Fat Products, Vol. 2: Edible Oil & Fat Products: Oils and Oil Seeds, 6th Edition*, John Wiley & Sons, Inc., New York.

Puumalainen T.J. et al. (2006), "Napins, 2S albumins, are major allergens in oilseed rape and turnip rape", *Journal of Allergy and Clinical Immunology*, Vol. 117, No. 2, pp. 426-432, http://dx.doi.org/10.1016/j.jaci.2005.10.004.

Racz, V. and D.A. Christensen (2004), *Whole Canola Seed Use and Value*, Prairie Feed Resource Centre, Saskatoon, Saskatchewan, Canada.

Seberry, D.E. et al. (2009), "Quality of Australian canola: 2008-09", *Australian Oilseeds Federation*, Vol. 15, NSW Department of Primary Industries, available at: www.australianoilseeds.com/__data/assets/pdf_file/0008/5849/2008_Quality_of_Australian_Canola_Book.pdf (accessed April 2010).

Seberry, D.E. et al. (2008), "Quality of Australian canola: 2007/08", *Australian Oilseeds Federation*, Vol. 14, NSW Department of Primary Industries, available at: www.australianoilseeds.com/__data/assets/pdf_file/0003/4197/2007_Book.pdf (accessed April 2010).

Seberry, D.E. et al. (2007), "Quality of Australian Canola: 2006/07", *Australian Oilseeds Federation*, Vol. 13, NSW Department of Primary Industries, available at: www.australianoilseeds.com/__data/assets/pdf_file/0004/2785/2006_Book.pdf (accessed April 2010).

Shearer, M. et al. (1996), "Chemistry, nutritional sources, tissue distribution and metabolism of vitamin K with special reference to bone health", *Journal of Nutrition*, Vol. 126, No. 4, Supplement, pp. 1 181S-1 186S.

Simbaya, J. et al. (1995), "Quality characteristics of yellow-seeded *Brassica* seed meals: Protein, carbohydrates, and dietary fiber components", *Journal of Agricultural and Food Chemistry*, Vol. 43, No. 8, pp. 2 062-2 066, http://dx.doi.org/10.1021/jf00056a020.

Suh, C.H. et al. (1998), "Oilseed rape allergy presented as occupational asthma in the grain industry", *Clinical and Experimental Allergy*, Vol. 28, No. 9, pp. 1 159-1 163.

Tan, S.H. et al. (2011), "Canola proteins for human consumption: Extraction, profile, and functional properties", *Journal of Food Science*, Vol. 76, Issue 1, pp. R16-R28, http://dx.doi.org/10.1111/j.1750-3841.2010.01930.x.

Trinidade, A. et al. (2010), "The prevalence of oilseed rape hypersensitivity in a mixed cereal farming population", *Clinical Otolaryngoly*, Vol. 35, No. 1, pp. 13-17, http://dx.doi.org/10.1111/j.1749-4486.2009.02062.x.

Tripathi, M.K and A.S. Mishra (2007), "Glucosinolates in animal nutrition: A review", *Animal Feed Science and Technology*, Vol. 132, Issues 1-2, pp. 1-27.

Unger, E.H. (1990), "Commercial processing of canola and rapeseed: Crushing and oil extraction", Chapter 14 in: Shahidi, F. (ed.), *Canola and Rapeseed: Production, Chemistry, Nutrition, and Processing Technology*, Van Nostrand Reinhold, New York, pp. 235-249.

USDA Agricultural Research Service (2011), *USDA Nutrient Database for Standard Reference, Release 23*, Nutrient Data Laboratory, United States Department of Agriculture, available at: www.ars.usda.gov/Services/docs.htm?docid=8964 (accessed October 2011).

USDA Foreign Agricultural Service (2011), *Oilseeds: World Markets and Trade*, Circular Series, FOP 10-11, October, available at: http://usda.mannlib.cornell.edu/usda/fas/oilseed-trade//2010s/2011/oilseed-trade-10-12-2011.pdf (accessed October 2011).

Van Barneveld, R. and R. Ed-King (2002), *Australian Canola Meal – A Valuable Component of Pig Feed*, Australian Pork Limited (APL) Technical Note.

Youngs, C.G. and L.R. Wetter (1969), "Processing of rapeseed for high quality meal. Rapeseed meal for livestock and poultry", *Rapeseed Association of Canada*, Publ. No. 3, pp. 2-3.

# *Chapter 9*

# Soybean (*Glycine max*)

*This chapter, prepared by the OECD Task Force for the Safety of Novel Foods and Feeds with the United States as the lead country, deals with the composition of soybean (*Glycine max*). It updates and revises the original publication on soybean composition issued in 2001. It contains elements that can be used in a comparative approach as part of a safety assessment of foods and feeds derived from new varieties. Background is given on soybean production, uses and processing, followed by appropriate varietal comparators and characteristics screened by breeders. Nutrients in soybean seed, oil, meal, hulls and forage, as well as other constituents (anti-nutrients and toxicants, other compounds, allergens), are then detailed. The final sections suggest key products and constituents for analysis of new varieties for food use and for feed use.*

# Background

## *Production*

The soybean[1] (*Glycine max* [L.] Merr.) is grown world-wide as an important staple and commercial crop. The soybean accounted for 56% of the main world oilseed crops production in 2011, being also the dominant species traded in international markets among all major oilseeds (American Soybean Association, 2012). The five major soybean producers in 2011 – the United States, Brazil, Argentina, the People's Republic of China (hereafter "China") and India – accounted for 90% of the total production (Table 9.1).

Table 9.1. **Production and export of soybeans in 2011**

Million metric tonnes

| Country/region | Production | Exports |
|---|---|---|
| United States | 83.2 | 34.7 |
| Brazil | 72.0 | 37.8 |
| Argentina | 48.0 | 8.9 |
| China (People's Republic of) | 13.5 | |
| India | 11.0 | |
| Canada | 4.2 | 2.9 |
| Paraguay | 6.4 | 5.0 |
| Others | 13.1 | 3.5 |
| Total | 251.5 | 92.8 |

*Source*: Adapted from American Soybean Association (2012).

## *Uses*

The major soybean commodity products are seeds, oil and meal. A bushel (27.2 kg) of soybeans yields about 21.8 kg of protein-rich meal and 5.0 kg of oil (American Soybean Association, 2012).

Unprocessed soybeans are not suitable for food and their use for animal feed remains limited because they contain anti-nutritional factors such as trypsin inhibitors and lectins. Adequate heat processing inactivates these factors.

Whole soybeans are utilised to produce soy sprouts, baked soybeans, roasted soybeans, full fat soy flour and the traditional soy foods (miso, soy milk, soy sauce and tofu). In addition to whole oil used for human consumption, refined soybean oil has many other technical and industrial applications. Glycerol, fatty acids, sterols and lecithin are all derived from soybean oil. Soy protein isolate is used as a source of amino acids in the production of infant food formula and other food products. Soybean meal is rich in essential amino acids, particularly lysine and tryptophan, which are required supplements in animal diets for optimum growth and health. Soybean meal is used in diets for poultry, swine, dairy cattle, beef cattle and pets.

Being rich in hydrocarbon, soybean oil is used for biodiesel fuel production (soy methyl esters). Approximately 4.8 kg of soybeans are required to produce 1 litre of biodiesel (American Soybean Association, 2012).

## *Processing*

Historically, the oil extraction process was conducted on a small scale using mechanical or hydraulic presses after the soybeans were rolled into flakes and properly conditioned by heat treatment. Gradually, the screw press (expeller) has replaced the hydraulic press; however, the hydraulic press is still efficiently used on small-scale individual farms for organic production and in developing countries.

Large-scale solvent extraction facilities produce the bulk of soybean oil (Johnson, 2008). The solvent hexane is used to extract the oil from flaked soybeans (Lusas, 2000).

The processing steps used to produce the various soy products are schematised in Figures 9.1 and 9.2.

Figure 9.1. **Whole soybean processing**

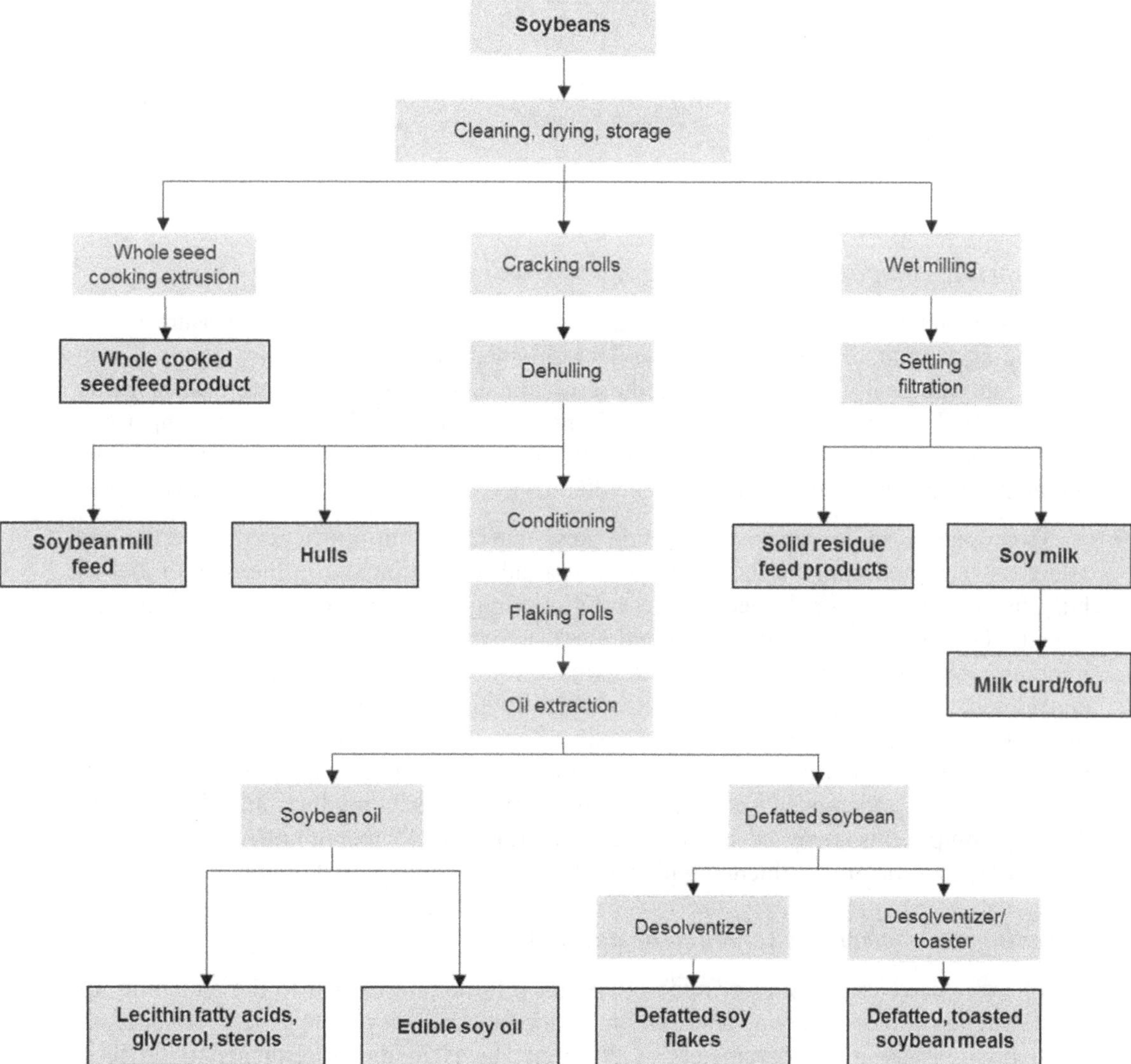


*Source*: Adapted from Waggle and Kolar (1979).

Figure 9.2. **Defatted soybean flakes processing**

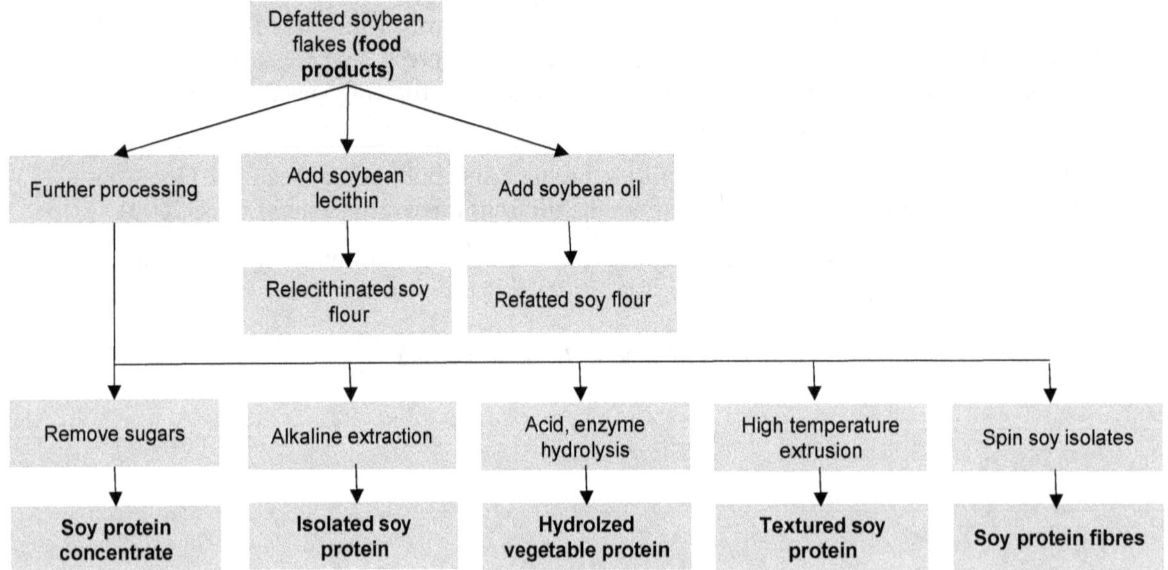


*Source*: Adapted from Sipos (1988).

## *Appropriate comparators for testing new varieties*

This chapter suggests parameters that soybean breeders should measure when developing new modified varieties. The data obtained in the analysis of a new soybean variety should ideally be compared to those obtained from an appropriate near isogenic non-modified variety, grown and harvested under the same conditions.[2] The comparison can also be made between values obtained from new varieties and data available in the literature, or chemical analytical data generated from other commercial soybean varieties.

Components to be analysed include key nutrients, anti-nutrients, toxicants and allergens. Key nutrients are those which have a substantial impact in the overall diet of humans (food) and animals (feed). These may be major constituents (fats, proteins, and structural and non-structural carbohydrates) or minor compounds (vitamins and minerals). Similarly, the levels of known anti-nutrients and allergens should be considered. Key toxicants are those toxicologically significant compounds known to be inherently present in the species, whose toxic potency and levels may impact human and animal health. Standardised analytical methods and appropriate types of material should be used, adequately adapted to the use of each product and by-product. The key components analysed are used as indicators of whether unintended effects of the genetic modification influencing plant metabolism have occurred or not.

## *Breeding characteristics screened by developers*

Phenotypic characteristics provide important information related to the suitability of new varieties for commercial distribution. Plant breeders developing new varieties of soybeans consider many parameters at different stages in the developmental process. In the early stages, breeders evaluate flower colour, plant standability, stand count, relative maturity, plant habit, pubescence colour, hila colour, pod wall colour, plant morphology, time of flowering, emergence, tolerance to low temperatures and general disease resistance. The latter disease screening depends on the maturity and area in which the seeds are being grown. Tolerance to low temperatures during flowering and early pod setting would be more important in developing genotypes for cooler soybean-growing regions.

Later on, as a new variety gets closer to commercialisation, breeders measure yield, first at one site then in larger plots and at increasing numbers of sites. Some of the factors considered in the evaluation process include maturity, height, lodging, flower colour, pubescence colour, pod wall colour, canopy width, leaf colour, hypocotyl elongation, emergence score, shattering score, seed size, seed quality, percent oil and percent protein.

Plants are also screened for resistance to various diseases.

In some cases, plants are modified for specific increases/decreases in certain components, and the plant breeder would be expected to analyse for such components. For plants modified for changes in specific compositional components, it is noted that careful consideration may be needed to determine an appropriate comparator.[3]

## Nutrients

Since the first issue of this document in 2001, several new sources of valuable information have become available.

Data from the International Life Sciences Institute (ILSI, 2010), the National Agricultural and Food Research Organization (NARO, 2001) and its National Food Research Institute (NFRI-NARO, 2011) in Japan, the *Food Composition and Nutrition Tables* (Souci et al., 2008), the *Danish Food Composition Databank* (revision 7.0, Saxholt et al., 2008), the Swedish National Food Administration's database (2011), Food Standards Australia New Zealand's *Nutrient Tables for Use in Australia* (NUTTAB, 2010) and the *USDA Nutrient Database for Standard Reference* (USDA Agricultural Research Service, 2008a) have been incorporated into this revised version of the original publication.[4]

### *Seeds*

Tables 9.2-9.6 provide data regarding the composition of soybean seed including proximates and fibre analysis, amino acids, fatty acids, minerals and vitamins.

It should be noted that soybean varieties that contain high levels of oleic acid have been developed, but corresponding data are not included in this chapter.

Some data sources only report fatty acid content based on the percent of the total fatty acids; data from these sources is reported under the "oil" section of this chapter.

Table 9.2. **Proximates and fibre analysis of soybean seed**

| Reference | ILSI | | NRC 2000 | NRC 2001 | Ensminger et al. | Souci et al.[1] | | USDA -ARS[2] | NFRI-NARO[3] | |
|---|---|---|---|---|---|---|---|---|---|---|
| | Mean | Range | Mean | Mean | Mean | Mean | Range | Mean | Mean | Range |
| | | | | | g/100 g fresh weight | | | | | |
| Moisture | 10.1 | 4.7-34.4 | 13.6 | 10.0 | 8.0 | 8.4 | | 8.5 | 11.1 | 9.2-13.7 |
| | | | | | g/100 g dry matter | | | | | |
| Crude protein | 39.5 | 33.2-45.5 | 40.3 | 39.2 | 41.7 | 41.7 | | 39.9 | 42.1 | 35.8-46.2 |
| Crude fat | 16.7 | 8.1-23.6 | 18.2 | 19.2 | 18.7 | 20.0 | 17.9-23.3 | 21.8 | 24.2 | 21.0-27.4 |
| Ash | 5.3 | 3.9-7.0 | 4.56 | 5.9 | 5.6 | 6.1 | | 5.3 | 5.6 | 5.0-6.5 |
| Acid detergent fibre | 12.0 | 7.8-18.6 | 11.1 | 13.1 | 11.0 | | | | | |
| Nuetral detergent fibre | 12.3 | 8.5-21.3 | 14.9 | 19.5 | | | | | | |
| Total dietary fibre | | | | | | 24.0 | | 10.2 | 18.7 | 15.9-22.9 |
| Crude fibre | 7.8 | | | | 5.8 | | | | | |
| Carbohydrates (by calculation) | 38.2 | 29.6-50.2 | | | 26.0[5] | | | 32.98 | 31.7 | 27.8-35.9[4] |
| Sugar (CHO-TDF) | | | | | | | | | 13.1 | 9.0-16.4 |

*Notes:* 1. Data converted from fresh weight to dry weight basis using given moisture level. 2. Data converted from fresh weight to dry weight basis using given moisture level. Data may include results from genetically engineered soybeans. 3. Data converted from fresh weight to dry weight basis using given moisture content. 4. Carbohydrate (by calculation) = 100% – (crude protein% + crude fat% + ash% + moisture%).

*Sources:* ILSI (2010); NRC (2000, 2001); Ensminger et al. (1990); Souci et al. (2008); USDA Agricultural Research Service (2008a); National Food Research Institute-NARO (2011).

Table 9.3. **Amino acid composition of soybean seed**

g/100 g dry matter

| Reference | ILSI | | NRC | Ensm. et al. | USDA-ARS[1] | Souci et al.[2] | | NFRI-NARO[2] | |
|---|---|---|---|---|---|---|---|---|---|
| | Mean | Range | Mean | Mean | Mean | Mean | Range | Mean | Range |
| Arginine | 2.84 | 2.28-3.4 | 2.95 | 2.86 | 3.45 | 2.58 | 2.19-2.91 | 3.04 | 2.44-3.62 |
| Cystine/cysteine | 0.59 | 0.37-0.81 | 0.57 | 0.45 | 0.72 | 0.64 | 0.57-0.72 | 0.66 | 0.52-0.73 |
| Histidine | 1.04 | 0.87-1.17 | 1.08 | 1.00 | 1.20 | 0.91 | 0.85-0.96 | 1.11 | 0.97-1.26 |
| Isoleucine | 1.81 | 1.53-2.07 | 1.73 | 1.76 | 2.16 | 1.94 | 1.72-2.16 | 1.85 | 1.59-2.06 |
| Leucine | 3.04 | 2.59-3.62 | 2.90 | 2.95 | 3.62 | 3.10 | | 3.17 | 2.69-3.49 |
| Lysine | 2.56 | 2.28-2.83 | 2.34 | 2.52 | 2.96 | 2.07 | 1.56-2.54 | 2.61 | 2.55-2.87 |
| Methionine | 0.55 | 0.43-0.68 | 0.58 | 0.52 | 0.60 | 0.63 | 0.53-0.74 | 0.57 | 0.49-0.62 |
| Phenylalanine | 1.98 | 1.63-2.34 | 1.96 | 1.91 | 2.32 | 2.15 | 2.00-2.35 | 2.11 | 1.72-2.45 |
| Threonine | 1.47 | 1.14-1.86 | 1.55 | 1.58 | 1.93 | 1.63 | 1.47-1.81 | 1.66 | 1.42-1.79 |
| Tryptophan | 0.43 | 0.36-0.50 | 0.51 | 0.61 | 0.65 | 0.49 | 0.44-0.56 | 0.55 | 0.49-0.63 |
| Valine | 1.91 | 1.59-2.20 | 1.84 | 1.75 | 2.22 | 1.92 | 1.55-2.12 | 1.94 | 1.70-2.19 |
| Glycine | 1.69 | 1.46-1.99 | | 1.55 | 2.06 | 1.55 | | 1.77 | 1.52-1.94 |
| Tyrosine | 1.32 | 1.01-1.61 | | 1.40 | 1.68 | 1.36 | 1.29-1.45 | 1.40 | 1.24-1.56 |
| Serine | 2.02 | 1.1-2.48 | | 2.16 | 2.58 | 1.84 | | 2.14 | 1.77-2.46 |
| Proline | 2.00 | 1.68-2.28 | | | 2.60 | 1.99 | | 2.12 | 1.78-2.41 |
| Alanine | 1.72 | 1.51-2.10 | | | 2.09 | 1.67 | | 1.78 | 1.59-1.95 |
| Aspartic acid | 4.49 | 3.81-5.12 | | | 5.59 | 4.36 | | 4.79 | 3.95-5.36 |
| Glutamic acid | 7.09 | 5.84-8.20 | | | 8.61 | 7.09 | | 7.73 | 6.21-8.60 |

*Notes:* 1. Data converted from fresh weight to dry weight basis using given moisture level. Data may include results from genetically engineered soybeans. 2. Data converted from fresh weight to dry weight basis using given moisture level.

*Sources:* ILSI (2010); NRC (2001); Ensminger et al. (1990); USDA Agricultural Research Service (2008a); Souci et al. (2008); National Food Research Institute-NARO (2011).

Table 9.4. **Fatty acid composition of soybean seed**

g/100 g dry matter

| Reference | | USDA-ARS[1] | ILSI[2] | | Souci et al.[2] | | NFRI-NARO[2] | |
|---|---|---|---|---|---|---|---|---|
| | | Mean | Mean | Range | Mean | Range | Mean | Range |
| Palmitic | C16:0 | 2.31 | 1.87 | 0.67-2.78 | 1.89 | 0.44-2.01 | 2.59 | 2.24-2.89 |
| Stearic | C18:0 | 0.78 | 0.68 | 0.28-1.13 | 0.63 | 0.46-1.30 | 0.76 | 0.42-1.18 |
| Oleic | C18:1 | 4.75 | 3.46 | 1.36-6.56 | 4.35 | 4.08-5.81 | 5.46 | 3.93-8.95 |
| Linoleic | C18:2 | 10.85 | 8.91 | 3.46-13.36 | 10.71 | 9.40-11.58 | 12.15 | 10.34-13.60 |
| Linolenic | C18:3 | 1.45 | 1.40 | 0.30-2.19 | 1.02 | 0.9-1.09 | 1.87 | 1.26-2.73 |
| Arachidic | C20:0 | | 0.06 | 0.02-0.11 | | 0.09-0.46 | 0.07 | 0.05-0.09 |

*Notes:* 1. Data converted from fresh weight to dry weight basis using given moisture level. Data may include results from genetically engineered soybeans. 2 Data converted from fresh weight to dry weight basis using given moisture level.

*Sources:* USDA Agricultural Research Service (2008a); ILSI (2010); Souci et al. (2008); National Food Research Institute-NARO (2011).

Table 9.5. **Mineral composition of soybean seed**

| Reference | USDA-ARS[1] | NRC 2000 | NRC 2001 | Ensminger et al. | ILSI[2] | | Souci et al.[2] | |
|---|---|---|---|---|---|---|---|---|
| | Mean | Mean | Mean | Mean | Mean | Range | Mean | Range |
| | | | | **g/100 g DM** | | | | |
| Calcium (Ca) | 0.30 | 0.27 | 0.32 | 0.27 | 0.22 | 0.12-0.31 | 0.22 | 0.22-0.24 |
| Phosphorus (P) | 0.77 | 0.65 | 0.60 | 0.65 | 0.71 | 0.50-0.94 | 0.60 | 0.52-0.70 |
| Magnesium (Mg) | 0.31 | 0.27 | 0.25 | 0.29 | 0.26 | 0.22-0.31 | 0.24 | 0.23-0.31 |
| Potassium (K) | 1.96 | 2.01 | 1.99 | 1.80 | 2.06 | 1.87-2.32 | 1.97 | 1.97-1.99 |
| | | | | **mg/100 g DM** | | | | |
| Iron (Fe) | 17.00 | 20.00 | 10.00 | 10.00 | 8.00 | 6.00-11.00 | 10.00 | |
| Sodium (Na) | 2.00 | 40.00 | 10.00 | 0.00 | | | 10.00 | |
| Selenium (Se) | 0.020 | | 0.028 | 0.012 | | | 0.02 | 0.01-0.08 |
| Manganese (Mn) | 2.75 | 34.5[3] | 2.9 | 3.96 | | | 2.95 | 0-5.90 |
| Copper (Cu) | 1.81 | 1.46 | 1.3 | 1.98 | | | 1.31 | 0.11-1.53 |
| Zinc (Zn) | 5.35 | 5.9 | 4.9 | 6.18 | | | 4.59 | 1.09-6.77 |

*Notes:* DM: dry matter.

1. Data converted from fresh weight to dry weight basis using given moisture level. Data may include results from genetically engineered soybeans. 2. Data converted from fresh weight to dry weight basis using given moisture level. 3. NRC (2000) indicates that manganese is present in soybean seeds at 345 mg/kg soybean seed. In contrast, NRC (1984) indicates that manganese is present in soybean seed at 39 mg/kg soybean seed (i.e. 3.9 mg Mn/100 g dry matter).

*Sources:* USDA Agricultural Research Service (2008a); NRC (2000, 2001); Ensminger et al. (1990); ILSI (2010); Souci et al. (2008).

Table 9.6. **Vitamin composition of soybean seed**

| Reference | Units/ 100 g DM | ILSI[1] | | Ensminger et al. | USDA-ARS[2] | Souci et al.[3] | | NFRI-NARO[4] | |
|---|---|---|---|---|---|---|---|---|---|
| | | Mean | Range | Mean | Mean | Mean | Range | Mean | Range |
| Folic acid | mg | 0.36 | 0.24-0.47 | | | 0.27 | | | |
| Vitamin A | IU | | | 160 | 24.05 | | | | |
| B-carotene | mg | | | | 0.014 | 0.42 | 0.37-0.44 | | |
| Vitamin B1 | mg | 0.20 | 0.10-0.25 | 0.12 | 0.96 | 1.12 | 0.95-1.30 | 0.99 | 0.79-1.31 |
| Vitamin B2 | mg | 0.27 | 0.19-0.32 | 0.32 | 0.95 | 0.50 | 0.25-1.42 | 0.33 | 0.27-0.51 |
| Vitamin E (α-tocopherol except when noted) | mg | 1.91 | 0.19-6.17 | 0.37 | 0.93 | 16.38[3] | | 3.97 | 1.25-10.75 |
| Vitamin K | mg | | | | 0.051 | 0.04 | 0.03-0.05 | 0.017 | 0.00-0.046 |
| Niacin | mg | | | 2.4 | 1.78 | 2.95 | 2.62-3.28 | 2.26 | 0.79-3.23 |
| Vitamin B6 | mg | | | 0.12 | 0.41 | 1.09 | 0.66-1.31 | 0.64 | 0.37-1.18 |

*Notes:* 1. www.cropcomposition.org (2010). 2. Data converted from fresh weight to dry weight basis using given moisture level. Data may include results from genetically engineered soybeans. 3. Data converted from fresh weight to dry weight basis using given moisture level. Vitamin E is reported as total tocopherol. 4. Data converted from fresh weight to dry weight basis using given moisture level.

*Sources:* ILSI (2010); Ensminger et al. (1990); USDA Agricultural Research Service (2008a); Souci et al. (2008); National Food Research Institute-NARO (2011).

## *Oil*

Triglycerides are the main constituents (99%) of soybean oil. Soybean oil is noted for its relatively high content of unsaturated fatty acids, oleic (C18:1), linoleic (C18:2) and linolenic (C18:3) acids. Soybean oil contains relatively lesser amounts of the saturated fatty acids, palmitic (C16:0) and stearic (C18:0) acids (Wang et al., 1997). Arachidic and behenic acid are also present, but at only low levels. The range of fatty acid composition in soybean oil is shown in Table 9.7.

In the human diet, soybean oil is considered a source of vitamins K and E, but not provitamin A. The vitamin composition of soybean oil has been described in several publications. Data suggest that there may be partial loss of vitamin K in vegetable oils due to refining processes (Gao and Ackman, 1995). The Food and Agriculture Organization and the World Health Organization of the United Nations (FAO/WHO, 2002) have noted that certain vegetable oils, including soybean oil, represent biologically available sources of vitamin K. Soybean oil is a source of vitamin E even though some of the vitamin E may be lost during the processing of soybean oil (Frankel, 1996). The vitamin K1 (phylloquinone) and vitamin E (reported as α-tocopherol) content of soybean oil is shown in Tables 9.8 and 9.9, respectively.

Table 9.7. **Fatty acid composition of soybean oil**

% of total fatty acids

| Reference | | USDA-ARS[1] | Codex[2] | Souci et al. | | MECSST[3] | *Danish Food Composit. Databank*[4] | Swedish National Food Adm. | NUTTAB[5] | Padgette et al. | |
|---|---|---|---|---|---|---|---|---|---|---|---|
| | | Mean | Range | Mean | Range | Mean | Mean | Mean | Mean | Mean | Range |
| Palmitic | C16:0 | 10.87 | 8.0-13.5 | 11.44 | 7.11-15.39 | 10.67 | 9.79 | 10.56 | 10.7 | 12.18 | 11.57-12.71 |
| Stearic | C18:0 | 4.61 | 2.0-5.4 | 3.98 | 0.53-9.45 | 4.31 | 3.65 | 4.28 | 3.9 | 4.45 | 4.19-4.95 |
| Oleic | C18:1 | 23.45 | 17.0-30.0 | 20.79 | 15.18-30.46 | 23.72 | 22.30 | 23.01 | 19.1 | 21.46 | 16.35-33.95 |
| Linoleic | C18:2 | 52.98 | 48.0-59.0 | 58.42 | 38.74-61.35 | 53.90 | 55.70 | 53.45 | 57.7 | 57.16 | 47.92-59.82 |
| Linolenic | C18:3 | 7.06 | 4.5-11.0 | 8.47 | 2.02-15.60 | 6.58 | 7.19 | 7.74 | 7.5 | 8.73 | 5.53-11.17 |
| Arachidic | C20:0 | 0.38 | 0.1-0.6 | 0.55 | 0.11-0.96 | 0.38 | 0.625 | 0.31 | 0.4 | 0.39 | 0.34-0.47 |

N*otes:* 1. Data converted from g/100 g to percentage of total fatty acid. Total fatty acid calculated as the sum of total saturated fatty acids, total monounsaturated fatty acids and total polyunsaturated fatty acids. Data may include results from genetically engineered soybeans. 2. Non-detects (ND) are ≤ 0.05 % total fatty acids. 3. Ministry of Education, Culture, Sports, Science and Technology (MECSST). Data converted from mg/100g to percentage of total fatty acids using given total fatty acid content. 4. Saxholt et al. (2008). 5. Food Standards Australia New Zealand (2010).

*Sources:* USDA Agricultural Research Service (2008a); Codex Alimentarius Commission (2009); Souci et al. (2008); Ministry of Education, Culture, Sports, Science and Technology (MECSST) (2005); Saxholt et al. (2008); Swedish National Food Administration (2011); Food Standards Australia New Zealand (2010); Padgette et al. (1996).

Table 9.8. **Vitamin K1 levels in commercially available soybean oil as measured by various types of HPLC-based analytical methodologies**

µg per 100 g of oil

| Reference | Sample type | **Vitamin K1 (phylloquinone)** | |
|---|---|---|---|
| | | Mean | Range |
| Ferland and Sadowski | Commercially available oil | 193 | 139-290 |
| Gao and Ackman | Retail expeller oil | 250 | |
| Shearer et al. | | 173 | |
| Piironen et al. | Refined oil | 145 | |
| Cook et al. | Commercially available oil | 114.2[1]<br>102.5[2] | |
| Bolton-Smith et al. | Aged soybean oil | 112 | |
| | Fresh soybean oil | 150 | |
| Kamao et al. | Retail oil | 234 | |
| USDA-ARS | Salad or cooking oil | 183.9 | |
| Booth and Suttie | | 193 | |

*Notes:* 1. Sample prepared by enzymatic digestion and extraction. 2. Sample prepared by direct extraction.

*Sources:* Ferland and Sadowski (1992); Gao and Ackman (1995); Shearer et al. (1996); Piironen et al. (1997); Cook et al. (1999); Bolton-Smith et al. (2000); Kamao et al. (2007); USDA Agricultural Research Service (2008a); Booth and Suttie (1998).

Table 9.9. **Vitamin E (α-tocopherol) levels in soybean oil as measured by different analytical methodologies**

mg/100 g oil

| Reference[1] | Sample type | Analytical method | α-tocopherol | |
|---|---|---|---|---|
| | | | Mean | Range |
| Shahidi | | HPLC | | 10.1-10.2 |
| Yuki and Ishikawa | Commercial refined, bleached and deodorised oil sample | TLC-GLC | 4.8 | |
| Codex Alimentarius Commission | | HPLC | | 0.9-35.2 |
| NUTTAB[2] | | | 8.3 | |
| USDA-ARS [3] | | GLC, HPLC | 8.2 | |
| Swedish National Food Administration | | | 12 | |
| Danish Food Composition Databank[4] | Refined oil | | 6.1 | 2.7-9.5 |
| Jung et al. | Crude oil | HPLC-UV | 4.6 | |
| | Degummed oil | | 5.3 | |
| | Refined oil | | 4.3 | |
| | Bleached oil | | 4.6 | |
| | Deodorised oil | | 4.0 | |

*Notes:* 1. Many of these references provide additional data on the β-tocopherol, γ-tocopherol, δ-tocopherol and total tocopherol content of the samples. 2. Food Standards Australia New Zealand (2010). 3. Value is converted to mg amounts based on the conversions of vitamin E in IU to mg as defined by the DRI report, 1 mg of α-tocopherol = IU of the RRR-α-tocopherol compound × 0.67, where RRR-α-tocopherol compound is natural vitamin E (Gebhardt and Holden, 2006). 4. Saxholt et al. (2008).

*Sources:* Shahidi (2002); Yuki and Ishikawa (1976); Codex Alimentarius Commission (2009); Food Standards Australia New Zealand (2010); USDA Agricultural Research Service (2008b); Swedish National Food Administration (2011); Saxholt et al. (2008); Jung et al. (1989).

## *Soybean meal*

Soybean meals, as present in the marketplace, are normally defatted and toasted to obtain a moisture content of approximately 9-11% (Table 9.10). Two types of meals are ordinarily produced. One is 44% crude protein on an as-is basis, with further addition of hulls. The other is a higher 49% (as-is basis) crude protein meal, without hulls. The reported ranges for protein, fat, ash, crude fibre, neutral detergent fibre, acid detergent fibre and carbohydrate (given as nitrogen-free extract [NFE]) content are shown in Table 9.10. The ranges in amino acid concentrations are shown in Table 9.11.

## *Hulls and forage*

Soybean hulls are generally removed from the beans before oil extraction. In animal feeds, hulls may be used as carriers and as a source of fibre. Soybean forage is usually harvested around the full seed (R6) stage. Soybean hay is produced from the harvested forage. The hay is allowed to sun-cure to about 11% moisture. The proximate nutrient content for soybean hulls, soybean forage and soybean hay are shown in Tables 9.12, 9.13 and 9.14, respectively.

Table 9.10. **Proximate and fibre content of soybean meal**

% of dry matter

| | 44% meal | | | | | 49% meal | | | | | Meal | Meal (dehulled, solvent extracted) |
|---|---|---|---|---|---|---|---|---|---|---|---|---|
| Reference | NRC 2000 (beef)[1] | NRC 2001 (dairy)[2] | NRC 1998 (swine)[3] | NRC 1994 (poultry)[4] | Ensminger et al.[5] | NRC 2000 (beef)[1] | NRC 2001 (dairy)[2] | NRC 1998 (swine)[3] | NRC 1994 (poultry)[4] | Ensminger et al.[5] | NARO[6] | NARO[6] |
| | Mean | Mean | Mean | Mean | Mean | Mean | Mean | Mean | Mean | Mean | Mean | Mean |
| Moisture | 10.9 | 10.9 | 11 | 11 | 11.0 | 10.1 | 10.5 | 10 | 10 | 10.0 | 11.7 | 9.8 |
| Crude protein | 49.9 | 49.9 | 49.21 | 49.44 | 49.8 | 54.0 | 53.8 | 52.78 | 53.89 | 54.6 | 52.2 | 56.2 |
| Crude fat (ether extract) | 1.6 | 1.6 | 1.69 | 0.90 | 1.7 | 1.1 | 1.1 | 3.33 | 1.11 | 1.4 | 1.5 | 1.3 |
| Ash | 7.2 | 6.6 | | | 7.2 | 6.7 | 6.4 | | | 6.8 | 6.7 | 7.0 |
| Crude fibre | | | | 7.87 | 7.0 | | | | 4.33 | 4.1 | 6.3 | 3.7 |
| Neutral detergent fibre | 14.9 | 14.9 | 14.94 | | 14.0 | 7.79 | 9.8 | 9.89 | | 7.4 | 14.3 | |
| Acid detergent fibre | 10.0 | 10.0 | 10.56 | | 10.0 | 6.10 | 6.2 | 6.00 | | 6.9 | 8.9 | |
| Nitrogen-free extract | | | | | 34.3 | | | | | 33.2 | 33.3 | 31.8 |

*Notes:* 1. 44% meal = seed, meal solvent extracted; 44% protein, 49% meal = seeds without hulls, meal. 2. 44% meal = meal, solvent, 44% CP; 49% meal = meal, solvent, 48% CP. 3. 44% meal = meal, solvent extracted; 49% meal = meal without hulls, solvent extracted. Data converted from fresh weight to dry weight basis using given moisture level. 4. 44% meal = seeds, meal solvent extracted; 49% meal = seeds without hulls, meal solvent extracted. Data converted from fresh weight to dry weight basis using given moisture level. 5. 44% meal = seeds, meal solvent extracted, 44% protein; 49% meal = seeds without hulls, meal solvent extracted, 49% protein. 6. The target amount of protein was not specified in this source.

*Sources:* NRC (2000, 2001, 1998, 1994); Ensminger et al. (1990); NARO (2001).

Table 9.11. **Amino acid composition of soybean meal**

% of dry matter

| | 44% meal | | | | 49% meal | | | | Meal | Meal (dehulled, solvent extracted) |
|---|---|---|---|---|---|---|---|---|---|---|
| Reference | NRC 2001 (dairy)[1] | NRC 1998 (swine)[2] | NRC 1994 (poultry)[3] | Ensminger et al.[4] | NRC 2001 (dairy)[1] | NRC 1998 (swine)[2] | NRC 1994 (poultry)[3] | Ensminger et al.[4] | NARO[5] | NARO[5] |
| | Mean | Mean | Mean | Mean | Mean | Mean | Mean | Mean | Mean | Mean |
| Arginine | 3.68 | 3.63 | 3.56 | 3.65 | 3.94 | 3.87 | 3.94 | 4.03 | 3.88 | 4.24 |
| Cystine | 0.76 | 0.79 | 0.75 | 0.75 | 0.81 | 0.82 | 0.81 | 0.83 | 0.80 | 0.85 |
| Histidine | 1.38 | 1.31 | 1.33 | 1.27 | 1.49 | 1.42 | 1.45 | 1.43 | 1.38 | 1.49 |
| Isoleucine | 2.28 | 2.24 | 2.22 | 2.38 | 2.45 | 2.40 | 2.40 | 2.60 | 2.39 | 2.61 |
| Leucine | 3.90 | 3.84 | 3.84 | 3.92 | 4.20 | 4.07 | 4.23 | 4.20 | 4.02 | 4.21 |
| Lysine | 3.13 | 3.18 | 3.05 | 3.20 | 3.38 | 3.36 | 3.35 | 3.44 | 3.21 | 3.53 |
| Methionine | 0.72 | 0.69 | 0.70 | 0.67 | 0.77 | 0.74 | 0.76 | 0.74 | 0.73 | 0.73 |
| Phenylalanine | 2.62 | 2.45 | 2.45 | 2.51 | 2.83 | 2.66 | 2.65 | 2.76 | 2.67 | 2.83 |
| Threonine | 1.99 | 1.94 | 1.95 | 2.03 | 2.13 | 2.06 | 2.12 | 2.23 | 2.07 | 2.21 |
| Tryptophan | 0.63 | 0.69 | 0.84 | 0.69 | 0.68 | 0.72 | 0.84 | 0.78 | 0.71 | 0.76 |
| Valine | 2.34 | 2.31 | 2.35 | 2.66 | 2.50 | 2.52 | 2.51 | 2.77 | 2.53 | 2.67 |
| Glycine | | | 2.15 | 2.36 | | | 2.32 | 2.66 | 2.26 | 2.34 |
| Serine | | | 2.60 | 2.66 | | | 2.81 | 3.08 | 2.68 | 2.89 |
| Tyrosine | | 1.90 | 2.17 | 1.80 | | 2.02 | 2.21 | 2.18 | 1.86 | 1.96 |
| Crude protein | 49.9 | 49.21 | 49.89 | 49.8 | 53.8 | 52.78 | 53.73 | 54.6 | 52.21 | 56.21 |

*Notes:* 1. 44% meal = meal, solvent, 44% CP; 49% meal = meal, solvent, 48% CP. 2. 44% meal = meal, solvent extracted; 49% meal = meal without hulls, solvent extracted. Data converted from fresh weight to dry weight basis using given moisture level. 3. 44% meal = seeds, meal solvent extracted; 49% meal = seeds without hulls, meal solvent extracted. Data converted from fresh weight to dry weight basis using given moisture level. 4. Seeds, meal solvent extracted, 44% protein; 49% meal = seeds without hulls, meal solvent extracted, 49% protein. 5. The target amount of protein was not specified in this source.

*Sources:* NRC (2001, 1998, 1994); Ensminger et al. (1990); NARO (2001).

Table 9.12. **Proximate and fibre content of soybean hulls**

| Reference | Ensminger et al.[1] | NARO | NRC 2001 | NRC 2000[2] |
|---|---|---|---|---|
| | Mean | Mean | Mean | Mean |
| | | g/100 g fresh weight | | |
| Moisture | 9.0 | 10.3 | 9.1 | 9.7 |
| | | g/100 g dry matter | | |
| Crude protein | 11.9 | 17.6 | 13.9 | 12.2 |
| Crude fat (ether extract) | 2.2 | 5.6 | 2.7 | 2.10 |
| Ash | 5.1 | 4.9 | 4.8 | 4.9 |
| Neutral detergent fibre | 65.6 | 54.4 | 60.3 | 66.3 |
| Acid detergent fibre | 46.8 | 41.8 | 44.6 | 49.0 |
| Nitrogen-free extract | 40.9 | 40.2 | | |

*Notes:* 1. Data listed as soybean "seed coats". 2. Data listed as soybean "seed coats".

*Sources:* Ensminger et al. (1990); NARO (2001); NRC (2001, 2000).

Table 9.13. **Proximate and fibre content of soybean forage**

| Reference | Ensminger et al.[1] | NARO | ILSI | |
|---|---|---|---|---|
| | Mean | Mean | Mean | Range |
| | | g/100 g fresh weight | | |
| Moisture | 77 | 77.1 | 77.0 | 73.5-81.6 |
| | | g/100 g dry matter | | |
| Crude protein | 17.9 | 18.8 | 19.38 | 14.37-24.71 |
| Crude fat (ether extract) | 4.0 | 2.2 | 3.14 | 1.30-5.13 |
| Ash | 10.5 | 8.3 | 9.04 | 6.71-10.78 |
| Crude fibre | 27.3 | 32.3 | 22.67 | 13.58-31.73 |
| Neutral detergent fibre | | 47.2 | | |
| Acid detergent fibre | | 37.6 | | |
| Nitrogen-free extract [1] | 40.3 | 38.4 | | |

*Note:* 1. Data reported as fresh forage.

*Sources:* Ensminger et al. (1990); NARO (2001); ILSI (2010).

Table 9.14. **Proximate and fibre content of soybean hay**

| Reference | Ensminger et al.[1] | NARO |
|---|---|---|
| | Mean | Mean |
| | g/100 g fresh weight | |
| Moisture | 11.0 | 12.8 |
| | g/100 g dry matter | |
| Crude protein | 15.8 | 18.1 |
| Crude fat (ether extract) | 2.5 | 1.6 |
| Ash | 8.0 | 6.8 |
| Neutral detergent fibre | | |
| Acid detergent fibre | 40.0 | |
| Nitrogen-free extract | 39.3 | 34.9 |

*Note:* 1. Data reported as sun-cured hay.

*Sources:* Ensminger et al. (1990); NARO (2001).

# Other constituents

## *Anti-nutrients and toxicants*

### *Oligosaccharides*

Soybean meal contains stachyose and raffinose oligosaccharides, which limit the energy availability for this co-product in swine and poultry. These two low molecular weight carbohydrates are considered anti-nutrients due to the gas production and resulting flatulence caused by their consumption (Rackis, 1974).

These compounds are present in defatted, toasted soybean meal as well as in raw soybeans (Padgette et al., 1996). Further processing of soybean meal into concentrate or isolate reduces or removes these oligosaccharides (Mounts et al., 1987). Data regarding the oligosaccharide content of soybean seed are shown in Table 9.15.

Table 9.15. **Oligosaccharide content of soybean seed (g/100 g dry matter)**

| Reference | ILSI | | Hymowitz et al. | | NFRI-NARO[1] | |
|---|---|---|---|---|---|---|
| | Mean | Range | Mean | Range | Mean | Range |
| Raffinose | 0.36 | 0.21-0.66 | 0.39 | 0.1-0.9 | 0.91 | 0.61-1.60 |
| Stachyose | 2.19 | 1.21-3.50 | 2.79 | 0.6-5.1 | 3.82 | 2.58-4.96 |

*Note:* 1. Data converted from fresh weight to dry weight basis using given moisture level.

*Sources:* ILSI (2010); Hymowitz et al. (1972); National Food Research Institute-NARO (2011).

### *Trypsin inhibitors*

Protease inhibitors, i.e. the Kunitz inhibitor and the Bowman-Birk inhibitor, are active against trypsin, while the latter is also active against chymotrypsin (Liener, 1994). These protease inhibitors interfere with the digestion of proteins resulting in decreased animal growth. The activity of these inhibitors is destroyed when the bean or meal is toasted or heated during processing. Data on trypsin inhibitor content from soybean seed are shown in Table 9.16.

### *Lectins*

Lectins are proteins that bind to carbohydrate-containing molecules. Lectins in raw soybeans can inhibit growth in animals (Liener, 1994). Soybean lectin is sometimes referred to as soybean hemagglutinin. Lectins are rapidly degraded upon heating. In one study, lectin levels dropped approximately 100-fold when the raw soybean was processed into defatted, toasted soybean meal (Padgette et al., 1996). However, Liu (1997) in his review found research to indicate that soy lectin is quite resistant to dry heat. Data regarding the lectin content of soybean seed are shown in Table 9.16.

### *Phytic acid*

Phytic acid (myo-Inositol 1,2,3,4,5,6-hexakis [dihydrogen phosphate]) is present in soybeans. Liener (2000) estimates that two-thirds of the phosphorus in soybeans is bound as phytate and is mostly unavailable to non-ruminant animals. This compound chelates mineral nutrients including calcium, magnesium, potassium, iron and zinc, rendering them unavailable to non-ruminant animals consuming the beans (NRC, 1998; Liener, 1994). Some processing steps such as boiling, steaming or fermenting may reduce the phytate content of soybeans (Reddy and Pierson, 1994). For example, during tempeh fermentation, the phytase action of *Rhizopus* sp. hydrolyzes phytate. Reddy and Pierson report that approximately 32.9-54.5% of the phytate in soybeans may be hydrolyzed during tempeh fermentation.

Phytic acid chelation of zinc present in corn-soybean meal diets used for growing swine requires supplements of zinc to avoid a parakeratosis condition (Smith et al., 1962). It is becoming common for feed formulators to add a phytic acid degrading enzyme, phytase, to swine and poultry diets to release phytin-bound phosphorus, so that the amount of this mineral added to the diet can be decreased, potentially reducing excess phosphorus in the environment.

Phytic acid naturally occurs in soybeans (and most soybean products) and can make up to 1-1.5 g per 100 g of the dry weight (Liener, 1994). Data on the phytic acid content of soybean seed are shown in Table 9.16.

Table 9.16. **Anti-nutrient content of soybean seed**

| Reference | ILSI | | NFRI-NARO | | Kakade et al.[3] | Liener |
|---|---|---|---|---|---|---|
| | Mean | Range | Mean | Range | Range | Range |
| Lectins | 1.72 HU[1]/ mg DM | 0.11-9.04 HU/ mg DM | | | 0.11-9.4 HU/ mg DM[4] | |
| Phytic acid, g/100 g DW | 1.12 | 0.63–1.96 | 1.63 | 0.8-2.5 | | 1.0-1.5 |
| Trypsin inhibitor | 48.33 TIU[2]/ mg DM | 19.59-118.68 TIU/ mg DM | 7.57 µg/ mg protein | 4.23-10.64 µg/mg protein | 100-184 TIU/ mg protein[4] | |

*Notes:* DW: dry weight; DM: dry matter.

1. HU: hemagglutination units. 2. TIU: trypsin inhibitor units. 3. Aqueous extractable proteins derived from defatted soybean seed extracts. 4. Activity reported is in the protein content of the crude defatted soybean seed extract.

*Sources:* ILSI (2010); National Food Research Institute-NARO (2011); Kakade et al. (1972); Liener (1994).

## *Other compounds*

### *Isoflavones*

Soybeans naturally contain a number of isoflavone compounds reported to possess biochemical activity, including estrogenic, anti-estrogenic and hypocholesterolemic effects, in mammalian species. These compounds have been implicated in adversely affecting reproduction in animals fed diets containing large amounts of soybean meal (Schutt, 1976). The effect of isoflavones in humans is an active area of research, including research on both the safety of isoflavones as well as the potential health benefits of isoflavones (Messina, 2010).

The isoflavones in soybeans and soy products have three basic types: diadzein, genistein and glycitein. Each of these three isomers, known as aglycones or free forms, can also exist in three conjugate forms: glucoside, acetylglucoside or malonylglucoside (Wang and Murphy, 1994a; Liu, 1997). Therefore, there are a total of 12 known isoflavone aglycones and glycosides in soybeans.

The isoflavone content of soybeans is greatly influenced by factors such as variety, growing locations, planting year, planting date and harvesting date (Wang and Murphy, 1994b; Aussenac et al, 1998; Murphy et al., 1998). For example, a study indicated that the total isoflavone content of a single variety, Vinton 81, ranged from 84.4 mg to 163.6 mg/100 g raw seeds among eight locations in 1995, and from 160.8 mg to 284.2 mg/100 g in 1996 (Hoeck et al., 2000). The total isoflavone content of raw soybean seeds of an individual variety has been shown to range three- to five-fold depending on location and year of growth (Wang and Murphy, 1994b; Hoeck et al., 2000, Aussenac et al., 2008).

Furthermore, differences in analytical methods and reporting conventions can also contribute significantly to variation in isoflavone values found in the literature. In some reports, total isoflavone is expressed as the sum of all 12 known isoflavone aglycones and glycosides (Wang and Murphy, 1994b). In other studies, only free (aglycones) or bound (conjugated) forms are tested and expressed (Coward et al., 1993; Taylor et al., 1996). Still, other sources describe that isoflavones are hydrolysed to their aglycone forms or the amount is normalised by molecular weight to the aglycone forms (Wang and Murphy, 1996). In the latter case, because the molecular weight of the glucosides is 1.6-1.9 times greater than the aglycones, the total isoflavone amount can be significantly less than the value from non-normalised data (Murphy et al., 1998).

Processing also significantly affects the retention and distribution of isoflavone isomers in the final products (Coward et al., 1993; Wang and Murphy, 1994a, 1996; Liu, 1997). Toasted soybean meal appears to have similar levels of phytoestrogens as the raw seed (Padgette et al., 1996; Wang and Murphy, 1996). Soybean sprouts have also been reported to contain coumestrol (Liener, 1994; Wang and Murphy, 1994a).

Table 9.17 contains data on various isoflavones found in soybean seeds.

Table 9.17. **Isoflavone content of soybean seed (mg/kg dry matter, unless noted)**

| Reference | ILSI | | Kim et al. | Lee et al. | Wang and Murphy[1] | Aussenac et al. | USDA-ARS[2] |
|---|---|---|---|---|---|---|---|
| | Mean | Range | Range | Range | Range | Range | Mean |
| Diadzin | | | 13.1-83.6 | 310.5-608.9 | 180-690 | 503.3-96 398.7 | |
| Malonyldaidzin | | | 61.9-558.1 | 1 204.3-1 806.3 | 241-300 | 1 104-3 889.7 | |
| Daidzein | | | 0.1-21.2 | 32.2-153.5 | 7-26 | 0.8-3.5 | |
| Total daidzein | 834.8 | 60.0-2 453.5 | | 1 568.0-2 568.8 | 240-600 | | 620.7 |
| Genistin | | | 11.7-143.0 | 493.2-773.5 | 394-852 | 378.6-957.7 | |
| Malonylgenistin | | | 135.5-603.4 | 153.5-1 981.3 | 738-743 | 1 407.7-3 752.4 | |
| Acetylgenistin | | | 0.1-21.0 | | 2-9 | | |
| Genistein | | | 0.5-22.6 | 9.3-31.5 | 17-29 | 1.1-5.5 | |
| Total genistein | 976.8 | 144.3-2 837.2 | | 1 751.2-2 646.6 | 648-954 | | 809.9 |
| Glycitin | | | 1.1-33.5 | 56.4-218.3 | 53-56 | 228.7-411.7 | |
| Malonylglycitin | | | 6.6-71.2 | 134.1-463.0 | 50-61 | 58.0-210.5 | |
| Glycitein | | | | 6.7-58.7 | 17-29 | | |
| Total glycitein | 156.6 | 15.3-310.4 | | 208.7-502.4 | 82-107 | | 149.9 |
| Total isoflavones | 2 123.8 | 678.7-3 688.9 | | 3 764.5-5 594.9 | 995-1 636 | 3 911.0-9 797.9 | 1 545.3 |

*Notes:* 1. Data converted from fresh weight to dry matter (DM) basis assuming average DM of 90%. 2. USDA Agricultural Research Service *Database for the Isoflavone Content of Selected Foods* (2008). Data source NBD No. 16108: Soybeans, mature seeds, raw (all sources). Data reported as mg/kg on a fresh weight basis.

*Sources:* ILSI (2010); Kim et al. (2005); Lee et al. (2003); Wang and Murphy (1994a); Aussenac et al. (1998); USDA Agricultural Research Service (2008b).

### *Phospholipids*

Phospholipids have been investigated for their medical and product stability characteristics (Hildebrand et al., 1984; O'Brien and Andrews, 1993). Soybean lecithin is known to contain the primary phospholipids identified as phosphatidylcholine, phosphatidylethanolamine, phosphatidylinositol and phosphatidic acid (Rydhag and Wilton, 1981). National Food Research Institute-NARO (2011) reported levels of these phospholipids and two other phospholipids, phosphatidylserine and phosphatidylglycerol, which are present in soybean seed at much lower levels.

These results are reported in Table 9.18.

Table 9.18. **Phosphopipids content of soybean seed**

g/100 g dry matter

| Reference | NFRI-NARO[1] | |
|---|---|---|
| | Mean | Range |
| Phosphatidylcholine | 0.41 | 0.21-0.55 |
| Phosphatidylethanolamine | 0.24 | 0.11-0.37 |
| Phosphatidylinositol | 0.28 | 0.19-0.42 |
| Phosphatidylserine | 0.04 | 0.02-0.08 |
| Phosphatidylglycerol | 0.05 | 0.01-0.10 |
| Phosphatidic acid | 0.2 | 0.08-0.29 |
| Lecithin | 1.40 | 1.0-1.79 |

*Note:* 1. Data converted from fresh weight to dry weight basis using given moisture levels.

*Source:* National Food Research Institute-NARO (2011).

## *Sterols*

Sterols are the other non-saponifiable components of vegetable oils besides tocopherols. Total sterols range from 180 mg to 450 mg/100 g of oil (Codex Alimentarius Commission, 2009). The proportions of major sterols are presented in Table 9.19.

Table 9.19. **Sterol levels in soybean oil**

% of total sterols

| Reference | Codex Alimentarius Commission[1] | Souci et al. |
|---|---|---|
| | Range | Mean |
| Cholesterol | 0.2-1.4 | 0.25 |
| Brassicasterol | ND-0.3 | |
| Campesterol | 15.8-24.2 | 10.00 |
| Stigmasterol | 14.9-19.1 | 10.92 |
| Beta-sitosterol | 47.0-60 | 29.85 |
| Delta-5-avenasterol | 1.5-3.7 | |
| Delta-7-stigmastenol | 1.4-5.2 | |
| Delta-7-avenasterol | 1.0-4.6 | |
| Others | ND–1.8 | 49.23[2] |

*Notes:* 1. Non-detects (ND) are ≤ 0.05 % total sterols. 2. Reported as "free sterols".

*Sources:* Codex Alimentarius Commission (2009); Souci et al. (2008).

## *Saponins*

Saponins are a diverse group of structurally related compounds containing a steroidal or triterpenoid aglycone linked to one or more oligosaccharides present in numerous plant families (Liener, 1994; Guclu-Ustundag and Mazza, 2007). Saponins from soybean have been shown to have no adverse effects when fed to laboratory animals at high levels (Liener, 1994). Unlike other plant saponins, soy saponins have only a weak effect on intestinal permeability and therefore have little impact on active nutrient transport (Liener, 1994).

Consequently, soybean saponins are not considered to be true anti-nutrients. Total saponin content of soybeans ranges from to 0.09-0.53 g/100 g dry matter (Anderson and Wolf, 1995).

### *Allergens*

Soybean is one of eight foods that account for 90% of all IgE-mediated food allergies (Taylor and Hefle, 2000). The prevalence of soybean allergy in the general population is reported to be between 0.3% and 0.7% (Becker et al., 2004; Roehr et al., 2004; Sicherer and Sampson, 2006; Zuidmeer et al., 2008; Boyce et al., 2010), with an increased prevalence reported in children with atopic eczema (Becker et al., 2004). Many cases of soy allergy are outgrown during childhood (Bock, 1987; Sampson and Scanlon, 1989; Host and Halken, 1990; Sicherer et al., 1998; Becker et al., 2004; Savage et al., 2010; Boyce et al., 2010). Allergic reactions resulting from soybean consumption are similar to those elicited by the other food allergens; however, the most severe allergic reactions, anaphylaxis and death, seem to be rare (Sicherer et al., 2000). Radioallergosorbent test (RAST) and skin prick test are both used in the diagnosis of soybean allergy; but these methods yield a far higher incidence of soybean allergy as compared with results from double-blind placebo controlled food challenges (Becker et al., 2004).

A number of immunological or immunochemical tests have been developed to examine allergenic proteins usually based on sera from sensitive subjects. There are a number of proteins in the soybean (see Table 9.20) that are considered potential allergens due to their IgE binding ability (L'Hocine and Boye, 2007; WHO/IUIS, 2011). These proteins are involved in storage, enzymatic and protective functions. Some of the proteins are associated with inhalation induced allergy, such as Gly m 1, Gly m 2 and Kunitz trypsin inhibitor. Other proteins are associated with food allergy and include P34, β-conglycinin and glycinin. IgE binding to all subunits of β-conglycinin and glycinin was recently demonstrated using sera from soybean allergic subjects (Holzhauser et al., 2009). The P34 allergen is considered as an immunodominant soybean allergen, i.e. responsible for a large percentage of the food allergy reactions to soybean (Wilson et al., 2005). Some soybean proteins are also known to cross-react with certain allergens present in legumes (e.g. peanut, green pea, green bean) (Herian et al., 1990). When compared to soybean seeds, sprouts exhibit similar ability to bind IgE from soy-allergic individuals (Herian et al., 1993; ILSI, 1999).

The effects of agronomic conditions, heating and processing on allergenicity of soybeans have been discussed in ILSI (1999) and Taylor and Lehrer (1996). Heating and other processing may increase or decrease the potency of soybean allergens (Taylor and Lehrer, 1996; Wilson et al., 2005).

Soybean products such as soybean oil and soybean lecithin contain low levels of soy protein. Soybean oils, particularly unrefined oils, may contain allergenic proteins (Bush et al., 1985; Paschke et al., 2001). Soy lecithin, which is largely composed of phospholipids, may also contain allergenic proteins (Porras et al., 1985; Awazuhara et al., 1998; Gu et al., 2001). Highly refined soybean oil has been studied in soy-allergic individuals; results from this study and other studies are consistent with the expectation that the amount of protein present in highly refined soybean oil does not elicit an allergic reaction in the overwhelming majority of these people (Bush et al., 1985; Awazuhara et al., 1998; Paschke et al., 2001).

Table 9.20. **Potential soybean allergens**

| IgE-binding proteins | Allergen nomenclature | Molecular weight (kDa) | Family |
|---|---|---|---|
| Hydrophobic proteins | Gly m 1[1] | 7.0-7.5 | Lipid transfer protein |
| Defensin | Gly m 2[1] | 8.0 | Storage protein |
| Profilin | Gly m 3[1] | 14 | Profilin |
| SAM22 | Gly m 4[1] | 16.6 | Pathogenesis related protein PR-10 |
| P34 | Gly m Bd 30 K | 34 | Protease |
| Unknown Asn-linked glycoprotein | Gly m Bd 28 K | 26 | Unknown |
| β-conglycinin (vicilin, 7S globulin) | Gly m 5[1] | 140-170 | Storage protein (with subunits) |
| Glycinin (legumin, 11S globulin) | Gly m 6[1] | 320-360 | Storage protein (with subunits) |
| 2S albumin | Not assigned | 12 | Prolamin |
| Lectin | Not assigned | 120 | Lectin |
| Lipoxygenase | Not assigned | 102 | Enzyme |
| Kunitz trypsin inhibitor | Not assigned | 21 | Protease inhibitor |
| Unknown | Not assigned | 39 | Unknown |
| Unknown | Not assigned | 50 | Homology to chlorophyll A-B binding protein |
| P22-25 | Not assigned | 22-25 | Unknown |

*Note:* 1. WHO/IUIS (2011) allergen nomenclature recognised by WHO and IUIS.

*Source*: Adapted from L'Hocine and Boye (2007); updated with information from WHO/IUIS (2011).

## Suggested constituents to be analysed related to food use

### *Key products consumed by humans*

Soybeans are consumed in non-fermented and fermented forms (IFIC, 2009; Liu, 2008). Non-fermented soy foods include dairy analogues (e.g. soymilk), meat analogues (e.g. "veggie burgers"), tofu, soy sprouts, yuba (soymilk film), okara (soy pulp), soy flours, soy protein (including isolates and concentrates), boiled soybeans (edamame) and baked soybeans ("soy nuts"). Fermented foods include soy sauce, miso, natto, tempeh, soy yogurt, sufu (fermented tofu) and fermented whole soybeans (Liu, 2008). Soybean oil, soy lecithin and soy protein isolate[5] are used in infant formulas. Soy protein products are also added to a number of meat, dairy, bakery and cereal products as protein extenders (Liu, 1997). Approximately 2% of soy protein is consumed by humans; the large majority of the remaining 98% is processed into soybean meal for livestock feed (Goldsmith, 2008). The daily intake of soy-based foods in Japan is generally estimated to be between 63.2 g and 70.2 g per person (Food Safety Commission, 2006). In Korea, consumption of soybean and soybean-based products, including tofu, soymilk, sprouts, soybean paste and other foods is estimated to be 21 g per person per day (Kim and Kwon, 2001).

Soybean oil is used in a wide variety of foods and is the predominant soybean-based product consumed by humans. Soybean oil makes up 94% of the soybean food ingredients consumed by humans. Yearly consumption of soybean oil per capita is 30 kg in Brazil, 4 kg in China and 27 kg in the United States (Goldsmith, 2008).

### *Suggested analysis for food use of new varieties*

Soybeans can be used in the diet to provide protein. Protein is evaluated in relationship to its biological value which is markedly influenced by the relative amounts of indispensable (essential) and dispensable (non-essential) amino acids and the form of nitrogen in the diet (WHO, 2007). WHO and the Institute of Medicine (IOM) list the following nine amino acids as indispensable, i.e. those that have carbon skeletons that cannot be synthesised to meet body needs from simpler molecules: histidine, isoleucine, leucine, lysine, methionine, phenylalanine, threonine, tryptophan and valine. Additionally, the IOM lists six amino acids as "conditionally indispensable", i.e. those amino acids requiring a dietary source when endogenous synthesis cannot meet metabolic needs: arginine, cysteine, glutamine, glycine, proline and tyrosine. However, WHO indicated that the requirement for indispensable amino acids is not an absolute value, and one must consider the total nitrogen content of the diet, including the dispensable amino acids particularly at lower levels of nitrogen consumption (National Academy of Sciences, 2005; WHO, 2007).

Soybeans are also used as a source of fat for human diets. Soybean oil is evaluated primarily for its fatty acid content, particularly for its unsaturated fatty acids: oleic, linoleic and linolenic acids. Linolenic and linoleic fatty acids have been recognised as essential, those that cannot be synthesised by the body (National Academy of Sciences, 2005). Soybean oil is also a source of vitamins E and K.

Soybeans contain several anti-nutrients that are relevant to nutrition and human health. For example, soybeans contain phytic acid, stachyose, raffinose, lectins and isoflavones.

The suggested key nutritional and anti-nutritional parameters to be analysed in soybean matrices for human food use are shown in Table 9.21. Fatty acids, vitamin E and vitamin K may be measured in seed or oil.

Table 9.21. **Suggested nutritional and compositional parameters to be analysed in soybean matrices for food use**

| Parameter | Seed | Soybean oil |
|---|---|---|
| Moisture[1] | X | |
| Crude protein[1] | X | |
| Crude fat (ether extract)[1] | X | |
| Crude fibre[1] | X | |
| Carbohydrate[2] | X | |
| Ash[1] | X | |
| Amino acids | X | |
| Fatty acids[3] | X | X |
| Vitamin E[3] (α-tocopherol) | X | X |
| Vitamin K1[3] | X | X |
| Phytic acid | X | |
| Stachyose | X | |
| Raffinose | X | |
| Lectins | X | |
| Isoflavones | X | |

*Notes:* 1. These components should be measured using a method suitable for the measurement of proximates. 2. Carbohydrate (by calculation) = 100% – (crude protein% + crude fat% + ash% + moisture%). 3. Measurement of this component can be conducted in either seed or oil.

## Suggested constituents to be analysed related to feed use

### *Key products consumed by animals*

Several whole and processed fractions of soybeans contribute to the total animal diet. Toasted soybeans (whole cooked seed feed product) are fed to cattle and swine on a limited basis, but the oil in toasted seeds causes the fat in pigs to take on an undesirable soft texture (Ensminger et al., 1990). Grummer and Rabelo (2000) suggest that whole cooked soybeans are a palatable protein and fat supplement that has the potential to increase lactation performance of dairy cattle when included at a rate of up to 24% of dry matter of the diet. Other methods of heating full-fat soybeans include jet-sploding, micronization and extrusion.

The main soybean product fed to animals is the defatted/toasted soybean meal (Thacker and Kirkwood, 1990). However, aspirated grain fractions, forage, hay, hulls, seed and silage are also fed to a limited extent, primarily to cattle. In some instances, bakery products containing soybean oil are also fed to cattle. It has been reported, however, that hay and hulls are also fed to poultry, and soybean aspirated grain fractions, hulls and seeds have been fed to swine (Ensminger et al., 1990).

The Environmental Protection Agency of the United States (US-EPA, 2008) has provided estimates on potential contribution of soybean products to the diets of high-producing beef cattle, dairy cattle, poultry and swine in the United States. The US-EPA provides these estimates as percentages of feedstuffs in livestock daily rations for mature and market animals based upon production data of livestock meat, milk and eggs for human consumption.

Estimated inclusion in animal diets for soybean fractions, based on the feedstuff's classification as roughage (R) or protein concentrate (PC), is shown in Table 9.22.

Table 9.22. **Estimated possible inclusion of soybean fractions to animal feeds**

| | | | Percent of diet (%) | | | |
|---|---|---|---|---|---|---|
| **Soybean fraction** | Classification[1] | Dry matter % | Feedlot beef | Mature, lactating dairy | Laying hen | Finishing swine |
| Seed | PC | 89 | 5 | 15 | 20 | 15 |
| Forage/silage[2] | R | 35 | NU[3] | 20 | NU | NU |
| Hay[2] | R | 89 | NU | 20 | NU | NU |
| Meal | PC | 92 | 5 | 10 | 25 | 15 |
| Hulls | R | 90 | 15 | 20 | NU | NU |

*Notes:* 1. Classification of the soybean fraction as roughage (R) or protein concentrate (PC). 2. Label restrictions about feeding may be allowed. 3. NU indicates the feedstuff is not used or is used at less than 5% of diet.

*Source*: Adapted from US-EPA (2008), Table 1 Feedstuffs (June 2008).

### *Suggested analysis for feed use of new varieties*

Soybean meal is fed to animals primarily as a source of protein, and is normally marketed as either a 44% protein product with hulls or a more refined 49% protein product with hulls removed. The amino acid profile of the two products is essentially the same, with the difference being that more fibre has been removed from the 49%

protein product. Soybean meal contains relatively high levels of certain essential amino acids that are deficient in many other common feedstuffs. However, addition of essential amino acids to the diet may still be needed to meet the essential amino acid requirements for swine and poultry (NRC, 1994, 1998).

Proximate analyses are commonly conducted on animal feedstuffs. The process determines amounts of nitrogen, ether extract, ash and crude fibre present in the feedstuff. Crude protein is calculated by multiplying the nitrogen content by 6.25, a conversion factor based on the average amount of nitrogen in protein. Fat is considered to be acid ether extractable material. For the ruminant animal, the traditional proximate analysis, crude fibre, is considered obsolete and has now been replaced by acid detergent fibre and neutral detergent fibre (NRC, 2001).

There is general agreement that the trypsin inhibitors are the primary soybean anti-nutrients that should be minimised in animal diets. However, the amount of lectins in the raw soybean and phytic acid levels are other important considerations. As previously discussed, toasting or heating reduces the content of trypsin inhibitors and lectin, and also will decrease urease concentrations.

The oligosaccharides, raffinose and stachyose, because of their adverse effect on energy availability in swine and poultry, may also be important. Isoflavones do not appear to be a concern when soybean meal is used in formulating livestock diets.

When considering the remainder of the soybean products that could be fed to animals, the composition of the soybean seed, soybean meal and the forage appear to be indicators of the safety and nutritional value of products derived from these matrices.

The suggested nutritional and compositional parameters to be analysed in soybean matrices for feed use are shown in Table 9.23. For all analytes, except fatty acids and lectins, analytes can be measured in seed or meal.

Table 9.23. **Suggested nutritional and compositional parameters to be analysed in soybean matrices for feed use**

| Parameter | Seed[1] | Meal[1] | Forage |
|---|---|---|---|
| Moisture[2] | X | X | X |
| Crude protein[2] | X | X | X |
| Crude fat (ether extract)[2] | X | X | X |
| Neutral detergent fibre | X | X | X |
| Acid detergent fibre | X | X | X |
| Carbohydrates[3] | X | X | X |
| Ash[2] | X | X | X |
| Amino acids | X | X | |
| Fatty acids | X | | |
| Calcium | X | X | |
| Phosphorus | X | X | |
| Stachyose | X | X | |
| Raffinose | X | X | |
| Phytic acid | X | X | |
| Trypsin inhibitors | X | X | |
| Lectins | X | | |

*Notes:* 1. Parameters that are shared between seed and meal can be measured in either matrix. 2. These components should be measured using a method suitable for the measurement of proximates. 3. Carbohydrate (by calculation) = 100% – (crude protein% + crude fat% + ash% + moisture%).

# Notes

1. For information on the environmental considerations for the safety assessment of soybean, see OECD (2000).
2. For additional discussion of appropriate comparators, see Codex Alimentarius Commission (2003: paragraphs 44 and 45).
3. For additional discussion of appropriate comparators for plants that have been modified for changes in specific compositional components, see Codex Alimentarius Commission (2003: paragraph 51).
4. On occasion, data from the original source may have been rounded to promote consistency in data presentation in the summary tables, and/or when specified, units were converted from a fresh weight to a dry weight basis.
5. The composition of soy protein isolate would, in effect, be considered when one considers the composition of the seed or meal from which the protein isolate would be derived. Any safety or nutritional issues associated with soy protein isolate would be expected to be detected from an analysis of the seed or meal. Additionally, the composition of the soy protein isolate could depend on the composition of the starting materials and the process used to obtain the isolate.

# *References*


American Soybean Association (2012), *Soy Stats: A Reference Guide to Important Soybean Facts and Figures*, St. Louis, Missouri.

Anderson, R.L. and W.J. Wolf (1995), "Compositional changes in trypsin inhibitors, phytic acid, saponins and isoflavones related to soybean processing", *Journal of Nutrition*, Vol. 125, No. 3, Supplement, pp. 581S-588S.

Aussenac, T. et al. (1998), "Quantification of isoflavones by capillary zone electrophoresis in soybean seeds: Effects of variety and environment", *American Journal of Clinical Nutrition*, Vol. 68, No. 6, Supplement, pp. 1 480S-1 485S.

Awazuhara, H. et al. (1998), "Antigenicity of the proteins in soy lecithin and soy oil in soybean allergy", *Clinical & Expimental Allergy*, Vol. 28, No. 12, pp. 1 559-1 564.

Becker, W. et al. (2004), "Opinion of the Scientific Panel on Dietetic Products, Nutrition and Allergies (NDA) on a request from the Commission relating to the evaluation of allergenic foods for labelling purposes", *EFSA Journal*, Vol. 32, pp. 1-197, http://dx.doi.org/10.2903/j.efsa.2004.32.

Bock, S.A. (1987), "Prospective appraisal of complaints of adverse reactions to foods in children during the first 3 years of life", *Pediatrics*, Vol. 79, No. 5, pp. 683-688.

Bolton-Smith, C. et al. (2000), "Compilation of a provisional UK Database for the phylloquinone (vitamin K1) content of foods", *British Journal of Nutrition*, Vol. 83, No. 4, pp. 389-399.

Booth, S.L. and J.W. Suttie (1998), "Dietary intake and adequacy of vitamin K1", *Journal of Nutrition*, Vol. 128, No. 5, pp. 785-788.

Boyce, J.A. et al. (2010), "Guidelines for the diagnosis and management of food allergy in the United States: Report of the NIAID-sponsored expert panel", *Journal of Allergy and Clinical Immunology*, Vol. 126, No. 6, Supplement, pp. S1-S58.

Bush, R.K. et al. (1985), "Soybean oil is not allergenic to soybean-sensitive individuals", *Journal of Allergy and Clinical Immunology*, Vol. 76, No. 2, Part 1, pp. 242-245.

Codex Alimentarius Commission (2009), "Codex Standard for Named Vegetable Oils", *Codex Standard Series No. 210-1999 (Revision 2001, Amendment 1003-2005)*, available at: www.codexalimentarius.net/download/standards/336/CXS_210e.pdf (accessed June 2012).

Codex Alimentarius Commission (2003; Annexes II and III adopted in 2008), *Guideline for the Conduct of Food Safety Assessment of Foods Derived from Recombinant DNA Plants*, CAC/GL 45/2003, available at: www.codexalimentarius.net/download/standards/10021/CXG_045e.pdf (accessed June 2012).

Cook, K.K. et al. (1999), "HPLC analysis for trans-vitamin K1 and dihydro-vitamin K1 in margarines and margarine-like products using the $C_{30}$ stationary phase", *Food Chemistry*, Vol. 67, No. 1, pp. 79-88.

Coward, L. et al. (1993), "Genistein, daidzein, and their b-glycoside conjugates: Antitumor isoflavones in soybean foods from American and Asian diets", *Journal of Agricultural and Food Chemistry*, Vol. 41, No. 11, pp. 1 961-1 967, http://dx.doi.org/10.1021/jf00035a027.

Ensminger, M.E. et al. (1990), *Feeds and Nutrition*, The Ensminger Publishing Company, Clovis, California.

FAO/WHO (2002), *Human Vitamin and Mineral Requirements: Chapter 10 Vitamin K*, Report of a Joint FAO/WHO Expert Consultation, Bangkok, Thailand, FAO and WHO, Rome, available at: www.fao.org/docrep/004/y2809e/y2809e00.htm.

Ferland, G. and J.A. Sadowski (1992), "Vitamin K1 (phylloquinone) content of edible oils: Effects of heating and light exposure", *Journal of Agricultural and Food Chemistry*, Vol. 40, No. 10, pp. 1 869-1 873, http://dx.doi.org/10.1021/jf00022a028.

Food Safety Commission, Novel Foods Expert Committee (2006), "Fundamental concepts in the safety assessment of foods containing soy isoflavones for the purpose of specified health use", Japan, May, available at: www.fsc.go.jp/english/evaluationreports/newfoods_sphealth/soy_isoflavones.pdf.

Food Standards Australia New Zealand (2010), "Australian food composition tables", *NUTTAB 2010*, www.foodstandards.gov.au/science/monitoringnutrients/nutrientables/pages/default.aspx.

Frankel, E.N. (1996), "Antioxidants in lipid foods and their impact on food quality", *Food Chemistry*, Vol. 57, Issue 1, September, pp. 51-55.

Gao, Z.H. and R.G. Ackman (1995), "Determination of vitamin K1 in canola oils by high performance liquid chromatography with menaquinone-4 as an internal standard", *Food Research International*, Vol. 28, Issue 1, No. 1, pp. 61-69.

Gebhardt, S.E. and J.M. Holden (2006), "Consequences of changes in the dietary reference intakes for nutrient databases", *Journal of Food Composition Analalysis*, Vol. 19, pp. 591-595.

Goldsmith, P.D. (2008), "Economics of soybean production, marketing, and utilization", Chapter 5 in: Johnson, L.A. et al. (eds.), *Soybeans: Chemistry, Production, Processing, and Utilization*, AOCS Press, Urbana, Illinois.

Grummer, R.R. and E. Rabelo (2000), *Soy in Animal Nutrition: Utilization of Whole Soybeans in Dairy Cattle Diets*, Federation of Animal Science Societies, Savoy, Illinois.

Gu, X. et al. (2001), "Identification of IgE-binding proteins in soy lecithin", *International Archives of Allergy and Immunology*, Vol. 126, No. 3, pp. 218-225.

Guclu-Ustundag, O. and G. Mazza (2007), "Saponins: Properties, applications and processing", *Critical Reviews of Food Science and Nutrition*, Vol. 47, No. 3, pp. 231-258.

Herian, A.M. et al. (1993), "Allergenic activity of various soybean products as determined by RAST inhibition", *Journal of Food Science*, Vol. 58, p. 385.

Herian, A.M. et al. (1990), "Identifcation of soybean allergens by immunoblotting with sera from soy-allergic adults", *International Archives of Allergy and Immunology*, Vol. 92, No. 2, pp. 193-198.

Hildebrand, D.S. et al. (1984), "Phospholipids plus tocopherols increase soybean oil stability", *Journal of the American Chemists' Society*, Vol. 61, No. 3, pp. 552-555.

Hoeck, J.A. et al. (2000), "Influence of genotype and environment on isoflavone contents of soybean", *Crop Science*, Vol. 40, No. 1, pp. 48-51, http://dx.doi.org/10.2135/cropsci2000.40148x.

Holzhauser, T. et al. (2009), "Soybean (*Glycine max*) allergy in Europe: Gly m 5 (β-conglycinin) and Gly m 6 (glycinin) are potential diagnostic markers for severe allergic reactions to soy", *Journal of Allergy and Clinical Immunology*, Vol. 123, No. 2, pp. 452-458, http://dx.doi.org/10.1016/j.jaci.2008.09.034.

Host, A. and S. Halken (1990), "A prospective study of cow milk allergy in Danish infants during the first 3 years of life", *Allergy*, Vol. 45, No. 8, pp. 587-596.

Hymowitz, T. et al. (1972), "Relationship between the content of oil, protein, and sugar in soybean seed", *Agronomy Journal*, Vol. 64, pp. 613-616.

IFIC (International Food Information Council) (2009), "Functional foods fact sheet: Soy", Food Insight, Washington, DC, available at: www.foodinsight.org/Resources/Detail.aspx?topic=Functional_Foods_Fact_Sheet_Soy, (accessed June 2012).

ILSI (2010), *ILSI Crop Composition Database*, Version 4.0, International Life Science Institute, Washington, DC, www.cropcomposition.org (accessed June 2012).

ILSI, North America Technical Committee on Food Components for Health Promotion (1999), "Safety assessment and potential health benefits of food components based on selected scientific criteria", *Critical Reviews of Food Science and Nutrition*, Vol. 39, No. 3, pp. 203-316.

Johnson, L.A. (2008), "Oil recovery from soybeans", Chapter 11 in: Johnson, L.A. et al. (eds.), *Soybeans: Chemistry, Production, Processing, and Utilization*, AOCS Press, Urbana, Illinois.

Jung, M.Y. et al. (1989) "Effects of processing stepson the contents of minor compounds and oxidation of soybean oil", *Journal of the American Oil Chemists' Society*, Vol. 66., pp. 118-120.

Kakade, M. et al. (1972), "Biochemical and nutritional assessment of different varieties of soybeans", *Journal of Agricultural and Food Chemistry*, Vol. 20, No. 1, pp. 87-90, http://dx.doi.org/10.1021/jf60179a024.

Kamao, M. et al. (2007), "Vitamin K content of foods and dietary vitamin K intake in Japanese young women", *Journal of Nutritional Science and Vitaminology*, Vol. 53, No. 6, pp. 464-470.

Kim, J.S. and C.S. Kwon (2001), "Estimated dietary isoflavone intake of Korean population based on National Nutrition Survey", *Nutrition Research*, Vol. 21, No. 7, pp. 947-53.

Kim, S. et al. (2005), "Analysis of isoflavone concentration and composition in soybean [*Glycine max* (L.)] seeds between the cropping year and storage for three years", *European Food Research and Technology*, Vol. 220, pp. 207-214, http://dx.doi.org/10.1007/s00217-004-1048-5.

Lee, S.J. et al. (2003), "Effects of year, site, genotype and their interactions on various soybean isoflavones", *Field Crops Research*, Vol. 81, Issue 2-3, February, pp. 181-192.

L'Hocine, L. and J.I. Boye (2007), "Allergenicity of soybean: New developments in identification of allergenic proteins, cross-reactivities and hypoallergenization technologies", *Critical Reviews of Food Science and Nutrition*, Vol. 47, No. 2, pp. 2 127-2 143.

Liener, I.E. (2000), "Non-nutritive factors and bioactive compounds in soy", in: Drackley, J.K. (ed.), *Soy in Animal Nutrition*, Federation of Animal Science Societies, Savoy, Illinois.

Liener, I.E. (1994), "Implications of antinutritional components in soybean foods", *Critical Reviews of Food Science and Nutrition*, Vol. 34, No. 1, pp. 31-67.

Liu, K. (2008), "Food use of whole soybeans", Chapter 13, in: Johnson, L.A. et al. (eds.), *Soybeans: Chemistry, Technology/Production, Processing, and Utilization*, Aspen Publishers, Inc., Gaithersburg, Maryland, AOCS Press, Urbana, Illinois, pp. 441-481.

Liu, K. (1997), "Chemistry and nutritional value of soybean components", Chapter 2, in *Soybeans: Chemistry, Technology, and Utilization*, Aspen Publishers, Inc. Frederick, Maryland, pp. 25-113.

Lusas, E.W. (2000), "Oilseeds and oil-bearing materials", in: Kulp, K. and J.G. Ponte, Jr. (eds.), *Handbook of Cereal Science and Technology*, Marcel Dekker, Inc., New York, pp. 297-362.

Messina, M. (2010), "A brief historical overview of the past two decades of soy and isoflavone research", *Journal of Nutrition*, Vol. 140, No. 7, pp.1 350S-1 354S, http://dx.doi.org/10.3945/jn.109.118315.

Ministry of Education, Culture, Sports, Science, and Technology, The Council for Science and Technology, Subdivision on Resources (2005), *Standard Tables of Food Composition in Japan*, Fifth Revised and Enlarged, Japan, available at: www.mext.go.jp/b_menu/shingi/gijyutu/gijyutu3/toushin/05031802/002.htm.

Mounts, T.L. et al. (1987), "Processing and utilization", in: Wilcox, J.R. (ed.), *Soybeans: Improvement, Production, and Uses, Agronomy Monographs*, No. 16, American Society of Agronomy, Madison, Wisconsin.

Murphy, P.A. et al. (1998), "Soy isoflavones in foods: Database development", Chapter 14 in: Shibamoto, T. et al. (eds.), *Functional Foods for Disease Prevention*, I. ACS Sym. Series 701, pp. 138-149.

NARO (2001), *Standard Tables of Feed Composition in Japan – Soybean,* National Agricultural Research Organization, Japan.

National Academy of Sciences (2005), *Dietary Reference Intakes for Energy, Carbohydrate, Fiber, Fat, Fatty Acids, Cholesterol, Protein, and Amino Acids (Macronutrients)*, Food and Nutrition Board, Institute of Medicine of the National Academies, National Academy of Sciences, Washington, DC, available at: www.nap.edu/catalog.php?record_id=10490.

National Food Research Institute-NARO (2011 – up to 2009 harvest data), "Soybean", *Food Composition Database for Safety Assessment of Genetically Modified Crops as Foods and Feeds*, Japan, available at: http://afdb.dc.affrc.go.jp/afdb/index_e.asp (accessed June 2012).

NRC (2001), *Nutrient Requirements of Dairy Cattle*, Seventh Revised Edition, National Research Council, National Academies Press, Washington, DC, available at: www.nap.edu/catalog/9825/nutrient-requirements-of-dairy-cattle-seventh-revised-edition-2001.

NRC (2000), *Nutrient Requirements of Beef Cattle*, Seventh Revised Edition: Update 2000, National Research Council, National Academies Press, Washington, DC, available at: www.nap.edu/catalog/9791/nutrient-requirements-of-beef-cattle-seventh-revised-edition-update-2000.

NRC (1998), *Nutrient Requirements of Swine*, Tenth Revised Edition, National Research Council, National Academies Press, Washington, DC, available at: www.nap.edu/catalog/6016/nutrient-requirements-of-swine-10th-revised-edition.

NRC (1994), *Nutrient Requirements of Poultry*, Ninth Revised Edition, National Research Council, National Academies Press, Washington, DC, available at: http://books.nap.edu/catalog/2114.html.

NRC (1984), *Nutrient Requirements of Beef, 6th Revised Edition*, Hale, W.H. et al. (eds.), National Academies Press, Washington, DC.

O'Brien, B.C. and V.G. Andrews (1993), "Influence of dietary egg and soybean phospholipids and triacylglycerols on human serum lipoproteins", *Lipids*, Vol. 28, Issue 1, pp. 7-12.

OECD (2000), "Consensus document on the biology of *Glycine max* (L.) Merr. (soybean)", *Series on Harmonisation of Regulatory Oversight in Biotechnology*, No. 15, OECD, Paris, available at: www.oecd.org/science/biotrack/46815668.pdf.

Padgette, S.R. et al. (1996), "The composition of glyphosate-tolerant soybean seeds is equivalent to that of conventional soybeans", *Journal of Nutrition*, Vol. 126, No. 3, pp. 702-716.

Paschke, A. et al. (2001), "Determination of the IgE-binding activity of soy lecithin and refined and non-refined soybean oils", *Journal of Chromatography B*, Vol. 756, Nos. 1-2, pp. 249-254.

Piironen, V. et al. (1997), "Determination of phylloquinone in oils, margarines, and butter by high-performance liquid chromatography with electrochemical detection", *Food Chemistry*, Vol. 59, No. 3, pp. 473-480.

Porras, O. et al. (1985), "Detection of soy protein in soy lectihin, margarine and occasionally, soy oil", *International Archives of Allergy and Applied Immunology*, Vol. 78, No. 1, pp. 30-32.

Rackis, J.J. (1974), "Biological and physiological factors in soybeans", *Journal of the American Oil Chemists' Society*, Vol. 51, Issue 1, pp. 161A-174A.

Reddy, N.R. and M.D. Pierson (1994), "Reduction in antinutritional and toxic components in plant foods by fermentation", *Food Research International*, Vol. 27, Issue 3, pp. 281-290.

Roehr, C.C. et al. (2004), "Food allergy and non-allergic food hypersensitivity in children and adolescents", *Clinical Experiences in Allergy*, Vol. 34, No. 10, pp. 1 534-1 542.

Rydhag, L. and I. Wilton (1981), "The function of phospholipids of soybean lecithin in emulsions", *Journal of the American Oil Chemists' Society*, Vol. 58, No. 8, pp. 830-837.

Sampson, H.A. and S.M. Scanlon (1989), "Natural history of food hypersensitivity in children with atopic dermatitis", *Journal of Pediatrics*, Vol. 115, No. 1, pp. 23-27.

Savage, J.H. et al. (2010), "The natural history of soy allergy", *Journal of Allergy and Clinical Immunology*, Vol. 125, No. 3, pp. 683-686, http://dx.doi.org/10.1016/j.jaci.2009.12.994.

Saxholt, E. et al. (2008), *Danish Food Composition Databank, Revision 7.1*, Department of Nutrition, National Food Institute, Technical University of Denmark, available at: www.foodcomp.dk (accessed 6 June 2011).

Schutt, D.A. (1976), "The effect of plant oesterogens on animal reproduction", *Endeavor*, Vol. 35, pp. 110-113.

Shahidi, F. (2002), "Phytochemicals in oilseeds", in: Meskin, M.S. et al. (eds.), *Phytochemicals in Nutrition and Health*, Omaye, CRC Press, Washington, DC, pp. 139-156.

Shearer, M.J. et al. (1996), "Chemistry, nutritional sources, tissue distribution and metabolism of vitamin K with special reference to bone health", *Journal of Nutrition*, Vol. 126, No. 4, Supplement, pp. 1 181S-1 186S.

Sicherer, S.H. and H.A. Sampson (2006), "Food allergy", *Journal of Allergy and Clinical Immunology*, Vol. 117, pp. S470-S475.

Sicherer, S.H. et al. (2000), "Peanut and soy allergy: A clinical and therapeutic dilemma", *Allergy*, Vol. 55, No. 6, pp. 515-521.

Sicherer, S.H. et al. (1998), "Clinical features of food protein-induced enterocolitis syndrome", *Journal of Pediatrics*, Vol. 133, No. 2, pp. 214-219.

Sipos, E.F. (1988), "Edible uses of soybean protein", in: McCana, L. (ed.), *Soybean Utilization Alternatives*, Center for Alternative Crops and Products, University of Minnesota, St. Paul, Minnesota, pp. 57-93.

Smith, W.H. et al. (1962), "Effect of source of protein on zinc requirement of the growing pig", *Journal of Animal Science*, Vol. 21, p. 399.

Souci, S.W. et al. (2008), *Food Composition and Nutrition Tables*, 7th revised and completed edition, MedPharm Scientific Publishers, Boca Raton, Florida.

Swedish National Food Administration (2011), *The National Food Administration's Food Database, Version 07/03/2011*, Uppsala, Sweden.

Taylor, S.L. and S.L. Helfe (2000), "Hidden triggers of adverse reactions to foods", *Canadian Journal of Allergy and Clinical Immunology*, Vol. 5, pp. 106-110.

Taylor, S.L. and S.B. Lehrer (1996), "Principles and characteristics of food allergens", *Critical Reviews of Food and Science Nutrition*, Vol. 36, Supplement, S91-118.

Taylor, N.B. et al. (1996), "Compositional analysis of glyphosate-tolerant soybeans treated with glyphosate", *Journal of Agricultural and Food Chemistry*, Vol. 47, No. 10, pp. 4 469-4 473.

Thacker, P.A. and R.N. Kirkwood (1990), *Non-traditional Feed Sources for Use in Swine Production*, Butterworth, Stoneham, Massachusetts, p. 441.

USDA Agricultural Research Service (2008a), *USDA Nutrient Database for Standard Reference, Release 22*, Washington, DC, available at: www.ars.usda.gov/Services/docs.htm?docid=8964 (accessed 8 September 2010).

USDA Agricultural Research Service (2008b) *USDA Database for the Isoflavone Content of Selected Foods*, Release 2.0. Washington, DC, available at: www.ars.usda.gov/SP2UserFiles/Place/12354500/Data/isoflav/Isoflav_R2.pdf (accessed 7 June 2011).

US-EPA (United States Environmental Protection Agency) (2008), *Memorandum: Revisions of Feedstuffs in Table 1 of OPPTS Test Guideline 860.1000 and Guidance on Constructing Maximum Reasonably Balanced Diets (MRBD). From Chemistry Science Advisory Council, Health Effects Division (7509P) to HED Residue Chemistry Staff*, Washington, DC, 30 June 2008.

Waggle, D.H. and C.W. Kolar (1979), "Types of soy protein products", in: Wilke, H.L. et al. (eds.), *Soy Protein and Human Nutrition*, Academic Press, New York, pp. 19-51.

Wang, H.J. and P.A. Murphy (1996), "Mass balance study of isoflavones during soybean processing", *Journal of Agricultural and Food Chemistry*, Vol. 44, No. 8, pp. 2 377-2 383, http://dx.doi.org/10.1021/jf950535p.

Wang, H.J. and P.A. Murphy (1994a), "Isoflavone content in commercial soybean foods", *Journal of Agricultural and Food Chemistry*, Vol. 42, No. 8, pp. 1 666-1 673, http://dx.doi.org/10.1021/jf00044a016.

Wang, H.J. and P.A. Murphy (1994b), "Isoflavone composition of American and Japanese soybeans in Iowa: Effects of variety, crop year and location", *Journal of Agricultural and Food Chemistry*, Vol. 42, No. 8, pp. 1 674-1 677, http://dx.doi.org/10.1021/jf00044a017.

Wang, T. et al. (1997), "Phospholipid fatty acid composition and stereospecific distribution of soybeans with a wide range of fatty acid composition", *Journal of the American Oil Chemists' Society*, Vol. 74, No. 12, pp. 1 587-1 594.

WHO (2007), "Protein and amino acid requirements in human nutrition", Report of a Joint WHO/FAO/UNU Expert Consultation, *WHO Technical Report Series No. 935*, World Health Organization, Geneva.

WHO/IUIS (WHO and International Union of Immunological Societies Allergen Nomenclature Sub-committee), www.allergen.org (accessed 1 December 2011).

Wilson, S. et al. (2005), "Allergenic proteins in soybean: Processing and reduction of P34 allergenicity", *Nutrition Reviews*, Vol. 63, No. 2, pp. 47-58.

Yuki, E. and Y. Ishikawa (1976), "Tocopherol contents of nine vegetable frying oils, and their changes under simulated deep-fat frying conditions", *Journal of the American Oil Chemists' Society*, Vol. 53, No. 11, pp. 673-676.

Zuidmeer, L. et al. (2008), "The prevalence of plant food allergies: A systemic review", *Journal of Allergy and Clinical Immunology*, Vol. 121, No. 5, pp. 1 210-1 218, http://dx.doi.org/10.1016/j.jaci.2008.02.019.

# *Chapter 10*

# Oyster mushroom (*Pleurotus ostreatus*)

*This chapter, prepared by the OECD Task Force for the Safety of Novel Foods and Feeds with Sweden as the lead country, deals with the composition of oyster mushroom (*Pleurotus ostreatus*). It contains elements that can be used in a comparative approach as part of a safety assessment of foods derived from new varieties. Background is given on oyster mushroom taxonomy, nomenclature and occurrence, cultivation, production, consumption and processing, followed by appropriate varietal comparators and characteristics screened by breeders. Nutrients in oyster mushroom, as well as other constituents (anti-nutrients, toxicants, allergens), are then detailed. The final sections suggest key products and constituents for analysis of new varieties for food use mainly, feed use of oyster mushroom remaining rare.*

# Background

## *Taxonomy, nomenclature and occurrence*

It is well recognised that the taxonomy of the genus *Pleurotus* is confusing as many species have been given several names (synonyms) and can be divided into subspecies. Guzmán (2000) reviewed the taxonomy of the *Pleurotus* genus, scrutinising more than 230 publications and concluded that it was necessary to revise several described species as well as describe new species based on modern methodology. Over the last 200 years, more than 1 000 names have been proposed in the genus but it is agreed that the number of species is much more limited. Furthermore, the species can be divided into several sections or subgenera. Further discussion on the taxonomy and natural distribution of the wild *Pleurotus* mushrooms can be found in the OECD consensus document on the biology of *Pleurotus* species (oyster mushrooms) (OECD, 2005).

Oyster mushrooms belong to the genus *Pleurotus* (Quel.) Fr., and today at least 70 species have been identified. *Pleurotus* was first recommended as a tribe within the genus *Agaricus* by Fries (1821), but was proposed as a separate genus by Quelet (1886). *Pleurotus ostreatus* (Jacq: Fr.) Kummer, the oyster mushroom, is the most cultivated species among oyster mushrooms and the type species of the genus *Pleurotus*. To clearly distinguish the type species from other oyster mushrooms it is called *P. ostreatus* in this chapter. Mating compatibility studies have shown that *P. columbinus*, *P. florida*, *P. salignus* and *P. spodoleucus* are synonyms or subspecies taxa for *P. ostreatus*.

*Pleurotus ostreatus* is in nature found in temperate zones of the northern hemisphere, such as Europe, North Africa, Asia and North America (Singer, 1986) because it forms fruit-bodies at relatively low temperature compared to other *Pleurotus* species. The macroscopic morphologic features of this wood-rotting fungus and the microscopic characteristics of the spores, basidia, cheilocystidia and pleurocystidia of the mushroom are summarised by OECD (2005). Figure 10.1 shows *P. ostreatus* growing on a tree.

Figure 10.1. **Macroscopic feature of *P. ostreatus***

*Source*: OECD (2005).

Other *Pleurotus* species growing in warmer climates have been found to be as easily cultivated as *P. ostreatus*. Therefore, species such as *P. sajor-caju*, *P. cystidiosus*, *P. eryngii*, *P.tuber-regium*, *P. pulmonarius*, *P. citrinopileatus/P. cornucopiae* and *P. djamor/P. flabellatus* are cultivated in various regions of the world, and there is scattered information available on the composition of these species.

In nature *P. ostreatus* can be found living on a large number of plants, including species of the genera *Abies*, *Acacia*, *Acer*, *Alnus*, *Betula*, *Carpinus*, *Carya*, *Castanea*, *Laurocersus*, *Liquidambar*, *Liriodendron*, *Lupinus*, *Magnolia*, *Malus*, *Morus*, *Nyssa*, *Ostrya*, *Pandanus*, *Picea*, *Pistacia*, *Populus*, *Pseudotsuga*, *Quercus*, *Salix*, *Tilia*, *Ulmus* and *Wisteria* (Farr et al., 1989). The broad host plant spectrum makes it easier to understand that the species can be cultivated on substrates containing different lignocellulosic materials. Although seen on dying trees, *P. ostreatus* is thought to be primarily a saprophyte, but behaves as a facultative parasite at the earliest opportunity. Occasionally, it grows on composting bales of straw and, for example, on the pulp residues from coffee production.

## *Cultivation of* **P. ostreatus** *and other oyster mushrooms*

*P. ostreatus* was first cultivated in the United States in 1900 and is now cultivated throughout the world. As indicated by the broad host plant spectrum, *P. ostreatus* and other oyster mushrooms can thrive in and on many lignocellulosic substrates, including, but not limited to, most hardwoods, wood by-products (e.g. sawdust, paper and pulp sludge), cereal straws, maize, maize cobs, coffee residues (e.g. coffee grounds), hulls, stalk and leaves, banana fronds and waste cotton, to mention a few (OECD, 2005). Oyster mushrooms (*Pleurotus* spp.) are now regarded as one of the three most important edible cultivated mushrooms together with the cultivated mushroom (*Agaricus bisporus*) and shiitake (*Lentinula edodes*).

Oyster mushrooms have many advantages as cultivated mushrooms: rapid mycelial growth, high ability for saprophytic colonisation, simple and inexpensive cultivation techniques, and several kinds of species available for cultivation under different climatic conditions. In addition, they are low in calories, sodium, fat and cholesterol, while rich in protein, carbohydrate, fibre, vitamins and minerals. These nutritional properties make these mushrooms good dietary foods. In addition, oyster mushrooms and products from them are consumed for medicinal purposes (Cohen et al., 2002; Kues and Liu, 2000). Owing to these attributes during recent years, the production and consumption of this mushroom has increased tremendously.

The oyster mushroom strain to be cultivated is aseptically inoculated onto a growing bed in glass jars or polypropylene plastic bags. The growing bed is prepared from a mix of water, lime and grain spawn (e.g. rye, wheat, sorghum), straw spawn (e.g. paddy rice straw, wheat straw) or other plant waste materials, and sterilised before use. Alternatively, the mushrooms can be grown by using cut wood logs. Different strains will be suitable depending on the climate and incubation conditions. The most commonly cultivated *Pleurotus* species is *P. ostreatus*. Other frequently cultivated species include grey oyster mushroom or phoenix-tail mushroom (*P. sajor-caju* [Fr.] Sing.), abalone mushroom (*P. cystidiosus* O.O. Miller), golden oyster mushroom (*P. citrinopileatus* Sing.), pink oyster mushroom (*P. flabellatus* [Berk. and Br.] Sacc.), black oyster mushroom (*P. sapidus* [Schulzer] Kalchbremer), *P. eryngii* (DC.: Fr.) Quel., *P. djamor* (Fr.) Boedjin *sensu* Lato and *P. tuberregium* (Fr.) Sing.

### *Production of oyster mushrooms*

The cultivation of oyster mushrooms world-wide has increased more than 25-fold during the last 30 years. The worldwide production of oyster mushrooms was 35 000 tonnes in 1981 and around 875 000 tonnes in 1997. In 1997, the *Pleurotus* spp. accounted for 14.2% of the world production of mushrooms (Chang and Miles, 2004). However, Chang and Miles (2004) also refer to a publication in Chinese, *The Market of Edible Fungi*, which states that *Pleurotus* spp. are the second most important cultivated mushroom in the People's Republic of China (hereafter "China") (26% of the market), and that in 2000 the production was more than 1 722 000 tonnes. No data on world production since 1997 have been found.

### *Consumption of* **Pleurotus ostreatus**

The oyster mushroom is generally consumed cooked or preserved. Older data on world-wide production of oyster mushrooms is available, along with some data on the extent of export and import. It should be noted that all mushrooms produced are not used as food. For example, these mushrooms are used for the production of enzymes and products for medicinal purposes. Data on the actual consumption of *P. ostreatus* has not been found.

### *Processing of* **Pleurotus ostreatus**

The harvested fresh mushroom has a relatively short shelf life. The oldest method of preserving *P. ostreatus* and other *Pleurotus* species is by air-drying cleaned samples. Mushrooms can also be preserved in brine or canned, but their texture is best fresh. If heat treatments are used during processing, it is acknowledged that flavour may be lost, particularly if the mushrooms are cooled too slowly. Water has been shown to remove less flavour than steam (Chang and Miles, 2004).

### *Appropriate comparators for testing new varieties*

This chapter suggests parameters that oyster mushroom breeders should measure when developing new modified varieties of *P. ostreatus*. The data obtained in the analysis of a new *P. ostreatus* variety should ideally be compared to those obtained from an appropriate near isogenic non-modified variety, grown and harvested under the same conditions.[1] The comparison can also be made between values obtained from new varieties and data available in the literature, or chemical analytical data generated from other commercial *P. ostreatus* varieties.

Components to be analysed include key nutrients, anti-nutrients, toxicants, allergens and other metabolites. Key nutrients are those which have a substantial impact in the overall diet of humans (food) and animals (feed). These may be major constituents (fats, proteins, and structural and non-structural carbohydrates) or minor compounds (vitamins and minerals). Similarly, the levels of known anti-nutrients and allergens should be considered. Key toxicants are those toxicologically significant compounds known to be inherently present in the species, whose toxic potency and levels may impact human and animal health. Standardised analytical methods and appropriate types of material should be used, adequately adapted to the use of each product and by-product. The key components analysed are used as indicators of whether unintended effects of the genetic modification influencing plant metabolism have occurred or not.

## *Genetic modification of* Pleurotus *mushrooms*

Several methods to introduce DNA into *Pleurotus* mushrooms have been studied, including those based on polyethylene glycol (Peng et al., 1993; Yanai et al., 1996; Jia et al., 1998; Kim et al., 1999; Honda et al., 2000; Amore et al., 2012), electroporation (Peng et al., 1992), restriction enzyme mediated integration (Irie et al., 2001; Joh et al., 2003; Fan et al., 2006), particle bombardment (Sunagawa and Magae, 2001) and *Agrobacterium tumefaciens* mediated transformation (Ding et al., 2011). Drawbacks, limiting the use of some of these methods, are the low transformation efficiency, heterogeneous integration into genomic loci and the need for using protoplasts, although improved procedures with enhanced efficiencies have been published (Li et al., 2006; Ding et al., 2011).

Marker or reporter genes successfully employed in identification and selection of mushroom transformants include those conferring antibiotic (hygromycin B) resistance (Peng et al., 1992; Irie et al., 2001), antimetabolite (5-fluoroindole, 5'-fluoro-orotic acid) resistance (Jia et al., 1998; Kim et al., 1999), metabolite (uracil) auxotrophy (Joh et al., 2003), fungicide (carboxin) resistance (Honda et al., 2000), herbicide (bialaphos) resistance (Yanai et al., 1996), as well as reporting successful transformation by expressing the green fluorescent protein (Li et al., 2006; Amore et al., 2012).

## *Traditional characteristics screened by developers of* Pleurotus *strains*

The development of breeding programmes for edible mushrooms such as the oyster mushroom relies on efficient methods to perform directed crosses between fungal strains. This requires in-depth understanding of the biology of the mushroom, including knowledge about the mating genes, genome structure and genetic breeding of higher mushrooms. In. *P. ostreatus*, the genes of two independent loci on one of the 11 chromosome pairs orchestrate the mating control system (Ramírez et al., 2000). The molecular map of the *P. ostreatus* genome is starting to become available. In 2000, Ramírez and co-workers had developed a map based on 196 RAPD (random amplified polymorphic DNA) and RFLP (restriction fragment length polymorphism) markers, as well as functional characters. Traits to be explored in the breeding programmes are being mapped into the species genome in order to facilitate future breeding. The first reviews on gene sequences of *Pleurotus* intracellular and secreted proteins were published by Whiteford and Thurston (2000).

Although commercial transgenic mushroom strains are not yet available, molecular breeding studies of mushrooms have been carried out world-wide. *Pleurotus ostreatus* is not only one of the most important cultivated mushrooms, but also a good model for understanding biochemical and physiological processes in mushrooms, including the production of enzymes and other biologically active compounds. Possible target genes to be introduced by genetic transformation include: genes producing sporeless strains with reduced ability to cause respiratory disease in mushroom cultivation workers (Baars et al., 2004), senescence genes to improve mushroom quality, substrate utilisation genes (especially with increased lignin degradation capability) to enhance yields (Ha et al., 2001) and developmental genes to control mushroom fruiting. There are numerous potential pest and disease resistance targets in strain development, including genes involved in response to fungal pathogens, toxicity to insects and natural pest resistance. In addition, transformations with mating type genes that regulate inter-strain compatibility can alter breeding behavior, whereas genes that influence the contents of essential nutrients may result in functional foods or foods with medicinal effects

(Sunagawa and Magae, 2001; Aida et al., 2009). Other genes contributing to efficient agro-industrial waste bioconversion (Cohen et al., 2002), toxic heavy metal biosorption activities (Pan et al., 2005), hydrophobicity (Ma et al., 2008), and production of live vaccines for animal feeds and human health might also be targets for strain development.

## Nutrients

### *Composition of the oyster mushroom (*Pleurotus ostreatus *L.)*

There is a considerable variation in the data published on nutritional parameters of the oyster mushrooms (*Pleurotus* spp). Most of this variation reflects *de facto* differences in chemical composition due to different strains having been investigated (e.g. Bautista Justo et al., 1998; Manzi et al., 1999; Rai et al., 1988), and different conditions having been present during mushroom cultivation. Factors during and after cultivation that would influence the level of individual constituents include chemical composition of the substrate (e.g. Bonatti et al., 2004; Papaspyridi et al., 2010; Shashirekha et al., 2005; Wang et al., 2001; Yildiz et al., 1998), temperature during cultivation (Pedneault et al., 2007), flushing cycle sampled (Mendez et al., 2005), as well as the conditions after harvest, including storage (Hammond, 1980; Mäkinen et al., 1978) and processing (Jaworska et al., 2011; Manzi et al., 2001). As constituent levels may also vary between different parts of the mushroom (e.g. Synytsya et al., 2008; Yilmaz et al., 2006), it is important to know which parts of the mushroom have been analysed. In part, differences in data may also reflect discrepancies in the analytical methods used. However, most studies referred to in this chapter have utilised standardised AOAC[2] methods, or similarly validated procedures.

This chapter only considers data on the chemical composition of fruiting bodies of various strains of *Pleurotus ostreatus*, including synonymous strain as defined above. The mushrooms analysed were either cultivated under controlled conditions or wild mushrooms.

#### *Proximates*

Representative data on proximate analysis of the *P. ostreatus* are presented in Table 10.1. All data originally reported on a fresh weight basis were recalculated and expressed on a dry weight basis in order to facilitate comparisons.

#### *Proteins*

The protein content is generally calculated from analytical measurements of total nitrogen content. As proteins contain about 16% nitrogen, a conversion factor of 6.25 (1/0.16) is commonly used in nutrition research to convert total nitrogen to protein content (FAO/WHO, 1991; Merrill and Watt, 1973). For mushrooms, this conversion factor is commonly adjusted to account for significant amounts of non-protein nitrogen present. Thus, a conversion factor of 4.38 (Bano and Rajarathnam, 1988; Crisan and Sands, 1978; USDA Agricultural Research Service, 2010) or a factor close to this value (Fujihara et al., 1995; Mattila et al., 2002b) has been suggested by several authors, while the traditional conversion factor of 6.25 has been used in other publications. To facilitate comparisons of results, protein values obtained from total nitrogen using conversion factors other than 4.38 have been recalculated using the conversion factor 4.38. These cases are indicated by a footnote in Table 10.1.

Table 10.1. **Proximate composition of *P. ostreatus***

| | Bautista Justo et al.[2] | Beluhan and Ranogajec[3] | Bonatti et al.[4] | Chirinang and Intarapichet[5,6] | Coli et al.[5] | Jaworska et al.[4,7] | Khanna and Garcha[5,8] | Mattila et al.[7,9] |
|---|---|---|---|---|---|---|---|---|
| | Range[1] | Mean | Range[1] | Mean | Range[1] | Mean | Range[1] | Mean |
| | | | | % fresh weight | | | | |
| Dry matter | 9.9-10.2 | 11.7 | 11.9-14.4 | .. | 10.7-12.0 | 8.8 | 8.5-10.8 | 8.0 |
| | | | | g/100 g dry weight | | | | |
| Carbohydrates | 70.4-73.2 | 61.9 | 71.2-74.5 | 78.0 | 54.0-67.4 | 70.9 | 59.2-65.1 | 62.5 |
| Protein | 17.3-20.0 | 24.9 | 13.1-16.9 | 15.3 | 13.6-23.7 | 16.7 | 19.2-26.1 | 24.6 |
| Total dietary fibre | 32.1-36.8 | .. | .. | 47.6 | .. | .. | .. | 30.0 |
| Fat | 1.1-1.9 | 2.1 | 6.0-6.3 | 0.6 | 3.4-3.9 | 5.5 | 2.3-3.7 | 4.4 |
| Ash | 7.7-8.8 | 7.6 | 5.6-6.1 | 6.1 | 5.0-6.4 | 6.7 | 11.0-13.4 | 8.0 |

| | Manzi et al.[9,10] | Rai et al.[4,8,9] | Shah et al.[2,6] | Sturion and Oetterer[11] | USDA-ARS[9] | Wang et al.[5] | Yang et al.[2] | Obodai and Apertorgbor[11] |
|---|---|---|---|---|---|---|---|---|
| | Mean | Range[a] | Mean | Range[a] | Mean | Range[a] | Mean | Mean |
| | | | | % fresh weight | | | | |
| Dry matter | 8.7 | 6.0-7.1 | 9.6 | 5.6-7.0 | 10.8 | .. | 11.4 | 9.1 |
| | | | | g/100 g dry weight | | | | |
| Carbohydrates | 67.0 | 63.6-64.6 | 75.7 | 66.3-73.5 | 56.3 | 51.7-57.9 | 66.4 | 70.4 |
| Protein | 18.6 | 25.8-26.2 | 15.9 | 17.4-24.1 | 30.6 | 29.1-37.4 | 23.9 | 20.0 |
| Total dietary fibre | 47.3 | .. | .. | .. | 21.3 | .. | .. | .. |
| Fat | 4.2 | 1.5-1.7 | 1.9 | 1.5-1.9 | 3.8 | 4.3-4.7 | 2.2 | 2.0 |
| Ash | 10.3 | 7.9-8.7 | 6.5 | 7.5-8.1 | 9.3 | 6.7-8.4 | 7.6 | 7.6 |

*Notes:* 1. Range of means due to different strains and/or substrates. 2. Original carbohydrate values not including fibre have been recalculated. 3. Wild mushrooms. 4. Carbohydrate value here presented as carbohydrates by difference. Value originally presented as analysed. 5. Protein value recalculated using the conversion factor 4.38. Carbohydrates by difference recalculated accordingly. 6. Data recalculated based on true dry matter content. 7. Protein value calculated by summing the amino acid residues. 8. Both *Pleurotus ostreatus* and *Pleurotus ostreatus* var. *florida*. 9. Original data given on fresh weight basis have been recalculated on dry weight basis. 10. Original carbohydrate value including ash. Value recalculated excluding ash. 11. Original data not including carbohydrates have been complemented with this data if sufficient information was available for such calculation.

*Sources:* Bautista Justo et al. (1998); Beluhan and Ranogajec (2011); Bonatti et al. (2004); Chirinang and Intarapichet (2009); Coli et al. (1988); Jaworska et al. (2011); Khanna and Garcha (1984); Mattila et al. (2002b); Manzi et al. (2001); Rai et al. (1988); Shah et al. (1997); Sturion and Oetterer (1995); USDA Agricultural Research Service (2010); Wang et al. (2001); Yang et al. (2001); Obodai and Apertorgbor (2008).

The protein content in *P. ostreatus* has been reported to be in the range 13.1-37.4 g/100 g dry weight (Table 10.1). The difference in protein content reported by various investigators has partly been linked to the problem of estimating the true total nitrogen content. It has also been linked to several other factors influencing protein quantities, including mushroom strain studied (Coli et al., 1988; Bautista Justo et al., 1998; Manzi et al., 1999), substrate used for cultivation (Wang et al., 2001), time of harvest (Mendez et al., 2005), storage and processing (Jaworska et al., 2011; Manzi et al., 2001).

Table 10.2 presents data on the total content of the various amino acids in *P. ostreatus*. The total amino acid composition includes free amino acids and those in proteins. Essential amino acids comprise 29-41% of the total amino acid content.

The most abundant amino acids are glutamine/glutamic acid and asparagine/aspartic acid, whereas the least abundant are cysteine, methionine and tryptophan.

Table 10.2. **Amino acid composition of *P. ostreatus***

g/100 g total amino acids

| | Bautista Justo et al. | Chirinang and Intarapichet[3] | Mattila et al.[4] | Manzi et al. | Shah et al. | USDA-ARS | Wang et al. | Range of mean values (g/100 g total a.a.) | Range of mean values (g/100 g DW[5]) |
|---|---|---|---|---|---|---|---|---|---|
| Alanine | 6.1-6.2 | 9.0 | 5.4 | 6.0-8.3 | 6.2 | 8.7 | 8.2 | 5.4-9.0 | 0.95-2.86 |
| Arginine | 6.6-8.5 | 15.5 | 7.8 | 7.0-11.5 | 6.2 | 6.6 | 7.9 | 6.2-15.5 | 0.95-2.76 |
| Aspartic acid[1] | 11.0-12.2 | 9.7 | 12.8 | 9.2-12.1 | 9.3 | 10.7 | 9.0 | 9.0-12.8 | 1.42-3.66 |
| Cysteineb | 1.5-1.7 | ND | 1.2 | 1.2-1.7 | 0.8 | 1.0 | 1.1 | 0.8-1.7 | 0.12-0.38 |
| Glutamic acid[1] | 18.4-22.6 | 23.7 | 15.9 | 13.1-16.6 | 17.7 | 22.9 | 15.3 | 13.1-23.7 | 2.71-5.84 |
| Glycine | 4.2-4.5 | 3.9 | 4.2 | 4.4-4.8 | 4.6 | 4.5 | 4.9 | 3.9-4.9 | 0.70-1.71 |
| Proline | 2.9-3.1 | 1.4 | 4.1 | 3.6-4.8 | 6.2 | 1.5 | 4.3 | 1.4-6.2 | 0.39-1.52 |
| Serine | 4.6-4.9 | 5.2 | 4.8 | 3.5-6.0 | 4.7 | 4.5 | 5.2 | 3.5-6.0 | 0.72-1.81 |
| Tyrosine | 3.3-3.5 | 2.7 | 9.6 | 3.6-4.6 | 3.5 | 3.0 | 3.8 | 2.7-9.6 | 0.54-2.74 |
| Histidine | 2.6-2.7 | 2.6 | 2.8 | 3.6-4.3 | 2.0 | 2.5 | 3.6 | 2.0-4.3 | 0.31-1.24 |
| Isoleucine | 3.7-4.1 | 2.9 | 3.6 | 3.9-4.7 | 5.8 | 4.1 | 4.7 | 2.9-5.8 | 0.71-1.62 |
| Leucine | 5.9-6.9 | 5.4 | 6.1 | 6.3-7.3 | 8.2 | 6.1 | 7.4 | 5.4-8.2 | 1.18-2.57 |
| Lysine | 6.7-7.1 | 3.4 | 5.5 | 5.4-6.4 | 7.2 | 4.5 | 6.6 | 3.4-7.2 | 1.10-2.29 |
| Methionine | 1.9-2.0 | 1.3 | 1.5 | 1.5-2.3 | 1.7 | 1.5 | 1.1 | 1.1-2.3 | 0.26-0.44 |
| Threonine | 4.9-5.1 | 4.8 | 4.7 | 4.7-5.3 | 4.8 | 5.1 | 4.9 | 4.7-5.3 | 0.73-1.71 |
| Tryptophan | 1.8-2.2 | 0.7 | NA | 1.1-1.6 | 1.5 | 1.5 | 1.4 | 0.7-2.2 | 0.23-0.48 |
| Phenylalanine | 3.5-4.8 | 3.5 | 4.9 | 3.8-4.7 | 4.3 | 4.1 | 4.4 | 3.5-4.9 | 0.66-1.52 |
| Valine | 4.5-4.9 | 4.3 | 4.9 | 4.3-5.2 | 5.0 | 7.1 | 6.0 | 4.3-7.1 | 0.77-2.10 |
| Non-essential | 59% | 71% | 66% | 66-70% | 59% | 63% | 60% | 59-71% | |
| Essential | 41% | 29% | 34% | 30-34% | 41% | 37% | 40% | 29-41% | |

*Notes:* DW: dry weight; NA: not analysed; ND: not detected; a.a.: amino acids.

1. The analytical method converts asparagine and glutamine to aspartic acid and glutamic acid, respectively. Values represent a total amount of both forms. 2. The analytical method converts cysteine to the dimer cystine, and is analysed as cysteic acid. Values represent a total amount of both forms. 3. Cysteine not included in total amino acids. 4. Tryptophan not included in total amino acids. 5. Data presented on a fresh weight basis have been recalculated on a dry weight basis.

*Sources:* Bautista Justo et al. (1999); Chirinang and Intarapichet (2009); Mattila et al. (2002b); Manzi et al. (1999); Shah et al. (1997); USDA Agricultural Research Service (2010); Wang et al. (2001).

Several studies have investigated the content of free amino acids in *P. ostreatus* (Abe et al., 1980; Beluhan and Ranogajec, 2011; Ginterova and Maxianová, 1975; Kazuno and Miura, 1985; Kim et al., 2009; Oka et al., 1984; Rai et al., 1988; Sato et al., 1985; Tsai et al., 2009; Yang et al., 2001). The reports differ in methodology used for free amino acids analysis and in the various amino acids analysed. Consequently, reported levels of total free amino acids differ considerably (0.4-16.1 g/100 g dry weight [dry matter]). The most common free amino acids in *P. ostreatus* are glutamine/glutamic acid and alanine. Several non-protein amino acids have been detected, with ornithine as the major constituent (Kazuno and Miura, 1985; Kim et al., 2009; Manzi et al., 1999; Oka et al., 1984; Sato et al., 1985). The occurrence of non-protein amino acids in *P. ostreatus* does not raise a safety concern.

### *Carbohydrates*

This chapter uses a common understanding of total carbohydrates in proximate analysis, that is, carbohydrate is calculated as the remaining component when crude protein, crude fat, ash and moisture have been determined and summed up and the total subtracted from 100%. By this definition, dietary fibre is included in total carbohydrates. Publications with no presented value for carbohydrates by difference, but with sufficient data to calculate it, have been complemented with a calculated value. Changes in presentation of original data have been highlighted by a footnote in Table 10.1.

The nature of determining the total carbohydrate content is reflected by the broad range 51.7-78.0 g/100 g dry weight reported in Table 10.1. As for other nutrients, the influence of external factors on the carbohydrate content has been studied, *inter alia* composition of the substrate for cultivation (Shashirekha et al., 2005; Wang et al., 2001), strain of *P. ostreatus* used (Kim et al., 2009; Rai et al., 1988), post-harvest storage (Hammond, 1980) and processing (Manzi et al., 2001) of the mushrooms.

Total dietary fibre comprises the carbohydrate elements (remnants of plant cells, polysaccharides, lignin and associated substances) resistant to hydrolysis (digestion) by human digestive enzymes. Standard analyses for total dietary fibre are based on enzymatic-gravimetric or enzymatic-chemical methods in accordance with AOAC recommendations. Both methods permit separation of an insoluble fraction and a fraction that is soluble in 80% alcohol. Reported amounts of total dietary fibre in *P. ostreatus* are in the range 21.3-47.6 g/100 g dry weight (Chirinang and Intarapichet, 2009; Bautista Justo et al., 1998; Manzi et al., 2001; Mattila et al., 2002b; USDA Agricultural Research Service, 2010). Of this quantity, soluble fibre is reported to constitute 5.1-21.4%, and insoluble fibre constitutes 78.6-94.9% (Lee et al., 2008; Manzi et al., 2001; Synytsya et al., 2008). The insoluble fibre fraction of carbohydrates is primarily made up of chitin from the cell walls, at 3.6-5.5 g/100 g dry weight (Manzi et al., 2001; Vetter, 2007).

The most important constituents in the soluble fibre fraction of oyster mushrooms are the β-glucans. β-glucans have been studied for their potential medical uses (Bobek et al., 2001; Dey et al., 2010; Gutiérrez et al., 1996; Karácsonyi and Kuniak, 1994; Lavi et al., 2006; Nosálóvá et al., 2001; Palacios et al., 2012; Patra et al., 2013; Rop et al., 2009; Rovenský et al., 2009; Sarangi et al., 2006; Sun and Liu, 2009; Tong et al, 2009; Yoshioka et al., 1975, 1985; Zhang et al., 2007). Most studies have concentrated on isolation and characterisation of specific β-glucans, and only a few have tried to quantify these polysaccharides. The chemical method used for analysis of these constituents in various types of mushroom has been shown to influence the quantity of β-glucans detected (Park et al., 2009). Using an enzymatic method, Manzi and co-workers determined the β-glucan content in *P. ostreatus* to be 0.14-0.38 g/100 g dry weight (Manzi and Pizzoferrato, 2000; Manzi et al., 2001), which is about 5% of the total dietary fibre. Two other studies, using a commercial kit for analysis, found levels in the range of 27.4-50.0 g/100 g dry matter (Papaspyridi et al., 2010; Synytsya et al., 2008).

To date, no investigators have been able to give a complete picture of the distribution of individual carbohydrate components in oyster mushrooms. Due to solubility and stability issues different studies for a given carbohydrate fraction are often contradictory. The soluble sugar portion of carbohydrates is usually extracted with 80% ethanol, and analysed after chromatographic separation of the mono- and oligo-saccharide components. Glucose (0.1-1.8 g/100g dry weight) and mannose (0.1-1.3 g/100g dry weight) are the most abundant monosaccharides found in oyster mushrooms (Kazuno and Miura, 1985; Yoshida et al., 1986; Yang et al., 2001; Kim et al., 2009; Tsai et al., 2009; Beluhan and Ranogajec, 2011), while low quantities of fructose (Yoshida et al., 1986); Reis et al., 2012a) and ribose (Kim et al., 2009) have also been reported. Most investigators have identified the disaccharide trehalose (two molecules of glucose) in *P. ostreatus* but the reported levels vary between 0.2 and 40.8 g/100g dry weight (Yoshida et al., 1986; Yang et al., 2001; Kim et al., 2009; Tsai et al., 2009; Beluhan and Ranogajec, 2011; Reis et al., 2012a). The most common sugar alcohol in the mushroom is mannitol, at quantities between 0.3 g and 5.0 g/100 g dry weight (Kazuno and Miura, 1985; Yoshida et al., 1986; Yang et al., 2001; Tsai et al., 2009; Beluhan and Ranogajec, 2011; Reis et al., 2012a). Other sugar alcohols occurring at lower amount include arabitol, sorbitol and myo-inositol (Kazuno and Miura, 1985; Yoshida et al., 1986; Yang et al., 2001; Tsai et al., 2009), although the latter is not a classical sugar alcohol.

*Lipids*

The crude fat content of *P. ostreatus* ranges from 0.6% to 6.3% of mushroom dry weight (Table 10.1). Individual fatty acids are generally analysed as methyl esters by gas-liquid chromatography or gas chromatography coupled to mass spectrometry. They are usually presented in relative terms, as a percentage of total fatty acids. This means that an accurate presentation requires approximately equal efficiency to identify and quantify the different fatty acids.

Table 10.3 presents literature data on profiles of major fatty acid constituents in *P. ostreatus*. Linoleic acid (C18:2) is the most common fatty acid, at 50-78% of the total fatty acids. Oleic acid (C18:1) and palmitic acid (C16:0) are the next most prominent fatty acids with ranges of 6-20% and 11-26% of total fatty acids, respectively. Also studies that only analysed for a few fatty acids found these fatty acids be the major ones (Bautista Justo et al., 1998; Hadar and Cohen-Arazi, 1986; Khanna and Garcha, 1981; Rashad et al., 2009; Shashirekha et al., 2005). Stearic acid (C18:0), palmitoleic acid (C16:1) and myristic acid (C14:0) occur in lesser quantities. Occasional data are available for other individual fatty acids. The most complete picture of the fatty acid profile has been reported by Pedneault et al. (2007). They noted that all saturated fatty acids not mentioned above with a chain length between 12 and 24 carbons (except fatty acids with a chain length of 19 and 21 carbon atoms), as well as the unsaturated fatty acids linolenic acid (18:3), gadoleic acid (C20:1), erucic acid (C22:1) and nervonic acid (C24:1) were minor fatty acids in the range 0.01-0.82% of total fatty acids. Stancher et al. (1992a) made similar observations. Small quantities of arachidic acid (C20:0) have been reported by Rashad et al. (2009) and Yilmaz et al. (2006). Only a few studies have presented data on absolute quantities of fatty acids; linoleic acid levels were in the range between 7.0 mg and 11.9 mg/g of dry weight oleic acid between 1.6 mg and 2.9 mg/g of dry weight, palmitic acid between 1.8 mg and 5.7 mg/g of dry weight and stearic acid between 0.3 mg and 0.5 mg/g of dry weight (Hadar and Cohen-Arazi, 1986; Bautista Justo et al., 1998; USDA Agricultural Research Service, 2010).

A few investigators have reported data on free fatty acids in *P. ostreatus* (Kazuno and Miura, 1985; Stancher et al., 1992b). Linoleic acid (C18:2) is not only the most common fatty acid in lipids but also the most common free fatty acid, at 61-64% of the total free fatty acids. The next most common free fatty acid is palmitic acid (C16:0) at around 21% of total fatty acids.

Table 10.3. **Fatty acid composition of *P. ostreatus***

g/100 g total fatty acids

| | Pedneault et al.[1,2] | Coli et al.[1] | Stancher et al.[2] | Reis et al. |
|---|---|---|---|---|
| C14:0 | 0.1-0.2 | ND-0.3 | 1.2 | |
| C16:0 | 11.6-12.8 | 19.0-25.8 | 21.5 | 11.2 |
| C16:1[3] | 0.4-0.5 | ND-0.2 | 0.9 | |
| C18:0 | 1.8-2.8 | 1.7-2.2 | 2.8 | 1.6 |
| C18:1[3] | 6.2-12.0 | 13.5-20.0 | 9.7 | 12.3 |
| C18:2 | 69.6-78.0 | 50.5-51.6 | 59.4 | 68.9 |
| C20:0 | | 1.3-1.6 | 0.14 | |
| Others | 1.9 | 5.5-6.1 | 1.3 | 5.9 |

*Note:* ND: not detected.

1. Range is due to data from different strains and cultivation conditions. 2. Original data separated in polar and non-polar fatty acids. Data recalculated into total fatty acids. 3. Data presented as undifferentiated by double-bond position and configuration.

*Sources:* Pedneault et al. (2007); Coli et al. (1988); Stancher et al. (1992a); Reis et al. (2012a).

## *Minerals*

Mushrooms are usually good at taking up minerals and heavy metals from soil. Several reports demonstrate that the content of minerals and heavy metals in the fruiting bodies of *P. ostreatus* mirrors the content in the substrate (Bressa et al., 1988; Favero et al., 1990a, 1990b; Sales-Campos et al., 2009). Consequently, there are considerable differences in mineral and heavy metal levels presented in studies on cultivated and wild oyster mushrooms dependent on substrate and environmental factors.

This chapter therefore separates mineral data on wild grown mushrooms and cultivated mushrooms. Still, substantial differences due to production areas are likely, and there is a great variability within the presented data. Data from studies on mushrooms collected from pronounced contaminated areas have been omitted.

Table 10.4 summarises data on the content of the most important minerals and heavy metals in the cultivated *P. ostreatus*.

Occasional data on other minerals or trace elements have been reported, i.e. Al, As, B, Ba, Li, Mo, Ni, Se, Sr, Ti, and V (Costa-Silva et al., 2011; Haldimann et al., 1995; Mattila et al., 2001; Petrovska, 1999; Procida and Pertoldi Marletta, 1995; Santoprete and Innocenti, 1984; USDA Agricultural Research Service, 2010; Vetter, 1989, 2005; Vetter et al., 2005).

Table 10.4. **Mineral content of the cultivated *P. ostreatus***

| | Çaglarirmak[1] | Çoli et al.[2] | Kawai et al.[3] | Manzi et al.[2] | Mattila et al. | Obodai and Apertorgbor | Petrovska[2] | Procida and Pertoldi Marletta[2] |
|---|---|---|---|---|---|---|---|---|
| | | | | mg/100 g dry weight | | | | |
| Calcium (Ca) | 110.1 | 40.2-42.0 | 4 | 23.5-48.6 | 1.0 | 43.1 | 36.0-48.0 | |
| Iron (Fe) | 20.1 | 11.4-23.1 | 8 | | 5.4 | 42.6 | 3.4-3.8 | 7.2-21.6 |
| Magnesium (Mg) | 301 | 129-151 | 156 | 161-203 | 200 | | 380-722 | |
| Phosphorus (P) | 1 355 | 697-1 027 | 1 061 | | 1 390 | 939 | 310-400 | |
| Potassium (K) | 3 019 | 967-2 503 | 2 720 | 2 185-3 444 | 3 730 | 3334 | | |
| Sodium (Na) | 1 049.8 | 1 400-1 485 | 74 | 25.2-136.0 | 13.0 | 56.2 | | |
| Zinc (Zn) | 15.2 | 6.2-10.4 | 10.8 | | 8.3 | | 1.7-2.6 | 11.3-14.2 |
| Copper (Cu) | | 8.8-11.4 | 1.6 | | 0.8 | | 1.0-2.4 | 0.5-4.6 |
| Manganese (Mn) | | | 1.5 | | 1.1 | | 0.3-1.1 | 0.8-1.0 |
| | | | | µg/100 g dry weight | | | | |
| Cobalt (Co) | | | | | | | ND-51 | |
| Chromium (Cr) | | | | | | | ND-22 | 35-47 |

| | Rashad and Abdou[2] | Sales-Campos et al.[2] | Sesli and Tüzen | Strmisková et al.[2] | Tshinyangu[2,5] | USDA-ARS[1] | Vetter | Vetter et al.[6] | Wang et al. | Yildiz et al.[2,7] |
|---|---|---|---|---|---|---|---|---|---|---|
| | | | | | mg/100 g dry weight | | | | | |
| Calcium (Ca) | 10.9-19.0 | 34.0-60.0 | | 12.8-17.5 | 101.7-108.3 | 28.0 | 89.0 | 82.0 | ND | 1.0-20.0 |
| Iron (Fe) | 42.9-209.9 | 11.6-15.1 | 5.8 | 5.7-14.2 | 6.7-8.2 | 12.3 | .. | 15.6 | 7.1 | 1.0-19.0 |
| Magnesium (Mg) | 136-166 | 157-250 | | 134-208 | 178-193 | 166 | 190 | 137 | 182 | .. |
| Phosphorus (P) | | 695-1 060 | | 942-1755 | 790-880 | 1 109 | 1 198 | 698 | 1 648 | .. |
| Potassium (K) | 632-2 020 | 3 683-4 218 | | 2 240-4 734 | 2 615-2 860 | 3 882 | 3 988 | 3 074 | 2 171 | 3 440-4 500 |
| Sodium (Na) | 433-654 | 15.4-19.4 | | 13.2-27.5 | 72.3-87.7 | 166.0 | | 26.9 | 21.9 | .. |
| Zinc (Zn) | 7.0-8.9 | 8.2-12.4 | 4.3 | 4.7-7.7 | 10.8-11.7 | 7.1 | 8.0 | 7.66 | 13.7 | 10.0-13.0 |
| Copper (Cu) | 6.0-11.9 | 0.9-1.2 | 0.7 | 0.8-2.7 | 1.5-2.0 | 2.3 | 2.2 | 1.87 | 2.5 | 3-30[h] |
| Manganese (Mn) | 1.6-3.5 | 1.6-2.3 | 1.3 | 0.5-1.0 | 1.0-1.4 | 1.0 | 1.1 | 0.96 | 1.6 | 2.0-4.0 |
| | | | | | µg/100 g dry weight | | | | | |
| Cobalt (Co) | | | 20 | | | | ND-19 | 4 | | |
| Chromium (Cr) | | | | 16-101 | | | ND-131 | 89 | | |

*Notes:* ND: not detected.

1. Original data given on fresh weight basis have been recalculated on dry weight basis. 2. Range of means due to different strains and/or substrates. 3. Average of 25 samples of different oyster mushroom cultivations on sawdust substrate. 4. *Pleurotus ostreatus* var *florida*. 5. *Pleurotus ostreatus* var *columbinus*. 6. Analysed part of the fruit body was the pileus. 7. *Pleurotus ostreatus* var *salignus*. 8. The value 30 mg/100 g is a suspected outlier. The other values are in the range 3-5 mg/100 g dry weight.

*Sources:* Çaglarirmak (2007); Çoli et al. (1988); Kawai et al. (1994); Manzi et al. (1999); Mattila et al. (2001); Obodai and Apertorgbor (2008); Petrovska (1999); Procida and Pertoldi Marletta (1995); Rashad and Abdou (2002); Sales-Campos et al. (2009); Sesli and Tüzen (1999); Strmisková et al. (1992); Tshinyangu (1996); USDA Agricultural Research Service (2010); Vetter (1989); Vetter et al.( 2005); Wang et al. (2001); Yildiz et al. (1998).

Numerous studies have determined the content of toxic heavy metals in the cultivated *P. ostreatus*. These data are presented in Table 10.5. Table 10.6 summarises a selection of studies on the content of the most important minerals and heavy metals in collected wild *P. ostreatus*.

Table 10.5. **Content of toxic heavy metals in cultivated *P. ostreatus***

μg/100 g dry weight

| Heavy metal | Range | References |
|---|---|---|
| Cadmium (Cd) | 20-294 | García et al. (2009); Haldimann et al. (1995); Kawai et al. (1994); Maihara et al. (2008); Mattila et al. (2001); Petrovska (1999); Procida and Pertoldi Marletta (1995); Reguła and Siwulski (2007); Santoprete and Innocenti (1984); Sesli and Tüzen (1999); Strmisková et al. (1992); Wahid et al. (1988); Vetter (1989); Vetter et al. (2005); Zurera-Cosano et al. (1987) |
| Lead (Pb) | ND-440 | García et al. (2009); Haldimann et al. (1995); Mattila et al. (2001); Petrovska (1999); Procida and Pertoldi Marletta (1995); Reguła and Siwulski (2007); Santoprete and Innocenti (1984); Sesli and Tüzen (1999); Strmisková et al. (1992); Wahid et al. (1988); Zurera-Cosano et al. (1987) |
| Mercury (Hg) | ND-110 | Haldimann et al. (1995); Kawai et al. (1994); Melgar et al. (2009); Reguła and Siwulski (2007); Santoprete and Innocenti (1984); Sesli and Tûzen (1999); Strmisková et al. (1992); Zurera-Cosano et al. (1988) |

*Note:* ND: not detected.

Table 10.6. **Mineral and toxic heavy metal content in wild *P. ostreatus***

mg/100 g dry weight

| Mineral/heavy metal | Range | References |
|---|---|---|
| Calcium (Ca) | 82-317 | Gençcelep et al. (2009); Vetter (1989) |
| Iron (Fe) | 9.9-68.2 | Gençcelep et al. (2009); Isildak et al. (2004); Tüzen et al. (1998); Vetter (1989); Zhu et al. (2011) |
| Magnesium (Mg) | 120-190 | Gençcelep et al. (2009) Vetter (1989) |
| Phosphorus (P) | 326-1 198 | Gençcelep et al. (2009); Vetter (1989) |
| Potassium (K) | 1 993-3 988 | Gençcelep et al. (2009); Vetter (1989) |
| Sodium (Na) | 19-153 | Gençcelep et al. (2009); Vetter (2003) |
| Zinc (Zn) | 1.9-14.2 | Alonso et al. (2003); Gençcelep et al. (2009); Isildak et al. (2004); Tüzen et al. (1998); Vetter (1989); Zhu et al. (2011) |
| Copper (Cu) | 0.5-4.7 | Alonso et al. (2003); Dogan et al. (2006); Gençcelep et al. (2009); Isildak et al. (2004); Tüzen et al. (1998); Vetter (1989); Zhu et al. (2011) |
| Manganese (Mn) | 0.7-3.7 | Dogan et al. (2006); Gençcelep et al. (2009); Isildak et al. (2004); Tüzen et al. (1998); Vetter (1989); Zhu et al. (2011) |
| Cadmium (Cd) | 0.023-0.30 | Dogan et al. (2006); Isildak et al. (2004); Tüzen et al. (1998); Vetter (1989); Zhu et al. (2011); Zurera-Cosano et al. (1987) |
| Lead (Pb) | 0.012-0.297 | Dogan et al. (2006); Tüzen et al. (1998); Zhu et al. (2011); Zurera-Cosano et al. (1987) |
| Chromium (Cr) | ND-4.1 | Dogan et al. (2006); Isildak et al. (2004); Vetter (1989); Zhu et al. (2011) |
| Mercury (Hg) | 0.002-0.142 | Nnorom et al. (2012); Tüzen and Soylak (2005); Tüzen et al. (1998); Vetter and Berta (1997) |

*Note:* ND: not detected.

*Vitamins*

Table 10.7 summarises data on the vitamin content of the oyster mushroom.

The level of β-carotene, the precursor of vitamin A, is reported to be very low (<3.1 mg/100g dry weight) and frequently below the limit of quantification.

Several studies have not been able to detect any vitamin C in oyster mushrooms (Okamura, 1998; USDA Agricultural Research Service, 2010; Yang et al., 2002), while others have presented quantities in the range 20.0-45.9 mg/100 g dry weight (Çaglarırmak, 2007; Bautista Justo et al., 1998; Li and Chang, 1985; Mattila et al., 2001; Rai et al., 1988), and one as high value as 113 mg/100 g dry weight (Bano and Rajarathnam, 1986). The latter observation was in a sample of *P. ostreatus* var. *florida*. These observations can partly be explained by the analytical method used, as Okamura (1998) showed that ascorbic acid occurs in the form of analogues (6-deoxyascorbic acid, erythroascorbic acid, 6-deoxy-5-O-(α-D-xylopyranosyl)-ascorbic acid, 6-deoxy-5-O-(α-D-glucopyranosyl)-ascorbic acid, 5-O-(α-D-glucopyranosyl)-erythroascorbic acid and 5-O-(α-D-xylopyranosyl)-erythroascorbic acid) rather than ascorbic acid itself in *P. ostreatus* and other mushrooms. The total level of the reduced and oxidised forms of these analogues, converted to ascorbic acid, was around 5 mg/100 g. Most of the analogues occurred in the reduced form (Okamura, 1998).

Also, reported vitamin E levels differ between investigators. Whereas one research team found α-tocopherol to be more common than γ-tocopherol and δ-tocopherol (Tsai et al., 2009), another research team made the opposite observation (Reis et al., 2012a).

The biosynthesis of vitamin D2 from ergosterol is ultraviolet light dependent, and its formation is influenced both by the amount of the precursor available, the moisture content of the mushroom, the supply of daylight and the temperature during exposure. Ergosterol has been reported to occur at quantities between 0.68 mg and 6.7 mg/g dry weight (Jasinghe and Perera, 2005; Koyama et al., 1984; Mattila et al., 2002a; Teichmann et al., 2007; Phillips et al., 2011).

Recently, Phillips et al. (2012) demonstrated that several mushrooms, including *P. ostreatus*, also contain vitamin D4, being produced in an ultraviolet light dependent process from the precursor ergosta-5,7-dienol (22,23-dihydroergosterol). The level of vitamin D4 in *P. ostreatus* was 18.3 μg/100 g dry weight, and the level of the precursor ergosta-5,7-dienol around 0.87 mg/g dry weight.

As no information on the cultivation conditions were available in the studies in Table 10.7 reporting vitamin D levels, it is not known whether the reported amounts fully describe the range in levels of these vitamins in *P. ostreatus*.

Table 10.7. **Vitamin content of the *P. ostreatus***

| Compound | Unit | Range | References |
|---|---|---|---|
| Vitamin C | mg/100 g d.w. | ND-113.0 | Bano and Rajarathnam (1986); Çaglarırmak (2007); Bautista Justo et al. (1998); Li and Chang (1985); Mattila et al. (2001); Okamura (1998); Rai et al. (1988); USDA Agricultural Research Service (2010); Wang et al. (2001); Yang et al. (2002) |
| Vitamin B1 | mg/100 g d.w. | 0.1-2.0 | Bano and Rajarathnam (1986); Çaglarırmak (2007); Bautista Justo et al. (1998); Mattila et al. (2001); USDA Agricultural Research Service (2010); Wang et al. (2001) |
| Vitamin B2 | mg/100 g d.w. | 2.3-7.9 | Bano and Rajarathnam (1986); Çaglarırmak (2007); Bautista Justo et al. (1998); Mattila et al. (2001); USDA Agricultural Research Service (2010); Wang et al. (2001) |
| Vitamin D2 | µg/100 g d.w. | 0.3-6.5 | Mattila et al. (2001); Teichmann et al. (2007); USDA Agricultural Research Service (2010); Phillips et al. (2011) |
| Vitamin D4 | µg/100 g d.w. | 18.3 | Phillips et al. (2012) |
| Folates | mg/100 g d.w. | 0.1-1.4 | Bano and Rajarathnam (1986); Çaglarırmak (2007); Lasota et al. (1983); Mattila et al. (2001); USDA Agricultural Research Service (2010) |
| Niacin | mg/100 g d.w. | 36.0-90.0 | Bano and Rajarathnam (1986); Çaglarırmak (2007); Bautista Justo et al. (1998); Lasota et al. (1983); Mattila et al. (2001); USDA Agricultural Research Service (2010); Wang et al. (2001) |
| Vitamin E | mg/100 g d.w. | ND-70 | Reis et al. (2012a); Tsai et al. (2009); USDA Agricultural Research Service (2010); Yang et al. (2002) |
| β-carotene | mg/100 g d.w. | ND-3.1 | Tsai et al. (2009); USDA Agricultural Research Service (2010); Yang et al. (2002) |

*Notes:* d.w.: dry weight; ND: not detected.

### *Other metabolites*

A few investigators have studied the composition of the flavour compounds of the oyster mushroom, volatile as well as soluble compounds. Tsai et al. (2009) and Zhang et al. (2008) studied these compounds in the fresh mushroom, and Misharina et al. (2009) studied them in cooked mushrooms. The volatile flavour compounds identified by Tsai et al. (2009) and Zhang et al. (2008) comprised six compounds with eight carbon atoms (1-octen-3-one, 1-octen-3-ol, 3-octanol, 3-octanone, 1-octanol and 2-octen-1-ol) and two aromatic compounds (benzaldehyde, benzyl alcohol), with 1-octene-3-ol, 3-octanone and 1-octen-3-one predominating. The aromatic compounds made up only about 1% of the volatile flavour compounds. Soluble flavours included several soluble sugars and polyols, free amino acids and 5'-nucleotides (Tsai et al., 2009).

Only limited data are available on the occurrence of other constituents in the oyster mushroom. Compounds that have been identified or quantified in *P. ostreatus* include organic acids (Yoshida et al., 1986), phenolic compounds (Del Signore et al., 1997; Rajarathnam et al., 2003; Reis et al., 2012b; Kim et al., 2008), indoles (Muszyńska et al., 2011), steroidal compounds (Chobot et al., 1997; Plemenitaš et al., 1999), glycoinositolphosphosphingolipids (Jennemann et al., 2001) and lovastatin, an inhibitor of 3-hydroxy-3-methylglutaryl-coenzyme A reductase (Gunde-Cimerman and Cimerman, 1995).

## Other constituents

### *Anti-nutrients*

Lectins are carbohydrate-binding proteins found in most vegetables and a broad range of mushrooms (Goldstein and Winter, 2007; Guillot and Konska, 1997). They are biologically active in higher animals by binding to cell-wall components in the gastro-intestinal tract, but their principal function in fungi has not been established (Goldstein and Winter, 2007; Guillot and Konska, 1997). One lectin protein, a glycoprotein containing 14% neutral carbohydrate, has been isolated (Conrad and Rüdiger, 1994; Kawagishi et al., 2000; Wang et al., 2000) and crystallised (Chattopadhyay et al., 1999) from *P. ostreatus*. Its molecular weight has been established to approximately 80-87 kDa using gel filtration and sodium dodecyl sulfate polyacrylamide gel electrophoresis (Conrad and Rüdiger, 1994; Kawagishi et al., 2000; Wang et al., 2000). Matrix-assisted laser desorption/ionisation-time of flight mass spectrometry confirmed the molecular weight to 81.6 kDa (Kawagishi et al., 2000).

Quantification of lectins in the mushroom is difficult as it is influenced by the efficacy of the purification. The lectin in the oyster mushroom belongs to the group of lectins causing erythrocyte agglutination and is frequently quantified by its haemagglutinating activity. Kawagishi et al. (2000) and Conrad and Rüdiger (1994) report a lectin content of 20 mg/100 g of fruiting bodies, whereas Wang et al. (2000) report a content of 7.9 mg/100 g fruiting bodies.

The toxicological data on the *P. ostreatus* lectin is limited and there are no incidents of human intoxication specifically related to the lectin in the oyster mushroom. However, decreased food intake has been reported in laboratory animals fed a diet including a powder of dried *P. ostreatus* (Kawagishi et al., 2000; Nieminen et al., 2009) whereas another study did not observe a similar effect (Bobek et al., 1991). Kawagishi et al. (2000) linked the reduced feed intake to the lectin fraction of the feed; the lower the lectin content of the oyster mushroom powder, the lower the influence on feed intake. Extrapolation of these experiments of repeated intake of relatively high doses of mushrooms pelleted into the animal's diet to the human situation has not yet been undertaken. Kawagishi et al. (2000) used a diet containing 5% mushroom powder and Nieminen et al. (2009) utilised a diet that resulted in a mushroom powder intake ranging from 1.8-5.4% of total feed. According to Nieminen et al. (2009), the latter range corresponds to an intake of 1.2-2.7 kg mushroom per day for a human weighing 70 kg.

### *Toxicants*

Two proteins with hemolytic and cytolytic properties (hemolysins) have been isolated from oyster mushroom, pleurotolysin (Bernheimer and Avigad, 1979) and ostreolysin (Berne et al., 2002). Pleurotolysin consists of two non-associated protein components with molecular weights of 17 kDa and 59 kDa, respectively (Sakurai et al., 2004; Tomita et al., 2004). These components co-operatively, in a larger assembled complex of 700 kDa, generate a pore structure in the cell membrane producing lysis. Ostreolysin is a membrane binding protein of 15-17 kDa interacting in particular with lipid membranes highly enriched in cholesterol and sphigomyelin, hypothetically involved in the development of fruit bodies (Berne et al., 2002; Skočaj et al., 2013). Besides showing hemolytic activity and increasing permeability of endothelial cell membranes in vitro (Berne et al., 2002; Maličev et al., 2007; Sepčić et al., 2003), ostreolysin contracts coronary blood vessels in laboratory animals supplied the protein intravenously (Juntes et al., 2009; Rebolj et al., 2007). An intravenous $LD_{50}$-value of 1 170 μg ostreolysin per kg body weight has been determined in the mouse (Zuzek et al., 2006). No data on quantities of pleurotolysin or ostreolysin in the fruit body of oyster mushroom is available in the literature.

There is no data on the toxicity of *P. ostreatus* hemolysins in oyster mushroom consumers. Oyster mushroom is considered a non-toxic mushroom and the presence of hemolysins does not influence this conclusion. Hemolysins are thermo-labile, and potential toxicity would be considered only for raw mushrooms. Furthermore, proteins are usually degraded in the gastrointestinal tract when ingested. Administration of aqueous extracts of *P. ostreatus* to mice demonstrated no acute effects (Al-Deen et al., 1987; Bedry et al., 2001) and repeated oral feeding of the mushroom to rodents revealed no histopathological changes of cardiac or hepatic tissue (Bobek et al., 1998; Nieminen et al., 2009).

### *Allergens*

Two types of allergy can be distinguished – food allergy manifested after consumption of allergenic mushrooms and respiratory allergy after inhalation of allergenic mycelia or basidiospores. The latter type of allergic disease may be due to the compost/cultivation conditions and is then frequently independent of the mushroom species cultivated. If it is due to mushroom tissues, usually spores, then it is frequently species dependent.

No case reports on individuals being allergic to oyster mushroom as food have been found in the literature.

Like many other cultivated mushrooms, *Pleurotus* species have been shown to give rise to mushroom grower's disease. Most likely mushroom grower's disease develops in workers that have worked in sheds in which spawning takes place and where the compost, spawn and organisms living in the media are mechanically mixed and where basidiospores are common. Characteristic symptoms include allergic rhinoconjunctivitis, asthma and hypersensitivity pneumonitis (Lehrer et al., 1994; Saikai et al., 2002; Helbling et al., 1999; Mori et al., 1998). All these symptoms of allergy have been described in workers cultivating *P. ostreatus* and its subspecies (Senti et al., 2000; Vereda et al., 2007), but hypersensitivity pneumonitis being particularly frequent (Zadrazil, 1973, 1974; Noster et al., 1976, 1978; Cox et al., 1988; Mori et al., 1998; Kamm et al., 1991). In addition, allergic contact dermatitis after exposure to *Pleurotus* mushrooms has been described. Symptoms appeared around harvest and included red scaly vesicular lesions on the hands, sometimes spreading to the upper and lower limbs, face and trunk (Rosina et al., 1995). The agent responsible for the contact dermatitis has not been identified.

## Suggested constituents to be analysed related to food use

### *Identification of* Pleurotus ostreatus *food products*

Oyster mushrooms stand for around 15% of the world production of cultivated mushrooms, and *P. ostreatus* is the most commonly cultivated oyster mushroom species. A significant part of the harvest is destined for human consumption. The mushrooms are either sold fresh or processed by industry for easy storage (dried, frozen, canned and freeze-dried mushrooms). Although mushrooms contain protein, vitamins and minerals, their main role in the human diet is to contribute flavours and enhance the total quality of the dish.

### *Recommendation of key components to be analysed related to food use*

The key constituents suggested to be analysed in new varieties of *P. ostreatus* using the appropriate methodology are shown in Table 10.8. In case minerals are also analysed, iron, phosphorus, potassium, zinc, copper, manganese and chromium are suggested. As all food products of the oyster mushroom used by consumers and the food industry are derived from the fresh fruit bodies of the mushrooms, it is considered sufficient, in most circumstances, to analyse key constituents only in the fresh mushrooms. It will not be necessary to perform separate analyses of key constituents in commodities such as dried, freeze-dried or canned fruit bodies of oyster mushroom.

Table 10.8. **Suggested constituents to be analysed in fresh fruit bodies of cultivated oyster mushroom, *P. ostreatus*, for food use**

| Constituents | Fruit bodies |
|---|---|
| Proximates | X |
| Amino acids | X |
| Fatty acids | X |
| Vitamins | X[1] |

*Note:* 1. The B-vitamins thiamine (B1), riboflavin (B2), niacin (B3) and folic acid (B9) are suggested.

## Suggested constituents to be analysed related to feed use

Mushrooms are not typically included as animal feed ingredients, and it is unlikely that mushrooms would become a large significant nutrient contributor in animal feed. In the rare cases when by-products of oyster mushroom cultivation and mushroom processing (mainly stipes) may be used as animal feed, it is probably locally in the neighbourhood of the mushroom farms.

Although it is unlikely that fresh or processed fruit bodies of *P. ostreatus* will be used as animal feed, fragments of the vegetative mycelium might be consumed. It has been observed that several agro-industrial by-products locally available in many nations, for example, cocoa pod husks, which when untreated have a limited value as animal feed due to high contents of lignin and non-starch polysaccharides such as cellulose, hemicelluloses and pectin, have improved nutritional utility after they have been chemically modified by being substrates during mushroom cultivation. However, in this case it is recognised that the main proportion of the animal feed ingredient will be the bio-converted agro-industrial byproduct, with mushroom mycelia being only a minor part. When included in this form, oyster mushroom would still be considered a very minor animal feed.

No specific studies on the chemical composition of by-products of oyster mushroom cultivation and processing are needed for considering these as animal feeds. Therefore, no specific requirements for constituents to be analysed for animal feed are recommended in this chapter. Any required compositional information can be obtained from the analysis of proximates performed as part of an assessment for food uses of new varieties of oyster mushroom.

# Notes

1. For additional discussion of appropriate comparators, see Codex Alimentarius Commission (2003: paragraphs 44 and 45).
2. "The Association Of Analytical Communities" AOAC INTERNATIONAL.

# *References*


Abe H., S. Gotoh and M. Aoyama (1980), "Comparison of Nature of Spontaneous and Cultivated Edible Mushrooms: Differences in Free and Combined Amino Acids in their Ethanolic Extracts, *Journal of Japanese Society of Food and Nutrition,* Vol. 33, pp. 177-184.

Aida, F.M.N.A. et al. (2009), "Mushroom as a potential source of prebiotics: A review", *Trends in Food Science & Technology*, Vol. 20, Issues 11-12, pp. 567-575.

Al-Deen, I.H.S. et al. (1987), "Toxicologic and histopathologic studies of Pleurotus ostreatus mushroom in mice", *Journal of Ethnopharmacology*, Vol. 21, No. 3, pp. 297-305.

Alonso, J. et al. (2003), "The concentrations and bioconcentration factors of copper and zinc in edible mushrooms", *Archives of Environmental Contamination and Toxicology*, Vol. 44, Issue 2, pp. 180-188, http://dx.doi.org/10.1007/s00244-002-2051-0.

Amore, A. et al. (2012), "Copper induction of enhanced green fluorescent protein expression in Pleurotus ostreatus driven by laccase poxa1b promoter", *FEMS Microbiology Letters*, Vol. 337, No. 2, pp. 155-163, http://dx.doi.org/10.1111/1574-6968.12023.

Baars, J.J.P. et al. (2004), "Prototype of a sporeless oyster mushroom", *Mushroom Science*, Vol. 16, pp. 139-147.

Bano, Z. and S. Rajarathnam (1988), "Pleurotus mushrooms. Part II: Chemical composition, nutritional value, post-harvest physiology, preservation, and role as human food", *Critical Reviews in Food Science and Nutrition*, Vol. 27, No. 2, pp. 87-158.

Bano, Z. and S. Rajarathnam (1986), "Vitamin values of Pleurotus mushrooms", *Plant Foods for Human Nutrition*, Vol. 36, Issue 1, pp. 11-15.

Bautista Justo, M.B. et al. (1999), "Calidad proteínica de tres cepas mexicanas de setas (*Pleurotus ostreatus*)", *Archivos Latinoamericanos de Nutrición*, Vol. 49, No. 1, pp. 81-85.

Bautista Justo, M.B. et al. (1998), "Composicion quimica de tres cepas mexicanas de setas (Pleurotus ostratus)", *Archivos Latinoamericanos de Nutrición*, Vol. 48, No. 4, pp. 359-363.

Bedry, R. et al. (2001), "Wild-mushroom intoxication as a cause of rhabdomyolysis", *New England Journal of Medicine*, Vol. 345, pp. 798-802, http://dx.doi.org/10.1056/NEJMoa010581.

Beluhan, S. and A. Ranogajec (2011), "Chemical composition and non-volatile components of Croatian wild edible mushrooms", *Food Chemistry*, Vol. 124, Issue 3, February, pp. 1 076-1 082.

Berne, S. et al. (2002), "Pleurotus and agrocybe hemolysins, new proteins hypothetically involved in fungal fruiting", *Biochimica et Biophysica Acta*, Vol. 1570, No. 3, pp. 153-159.

Bernheimer, A.W. and L.S. Avigad (1979), "A cytolytic protein from the edible mushroom, Pleurotus ostreatus", *Biochimica et Biophysica Acta*, Vol. 585, No. 3, pp. 451-461.

Bobek, P. et al. (2001), "Effect pleuran (beta-glucan from Pleurotus ostreatus) in diet or drinking fluid on colitis in rats", *Nahrung/Food*, Vol. 45, No. 5, pp. 360-363.

Bobek, P. et al. (1998), "Dose- and time-dependent hypocholesterolemic effect of oyster mushroom (Pleurotus ostreatus) in rats", *Nutrition*, Vol. 14, No. 3, pp. 282-286.

Bobek, P. et al. (1991), "Effect of mushroom Pleurotus ostreatus and isolated fungal polysaccharide on serum and liver lipids in Syrian hamsters with hyperlipoproteinemia", *Nutrition*, Vol. 7, No. 2, pp. 105-108.

Bonatti, M. et al. (2004), "Evaluation of Pleurotus ostreatus and P. sajor-caju nutritional characteristics when cultivated in different lignocellulosic wastes", *Food Chemistry*, Vol. 88, No. 3, pp. 425-428, http://dx.doi.org/10.1016/j.foodchem.2004.01.050.

Bressa, G. et al. (1988), "Bioaccumulation of Hg in the mushroom Pleurotus ostreatus", *Ecotoxicology and Environmental Safety*, Vol. 16, No. 2, pp. 85-89.

Çaglarırmak, N. (2007), "The nutrients of exotic mushrooms (*Lentinula edodes* and *Pleurotus* species) and an estimated approach to the volatile compounds", *Food Chemistry*, Vol. 105, Issue 3, pp. 1 188-1 194.

Chang, S.-T. and P.G. Miles (2004), *Mushrooms. Cultivation, Nutritional Value, Medicinal Effect, and Environmental Impact*, CRC Press, Boca Raton, Florida.

Chattopadhyay, T.K. et al. (1999), "Crystallization of Pleurotus ostreatus (oyster mushroom) lectin", *Acta Crystallographica Section D*, Vol. 55, pp. 1 589-1 590.

Chirinang, P. and K.-O. Intarapichet (2009), "Amino acids and antioxidant properties of the oyster mushrooms Pleurotus ostreatus and Pleurotus sajor-caju", *ScienceAsia*, Vol. 35, pp. 326-331, http://dx.doi.org/10.2306/scienceasia1513-1874.2009.35.326.

Chobot, V. et al. (1997), "Ergosta-4,6,8,22-tetraen-3-one from the edible fungus Pleurotus ostreatus (oyster fungus)", *Phytochemistry*, Vol. 45, Issue 8, pp. 1 669-1 671.

Codex (Codex Alimentarius Commission) (2003; with Annexes II and III adopted in 2008), *Guideline for the Conduct of Food Safety Assessment of Foods Derived from Recombinant DNA Plants*, CAC/GL 45/2003, available at: www.codexalimentarius.net/download/standards/10021/CXG_045e.pdf.

Cohen R. et al. (2002), "Biotechnological applications and potential of wood-degrading mushrooms of the genus Pleurotus", *Applied Microbiology and Biotechnology*, Vol. 58, No. 5, pp. 582-594.

Coli, R. et al. (1988), "Composizione chimica e valore nutritivo di alcuni ceppi di pleurotus eryngii, P. nebrodensis, e P. ostreatus coltivati in serra", *Annali della Facultà di Agraria di Università degli studi di Perugia*, Vol. 42, pp. 847-859.

Conrad, F. and H. Rüdiger (1994), "The lectin from Pleurotus ostreatus: Purification, characterization and interaction with a phosphatase", *Phytochemistry*, Vol. 36, Issue 2, pp. 277-283.

Costa-Silva, F. et al. (2011), "Selenium contents of Portuguese commercial and wild edible mushrooms", *Food Chemistry*, Vol. 126, Issue 1, pp. 91-96.

Cox, A. et al. (1988), "Extrinsic allergic alveolitis caused by spores of the oyster mushroom Pleurotus ostreatus", *European Respiratory Journal*, Vol. 1, No. 5, pp. 466-468.

Crisan, E.V. and A. Sands (1978), "Nutritional value", in: Chang, S.T. and W.A. Hayes (eds.), *The Biology and Cultivation of Edible Mushrooms*, Academic Press, New York.

Del Signore, A. et al. (1997), "Content of phenolic substances in basidiomycetes", *Mycological Research*, Vol. 101, No. 5, pp. 552-556.

Dey, B. et al. (2010), "Chemical analysis of an immuno-enhancing water-soluble polysaccharide of an edible mushroom, Pleurotus florida blue variant", *Carbohydrate Research*, Vol. 345, No. 18, pp. 2 736-2 741, http://dx.doi.org/10.1016/j.carres.2010.09.032.

Ding, Y. et al. (2011), "Agrobacterium tumefaciens mediated fused egfp-hph gene expression under the control of gpd promoter in Pleurotus ostreatus", *Microbiological Research*, Vol. 166, No. 4, pp. 314-322.

Dogan, H.H. et al. (2006), "Contents of metals in some wild mushrooms: Its impact in human health", *Biological Trace Element Research*, Vol. 110, No. 1, pp. 79-94.

Fan, L. et al. (2006), "Advances in mushroom research in the last decade", *Food Technology and Biotechnology*, Vol. 44, pp. 303-311.

FAO/WHO (1991), "Protein quality evaluation", *Food and Nutrition Paper*, No. 51, Food and Agriculture Organization, Rome.

Farr, D.F. et al. (1989), *Fungi on Plants and Plant Products in the United States*, APS Press, St. Paul, Minnesota.

Favero, N. et al. (1990a), "Response of Pleurotus ostreatus to cadmium exposure", *Ecotoxicology and Environmental Safety*, Vol. 20, No. 1, pp. 1-6.

Favero, N. et al. (1990b), "Role of copper in cadmium metabolism in the basidiomycetes *Pleurotus ostreatus*", *Comparative Biochemistry and Physiology-Part C: Comparative Pharmacology*, Vol. 97, Issue 2, pp. 297-303.

Fries, E. (1821), *Systema mycologicum*, Vol. 1, Lund.

Fujihara, S. et al. (1995), "Nitrogen-to-protein conversion factors for some common edible mushrooms", *Journal of Food Science*, Vol. 60, Issue 5, pp. 1 045-1 047, http://dx.doi.org/10.1111/j.1365-2621.1995.tb06289.x.

García, M.Á. et al. (2009), "Lead in edible mushrooms: Levels and bioaccumulation factors", *Journal of Hazardous Materials*, Vol. 167, Issues 1-3, pp. 777-783.

Gençcelep, H. et al. (2009), "Determination of mineral contents of wild-grown edible mushrooms", *Food Chemistry*, Vol. 113, Issue 4, pp. 1 033-1 036.

Ginterová, A. and A. Maxianova (1975), "The balance of nitrogen and composition of proteins in Pleurotus ostreatus grown on natural substrates", *Folia Microbiol (Praha)*, Vol. 20, No. 3, pp. 246-250.

Goldstein, I.J. and H.C. Winter (2007), "Mushroom lectins", in: Johannis, P.K., *Comprehensive Glycoscience*, Elsevier, Oxford, pp. 601-621.

Guillot, J. and G. Konska (1997), "Lectins in higher fungi", *Biochemical Systematics & Ecology*, Vol. 25, Issue 3, pp. 203 230.

Gunde-Cimerman, N. and A. Cimerman (1995), "Pleurotus fruiting bodies contain the inhibitor of 3-hydroxy-3-methylglutaryl-coenzyme A reductase – lovastatin", *Experimental Mycology*, Vol. 19, No. 1, pp. 1-6.

Gutiérrez, A. et al. (1996), "Structural characterization of extracellular polysaccharides produced by fungi from the genus Pleurotus", *Carbohydrate Research*, Vol. 281, No. 1, pp. 143-154.

Guzmán, G. (2000), "Genus *Pleurotus* (Jacq.:Fr.) P. Kumm. (Agaricomycetideae): Diversity, taxonomic problems, and cultural and traditional medicinal uses", *International Journal of Medicinal Mushrooms*, Vol. 2, pp. 95 123, http://dx.doi.org/10.1615/IntJMedMushr.v2.i2.10.

Ha, H.C. et al. (2001), "Production of manganese peroxidase by pellet culture of the lignin-degrading basidiomycete, Pleurotus ostreatus", *Applied Microbiology and Biotechnology*, Vol. 55, No. 6, pp. 704-711.

Hadar, Y. and E. Cohen-Arazi (1986), "Chemical composition of the edible mushroom Pleurotus ostreatus produced by fermentation", *Applied and Environmental Microbiology*, Vol. 51, No. 6, pp. 1 352-1 354.

Haldimann, M. et al. (1995), "Vorkommen von Arsen, Blei, Cadmium, Quecksilber und Selen in Zuchtpilzen", *Mitteilungen aus dem Gebiete der Lebensmitteluntersuchung und Hygiene*, Vol. 86, pp. 463-484.

Hammond, J.B.W. (1980), "The composition of fresh and stored oyster mushrooms (Pleurotus ostreatus)", *Phytochemistry*, Vol. 19, No. 12, pp. 2 565-2 568, http://dx.doi.org/10.1016/S0031-9422(00)83919-9.

Helbling, A. et al. (1999), "Respiratory allergy to mushroom spores: Not well recognised, but relevant", *Annales of Allergy, Asthma & Immunology*, Vol. 83, No. 1, pp. 17-19.

Honda, Y. et al. (2000), "Carboxin resistance transformation of the homobasidiomycete fungus Pleurotus ostreatus", *Current Genetics*, Vol. 37, No. 3, pp. 209-212.

Irie, T. et al. (2001), "Stable transformation of Pleurotus ostreatus to hygromycin B resistance using Lentinus edodes gpd expression signals", *Applied Microbiology and Biotechnology*, Vol. 56, Issues 5-6, pp. 707-709.

Isildak, Ö. et al. (2004), "Analysis of heavy metals in some wild-grown edible mushrooms from the Middle Black Sea Region, Turkey", *Food Chemistry*, Vol. 86, Issue 4, pp. 547-552.

Jasinghe, V.J. and C.O. Perera (2005), "Distribution of ergosterol in different tissues of mushrooms and its effect on the conversion of ergosterol to vitamin D2 by UV irradiation", *Food Chemistry*, Vol. 92, Issue 3, September, pp. 541-546.

Jaworska, G. et al. (2011), "Effect of production process on the amino acid content of frozen and canned Pleurotus ostreatus mushrooms", *Food Chemistry*, Vol. 125, Issue 3, April, pp. 936-943.

Jennemann, R. et al. (2001), "Glycoinositolphosphosphingolipids (basidiolipids) of higher mushrooms", *European Journal of Biochemistry*, Vol. 268, No. 5, pp. 1 190-1 205.

Jia, J.-H. et al. (1998), "Transformation of the edible fungi, Pleurotus ostreatus and volvariella volvacea", *Mycological Research*, Vol. 102, Issue 7, July, pp. 876-880.

Joh, J.-H. et al. (2003), "The efficient transformation of Pleurotus ostreatus using remi method", *Mycobiology*, Vol. 31, No. 1, pp. 32-35, http://dx.doi.org/10.4489/MYCO.2003.31.1.032.

Juntes, P. et al. (2009), "Ostreolysin induces sustained contraction of porcine coronary arteries and endothelial dysfunction in middle- and large-sized vessels", *Toxicon*, Vol. 54, No. 6, pp. 784-792, http://dx.doi.org/10.1016/j.toxicon.2009.06.005.

Kamm, Y.J. et al. (1991), "Provocation tests in extrinsic allergic alveolitis in mushroom workers", *Netherlands Journal of Medicine*, Vol. 38, No. 1-2, pp. 59-64.

Karácsonyi, S. and L. Kuniak (1994), "Polysaccharides of Pleurotus ostreatus: Isolation and structure of pleuran, an alkali-insoluble [beta]-d-glucan", *Carbohydrate Polymers*, Vol. 24, Issue 2, pp. 107-111.

Kawagishi, H. et al. (2000), "A lectin from an edible mushroom Pleurotus ostreatus as a food intake-suppressing substance", *Biochimica et Biophysica Acta*, Vol. 1474, No. 3, pp. 299-308.

Kawai, H. et al. (1994), "Relationship between fruiting bodies compositions and substrate in hiratake and maitake mushrooms cultivated on sawdust substrate beds", *Nippon Shokuhin Kogyo Gakkaishi*, Vol. 41, No. 6, pp. 419-424, http://dx.doi.org/10.3136/nskkk1962.41.419.

Kazuno, C. and H. Miura (1985), "Chemical constituents of Pleurotus ostreatus: Studies on constituents of edible fungi Part II", *Nippon Shokuhin Kogyo Gakkaishi*, Vol. 32, No. 5, pp. 338-343, http://dx.doi.org/10.3136/nskkk1962.32.338.

Khanna, P.K. and H.S. Garcha (1984), "Pleurotus mushroom: A source of food protein", *Mushroom Newsletter for the Tropics*, No. 4, pp. 9-15.

Khanna, P.S. and H.S. Garcha (1981), "Nutritive value of mushroom Pleurotus florida", *Mushroom Science*, Vol. 11, Part 2, Article 46, Proceedings of the 11th International Scientific Congress on the Cultivation of Edible Fungi, pp. 561-572.

Kim, B.G. et al. (1999), "Isolation and transformation of uracil auxoprophs of the edible basidomycete Pleurotus ostreatus", *FEMS Microbiology Letters*, Vol. 181, No. 2, pp. 225-228.

Kim, M.Y. et al. (2009), "Comparison of free amino acid, carbohydrates concentrations in Korean edible and medicinal mushrooms", *Food Chemistry*, Vol. 113, No. 2, March, pp. 386-393.

Kim, M.Y. et al. (2008), "Phenolic compound concentration and antioxidant activities of edible and medicinal mushrooms from Korea", *Journal of Agricultural and Food Chemistry*, Vol. 56, No. 16, pp. 7 265-7 270, http://dx.doi.org/10.1021/jf8008553.

Koyama, N. et al. (1984), "Fatty acid composition and ergosterol contents of edible mushrooms", *Nippon Shokuhin Kogyo Gakkaishi*, Vol. 31, No. 11, pp. 732-738, http://dx.doi.org/10.3136/nskkk1962.31.11_732.

Kues, U. and Y. Liu (2000), "Fruiting body production in basidiomycetes", *Applied Microbiology and Biotechnology*, Vol. 54, Issue 2, pp. 141-152.

Lasota, W. et al. (1983), "Poziom Wybranych Witaminy z Grupy B w Niektórych Suszach Grzybów Wielkoowocnikowych", *Bromatologia i Chemia Toksykologiczna*, Vol. 16, pp. 177-180.

Lavi, I. et al. (2006), "An aqueous polysaccharide extract from the edible mushroom Pleurotus ostreatus induces anti-proliferative and pro-apoptotic effects on HT-29 colon cancer cells", *Cancer Letters*, Vol. 244, No. 1, pp. 61-70.

Lee, Y. et al. (2008), "Analytical dietary fiber database for the National Health and Nutrition Survey in Korea", *Journal of Food Composition and Analysis*, Vol. 21, Supplement 1, pp. S35-S42.

Lehrer, S.B. et al. (1994), "Prevalence of basidiomycete allergy in the USA and Europe and its relationship to allergic respiratory symptoms", *Allergy*, Vol. 49, No. 6, pp. 460-465.

Li, G. et al. (2006), "A highly efficient polyethylene glycol-mediated transformation method for mushrooms", *FEMS Microbiology Letters*, Vol. 256, No. 2, pp. 203-208.

Li, G.S.F. and S.T. Chang (1985), "Determination of vitamin C (ascorbic acid) in some edible mushrooms by differential pulse polarography", *Mushroom Newsletter for the Tropics*, Vol. 5, pp. 11-16.

Ma, A.M. et al. (2008), "Partial characterization of a hydrophobin protein Po. HYD1 purified from the oyster mushroom Pleurotus ostreatus", *World Journal of Microbiology and Biotechnology*, Vol. 24, No. 4, pp. 501-507.

Maihara, V.A. et al. (2008), "Arsenic and cadmium content in edible mushrooms from São Paulo, Brazil determined by INAA and GF AAS", *Journal of Radioanalytical and Nuclear Chemistry*, Vol. 278, Issue 2, pp. 395-397, http://dx.doi.org/10.1007/s10967-008-0807-3.

Mäkinen, S. et al. (1978), "On the thiamine content of some edible mushrooms", *Karstenia*, Vol. 18, pp. 29-32.

Maličev, E. et al. (2007), "Effect of ostreolysin, an Asp-hemolysin isoform, on human chondrocytes and osteoblasts, and possible role of Asp-hemolysin in pathogenesis", *Medical Mycology*, Vol. 45, No. 2, pp. 123-130.

Manzi, P. and L. Pizzoferrato (2000), "Beta-glucans in edible mushrooms", *Food Chemistry*, Vol. 68, Issue 3, February, pp. 315-318.

Manzi, P. et al. (2001), "Nutritional value of mushrooms widely consumed in Italy", *Food Chemistry*, Vol. 73, Issue 3, May, pp. 321-325.

Manzi, P. et al. (1999), "Nutrients in edible mushrooms: An inter-species comparative study", *Food Chemistry*, Vol. 65, Issue 4, June, pp. 477-482.

Mattila, P. et al. (2002a), "Sterol and vitamin D2 contents in some wild and cultivated mushrooms", *Food Chemistry*, Vol. 76, Issue 3, March, pp. 293-298.

Mattila, P. et al. (2002b), "Basic composition and amino acid contents of mushrooms cultivated in Finland", *Journal of Agricultural and Food Chemistry*, Vol. 50, No. 22, pp. 6 419-6 422.

Mattila, P. et al. (2001), "Contents of vitamins, mineral elements, and some phenolic compounds in cultivated mushrooms", *Journal of Agricultural and Food Chemistry*, Vol. 49, No. 5, pp. 2 343-2 348.

Melgar, M.J. et al. (2009), "Mercury in edible mushrooms and underlying soil: Bioconcentration factors and toxicological risk", *Science of the Total Environment*, Vol. 407, No. 20, October, pp. 5 328-5 334.

Mendez, L.A. et al. (2005), "Effect of substrate and harvest on the amino acid profile of oyster mushroom (Pleurotus ostreatus)", *Journal of Food Composition and Analysis*, Vol. 18, Issue 5, August, pp. 447-450.

Merrill, A.L. and B.K. Watt (1973), *Energy Value of Foods [Electronic Resource]: Basis and Derivation*, Human Nutrition Research Branch, Agricultural Research Service, United States Department of Agriculture, Washington, DC.

Misharina, T.A. et al. (2009), "The composition of volatile components of cepe (Boletus edulis) and oyster mushrooms (Pleurotus ostreatus)", *Applied Biochemistry and Microbiology*, Vol. 45, No. 2, pp. 207-213.

Mori, S. et al. (1998), "Mushroom worker's lung resulting from indoor cultivation of Pleurotus ostreatus", *Occupational Medicine (Lond)*, Vol. 48, No. 7, pp. 465-468, available at: http://occmed.oxfordjournals.org/content/48/7/465.full.pdf.

Muszynska, B. et al. (2011), "Indole compounds in some culinary-medicinal higher basidiomycetes from Poland", *International Journal of Medicinal Mushrooms*, Vol. 13, No. 5, pp. 449-454.

Nieminen, P. et al. (2009), "Myo- and hepatotoxic effects of cultivated mushrooms in mice", *Food and Chemical Toxicology*, Vol. 47, No. 1, pp. 70-74, http://dx.doi.org/10.1016/j.fct.2008.10.009.

Nnorom, I.C. et al. (2012), "Occurrence and accumulation of mercury in two species of wild grown Pleurotus mushrooms from southeastern Nigeria", *Ecotoxicological and Environmental Safety*, Vol. 84, pp. 78-83, http://dx.doi.org/10.1016/j.ecoenv.2012.06.024.

Nosálóvá, V. et al. (2001), "Effects of pleuran (β-glucan isolated from Pleurotus ostreatus) on experimental colitis in rats", *Physiological Research*, Vol. 50, No. 6, pp. 575-581.

Noster, U. et al. (1978), "Immunofluorescence test in the diagnosis of mushroom worker's lung", *Deutsche Medizinische Wochenschrift*, Vol. 103, No. 15, pp. 655-657.

Noster, U. et al. (1976), "Mushroom worker's lung caused by inhalation of spores of the edible fungus Pleurotus florida (oyster mushroom)", *Deutsche Medizinische Wochenschrift*, Vol. 101, No. 34, pp. 1 241-1 245.

Obodai, M. and M. Apertorgbor (2008), "Proximate composition and nutrient content of some wild and cultivated mushrooms of Ghana", *Journal of Ghana Science Association*, Vol. 10, No. 2, pp. 139-144, http://dx.doi.org/10.4314/jgsa.v10i2.18050.

OECD (2007), "Consensus document on compositional considerations for new varieties of the cultivated mushroom Agaricus bisporus: Key food and feed nutrients, anti-nutrients and toxicants", *Series on the Safety of Novel Foods and Feeds*, No. 15, OECD, Paris, available at: www.oecd.org/env/ehs/biotrack/46815276.pdf.

OECD (2005), "Consensus document on the biology of Pleurotus spp. (oyster mushroom)", *Series on Harmonisation of Regulatory Oversight in Biotechnology*, No. 34, OECD, Paris, available at: www.oecd.org/env/ehs/biotrack/46815828.pdf.

OECD (2000), "Report of the Task Force for the Safety of Novel Foods and Feeds", prepared for the G8 Summit held in Okinawa, Japan on 21-23 July 2000, C(2000)86/ADD1, OECD, Paris, available at: www.oecd.org/chemicalsafety/biotrack/REPORT-OF-THE-TASK-FORCE-FOR-THE-SAFETY-OF-NOVEL.pdf.

OECD (1997), "Report of the OECD Workshop on the toxicological and nutritional testing of novel foods", held in Aussois, France, 5-8 March 1997, final text issued in February 2002 as SG/ICGB(98)1/FINAL, OECD, Paris, available at: www.oecd.org/env/ehs/biotrack/SG-ICGB(98)1-FINAL-ENG%20-Novel-Foods-Aussois-report.pdf.

Oka, Y. et al. (1984), "First evidence for the occurrence of N delta-acetyl-L-ornithine and quantification of the free amino acids in the cultivated mushroom Pleurotus ostreatus", *Journal of Nutritional Science and Vitaminology (Tokyo)*, Vol. 30, No. 1, pp. 27-35.

Okamura, M. (1998), "Separative determination of ascorbic acid analogs contained in mushrooms by high-performance liquid chromatography", *Journal of Nutritional Science and Vitaminology*, Vol. 44, No. 1, pp. 25-35.

Palacios, I. et al. (2012), "Novel isolation of water-soluble polysaccharides from the fruiting bodies of Pleurotus ostreatus mushrooms", *Carbohydrate Research*, Vol. 358, pp. 72-77, http://dx.doi.org/10.1016/j.carres.2012.06.016.

Pan, X.L. et al. (2005), "Biosorption of Pb (II) by Pleurotus ostreatus immobilized in calcium alginate gel", *Process Biochemistry*, Vol. 40, Issue 8, July, pp. 2 799-2 803.

Papaspyridi, L.M. et al. (2010), "Optimization of biomass production with enhanced glucan and dietary fibres content by Pleurotus ostreatus ATHUM 4438 under submerged culture", *Biochemical Engineering Journal*, Vol. 50, Issue 3, July, pp. 131-138.

Park, H.G. et al. (2009), "New method development for nanoparticle extraction of water-soluble β-(1→3)-d-glucan from edible mushrooms Sparassis crispa and Phellinus linteus", *Journal of Agricultural and Food Chemistry*, Vol. 57, No. 6, pp. 2 147-2 154, http://dx.doi.org/10.1021/jf802940x.

Patra, S. et al. (2013), "A heteroglycan from the mycelia of Pleurotus ostreatus: Structure determination and study of antioxidant properties", *Carbohydrate Research*, Vol. 368, pp. 16-21, http://dx.doi.org/10.1016/j.carres.2012.12.003.

Pedneault, K. et al. (2007), "Fatty acid profiles of polar and non-polar lipids of Pleurotus ostreatus and P. cornucopiae var. 'citrino-pileatus' grown at different temperatures", *Mycological Research*, Vol. 111, Part 10, pp. 1 228-1 234.

Peng, M. et al. (1993), "Improved conditions for protoplast formation and transformation of Pleurotus ostreatus", *Applied Microbiology and Biotechnology*, Vol. 40, Issue 1, pp. 101-106.

Peng, M. et al. (1992), "Recovery of recombinant plasmids from Pleurotus ostreatus transformants", *Current Genetics*, Vol. 22, No. 1, pp. 53-59.

Petrovska, B.B. (1999), "Mineral composition of some macedonian edible mushrooms", *Acta Pharmaceutica*, Vol. 49, No. 1, pp. 59-64.

Phillips, K.M. et al. (2012), "Vitamin D4 in mushrooms", *PLoS ONE*, Vol. 7, No. 8, pp. 1-10, http://dx.doi.org/10.1371/journal.pone.0040702.

Phillips, K.M. et al. (2011), "Vitamin D and sterol composition of 10 types of mushrooms from retail suppliers in the United States", *Journal of Agricultural and Food Chemistry*, Vol. 59, No. 14, pp. 7 841-7 853, http://dx.doi.org/10.1021/jf104246z.

Plemenitaš, A. et al. (1999), "Steroidogenesis in the fungus Pleurotus ostreatus", *Comparative Biochemistry and Physiology B*, Vol. 123, No. 2, June, pp. 175-179.

Procida, G. and G. Pertoldi Marletta (1995), "Presenza di elementi in tracce in funghi coltivati e spontanei", *La Rivista di Scienza dell'Alimentazione*, Vol. 24, No. 4, pp. 535-542.

Quelet, L. (1886), *Enchiridion Fungorum in Europa Media et Praesertim in Gallia Vigentium*, Lutetiae.

Rai, R.D. et al. (1988), "Comparative nutritional value of various Pleurotus species grown under identical conditions", *Mushroom Journal for the Tropics*, Vol. 8, pp. 93-98.

Rajarathnam, S. et al. (2003), "Biochemical changes associated with mushroom browning in Agaricus bisporus (Lange) Imbach and Pleurotus florida (Block & Tsao): Commercial implications", *Journal of the Science of Food and Agriculture*, Vol. 83, No. 14, pp. 1 531-1 537, http://dx.doi.org/10.1002/jsfa.1562.

Ramírez, L. et al. (2000), "Molecular tools for breeding basidiomycetes", *International Microbiology*, Vol. 3, No. 3, pp. 147-152.

Rashad, M.M. and H.M. Abdou (2002), "Production and evaluation of Pleurotus ostreatus mushroom cultivated on some food processing wastes", *Advances in Food Sciences*, Vol. 24, No. 2, pp. 79-84.

Rashad, M.M. et al. (2009), "Nutritional analysis and enzyme activities of Pleurotus ostreatus cultivated on citrus limonium and carica papaya wastes", *Australian Journal of Basic and Applied Sciences*, Vol. 3, pp. 3 352-3 360.

Rebolj, K. et al. (2007), "Ostreolysin affects rat aorta ring tension and endothelial cell viability in vitro", *Toxicon*, Vol. 49, No. 8, pp. 1 211-1 213.

Reguła, J. and M. Siwulski (2007), "Dried shiitake (Lentinulla edodes) and oyster (Pleurotus ostreatus) mushrooms as a good source of nutrients", *Scientiarum Polonorum Technologia Alimentaria*, Vol. 6, No. 4, pp. 135-142, available at: www.food.actapol.net/pub/13_4_2007.pdf.

Reis, F.S. et al. (2012a), "Chemical composition and nutritional value of the most widely appreciated cultivated mushrooms: An inter-species comparative study", *Food and Chemical Toxicology*, Vol. 50, No. 2, pp. 191-197, http://dx.doi.org/10.1016/j.fct.2011.10.056.

Reis, F.S. et al. (2012b), "Antioxidant properties and phenolic profile of the most widely appreciated cultivated mushrooms: A comparative study between in vivo and in vitro samples", *Food and Chemical Toxicology*, Vol. 50, No. 5, pp. 1 201-1 207, http://dx.doi.org/10.1016/j.fct.2012.02.013.

Rop, O. et al. (2009), "Beta-glucans in higher gungi and their health effects", *Nutrition Reviews*, Vol. 67, No. 11, pp. 624-631, http://dx.doi.org/10.1111/j.1753-4887.2009.00230.x.

Rosina, P. et al. (1995), "Allergic contact dermatitis from Pleurotus mushroom", *Contact Dermatitis*, Vol. 33, No. 4, pp. 277-278, http://dx.doi.org/10.1111/j.1600-0536.1995.tb00491.x.

Rovenský, J. et al. (2009), "The effects of β-glucan isolated from Pleurotus ostreatus on methotrexate treatment in rats with adjuvant arthritis", *Rheumatology International*, Vol. 31, No. 4, pp. 507-511, http://dx.doi.org/10.1007/s00296-009-1258-z.

Saikai, T. et al. (2002), "Hypersensitivity pneumonitis induced by the spore of Pleurotus eryngii (eringi)", *Internal Medicine*, Vol. 41, No. 7, pp. 571-573.

Sakurai, N. et al. (2004), "Cloning, expression and pore-forming properties of mature and precursor forms of pleurotolysin, a sphingomyelin-specific two-component cytolysin from the edible mushroom Pleurotus ostreatus", *Biochimica et Biophysica Acta (BBA) – Gene Structure and Expression*, Vol. 1679, No. 1, pp. 65-73.

Sales-Campos, C. et al. (2009), "Mineral composition of raw material, substrate and fruit body of Pleurotus ostreatus in culture", *Interciencia*, Vol. 34, No. 6, pp. 432-436.

Santoprete, G. and G. Innocenti (1984), "Indagini sperimentali sul contenuto di oligoelementi nei funghi del Bolognese e di altre provenienze", *Micologia italiana*, Vol. 13, No. 1, pp. 11-28.

Sarangi, I. et al. (2006), "Anti-tumor and immunomodulating effects of Pleurotus ostreatus mycelia-derived proteoglycans", *International Immunopharmacology*, Vol. 6, No. 8, pp. 1 287-1 297.

Sato, E. et al. (1985), "Contents of free amino acids in mushrooms", *Nippon Shokuhin Kogyo Gakkaishi*, Vol. 32, No. 7, pp. 509-521, http://dx.doi.org/10.3136/nskkk1962.32.7_509.

Senti, G. et al. (2000), "Allergic asthma to shiitake and oyster mushroom", *Allergy*, Vol. 55, No. 10, pp. 975-976.

Sepčić, K. et al. (2003), "Interaction of ostreolysin, a cytolytic protein from the edible mushroom Pleurotus ostreatus, with lipid membranes and modulation by lysophospholipids", *European Journal of Biochemistry*, Vol. 270, No. 6, pp. 1 199-1 210.

Sesli, E. and M. Tüzen (1999), "Levels of trace elements in the fruiting bodies of macrofungi growing in the East Black Sea Region of Turkey", *Food Chemistry*, Vol. 65, Issue 4, June, pp. 453-460.

Shah, H. et al. (1997), "Nutritional composition and protein quality of Pleurotus mushroom", *Sarhad Journal of Agriculture*, Vol. 13, pp. 621-626.

Shashirekha, M.N. et al. (2005), "Effects of supplementing rice straw growth substrate with cotton seeds on the analytical characteristics of the mushroom Pleurotus florida (Block & Tsao)", *Food Chemistry*, Vol. 92, Issue 2, September, pp. 255-259.

Singer, R. (1986), *The Agaricales in Modern Taxonomy-4th Edition*, Koeltz Scientific Books. Germany, pp. 174-179.

Skočaj, M. et al. (2013), "The sensing of membrane microdomains based on pore-forming toxins", *Current Medicinal Chemistry*, Vol. 20, No. 4, pp. 491-501.

Stancher, B. et al. (1992a), "Caratterizzazione Merceologica dei Più Comuni Funghi Cotivati in Italia. Nota III – La Frazione Lipidica: Determinazione del Centenuto in Acidi Grassi", *Industrie Alimentari*, Vol. 31, pp. 431-434, 438.

Stancher, B. et al. (1992b), "Caratterizzazione Merceologica dei Più Comuni funghi Cotivati in Italia. Nota IV – La Frazione Lipidica: Determinazione del Centenuto in Acidi Grassi Liberi e Combinati", *Industrie Alimentari*, Vol. 31, pp. 744-747, 750.

Strmisková, G. et al. (1992), "Mineral composition of oyster mushroom", *Die Nahrung*, Vol. 36, No. 2, pp. 210-212, http://dx.doi.org/10.1002/food.19920360218.

Sturion, G. L. and M. Oetterer (1995), "Composicao quimica de cogumelos comestiveis (Pleurotus spp.) originados de cultivos em diferentes substrates", *Ciencia e Tecnologia de Alimentos*, Vol. 15, pp. 189-193.

Sun, Y. and J. Liu (2009), "Purification, structure and immunobiological activity of a water-soluble polysaccharide from the fruiting body of Pleurotus ostreatus", *Bioresource Technology*, Vol. 100, No. 2, pp. 983-986, http://dx.doi.org/10.1016/j.biortech.2008.06.036.

Sunagawa, M. and Y. Magae (2001), "Transformation of the edible mushroom Pleurotus ostreatus by particle bombardment", *FEMS Microbiology Letters*, Vol. 211, No. 2, pp. 143-146.

Synytsya, A. et al. (2008), "Mushrooms of genus Pleurotus as a source of dietary fibres and glucans for food supplements", *Czech Journal of Food Science*, Vol. 26, pp. 441-446, available at: http://www.agriculturejournals.cz/publicFiles/03098.pdf.

Teichmann, A. et al. (2007), "Sterol and vitamin D2 concentrations in cultivated and wild grown mushrooms: Effects of UV irradiation", *LWT – Food Science and Technology*, Vol. 40, No. 5, June, pp. 815-822.

Tomita, T. et al. (2004), "Pleurotolysin, a novel sphingomyelin-specific two-component cytolysin from the edible mushroom Pleurotus ostreatus, assembles into a transmembrane pore complex", *Journal of Biological Chemistry*, Vol. 279, No. 26, pp. 26 975-26 982.

Tong, H. et al. (2009), "Structural characterization and in vitro antitumor activity of a novel polysaccharide isolated from the fruiting bodies of Pleurotus ostreatus", *Bioresource Technology*, Vol. 100, No. 4, pp. 1 682-1 686, http://dx.doi.org/10.1016/j.biortech.2008.09.004.

Tsai, S.-Y. et al. (2009), "Flavour components and antiocidant properties of several cultivated mushrooms", *Food Chemistry*, Vol. 113, Issue 2, March, pp. 578-584.

Tshinyangu, K.K. (1996), "Effect of grass hay substrate on nutritional value of Pleurotus ostreatus var. columbinus", *Food/Nahrung*, Vol. 40, Issue 2, pp. 79-83, http://dx.doi.org/10.1002/food.19960400207.

Tüzen, M. and M. Soylak (2005), "Mercury contamination in mushroom samples from Tokat, Turkey", *Bulletin of Environmental Contamination and Toxicology*, Vol. 74, No. 5, pp. 968-972.

Tüzen, M. et al. (1998), "Study of heavy metals in some cultivated and uncultivated mushrooms of Turkish origin", *Food Chemistry*, Vol. 63, Issue 2, October, pp. 247-251.

USDA Agricultural Research Service (2010), *USDA National Nutrient Database for Standard Reference*, Release 23, Nutrient Data Laboratory, available at: www.ars.usda.gov/Services/docs.htm?docid=22115 (accessed July 2013).

Vereda, A. et al. (2007), "Occupational asthma due to spores of Pleurotus ostreatus", *Allergy Net*, Vol. 20, pp. 211-212.

Vetter, J. (2007), "Chitin content of cultivated mushrooms Agaricus bisporus, Pleurotus ostreatus and Lentinula edodes", *Food Chemistry*, Vol. 102, No. 1, pp. 6-9.

Vetter, J. (2005), "Lithium content of some common edible wild-growing mushrooms", *Food Chemistry*, Vol. 90, Issue 1-2, March-April, pp. 31-37.

Vetter, J. (2003), "Data on sodium content of common edible mushrooms", *Food Chemistry*, Vol. 81, Issue 4, June, pp. 589-593.

Vetter, J. (1989), "Vergleichende Untersuchung des Mineralstoffgehaltes der Gattungen Agaricus (Champignon) und Pleurotus (Austernseitling)", *Zeitschrift für Lebensmitteluntersuchung und Forschung A*, Vol. 189, pp. 346-350.

Vetter, J. and E. Berta (1997), "Mercury content of some wild edible mushrooms", *Zeitschrift für Lebensmittel¬untersuchung und -Forschung A*, Vol. 205, No. 4, pp. 316-320, http://dx.doi.org/10.1007/s002170050172.

Vetter, J. et al. (2005), "Mineral composition of the cultivated mushrooms Agaricus bisporus, Pleurotus ostreatus and Lentinula edodes", *Acta Alimentaria*, Vol. 34, No. 4, pp. 441-451, http://dx.doi.org/10.1556/AAlim.34.2005.4.11.

Wahid, M. et al. (1988), "Composition of wild and cultivated mushrooms of Pakistan", *Mushroom Journal for the Tropics*, Vol. 8, pp. 47-51.

Wang, D. et al. (2001), "Biological efficiency and nutritional value of Pleurotus ostreatus cultivated on spent beer grain", *Bioresourse Technology*, Vol. 78, Issue 3, July, pp. 293-300.

Wang, H. et al. (2000), "A new lectin with highly potent antihepatoma and antisarcoma activities from the oyster mushroom Pleurotus ostreatus", *Biochemical and Biophysical Research Communications*, Vol. 275, No. 3, pp. 810-816.

Whiteford, J.R. and C.F. Thurston (2000), "The molecular genetics of cultivated mushrooms", *Advances in Microbial Physiology*, Vol. 42, pp. 1-23.

Yanai, K. et al. (1996), "The integrative transformation of Pleurotus ostreatus using bialaphos resistance as a dominant selectable marker", *Bioscience, Biotechnology and Biochemistry*, Vol. 60, No. 3, pp. 472-475.

Yang, J.-H. et al. (2002), "Antioxidant properties of several commercial mushrooms", *Food Chemistry*, Vol. 77, Issue 2, May, pp. 229-235.

Yang, J.-H. et al. (2001), "Non-volatile taste components of several commercial mushrooms", *Food Chemistry*, Vol. 72, Issue 4, March, pp. 465-471.

Yildiz, A. et al. (1998), "Studies on Pleurotus ostreatus (Jacq. ex Fr.) Kum. var. salignus (Pers. ex Fr.) Konr. et Maubl.: Cultivation, proximate composition, organic and mineral composition of carpophores", *Food Chemistry*, Vol. 61, Issues 1-2, January, pp. 127-130.

Yilmaz, N. et al. (2006), "Fatty acid composition in some wild edible mushrooms growing in the Middle Black Sea Region of Turkey", *Food Chemistry*, Vol. 99, Issue 1, pp. 168-174.

Yoshida, H. et al. (1986), "Changes in carbohydrates and organic acids during development of mycelium and fruit-bodies of hiratake mushroom (Pleurotus ostreatus)", *Nippon Shokuhin Kogyo Gakkaishi*, Vol. 33, pp. 519-528.

Yoshioka, Y. et al. (1985), "Antitumor polysaccharides from P. ostreatus (Fr.) Quél.: Isolation and structure of a [beta]-glucan", *Carbohydrate Research*, Vol. 140, No. 1, pp. 93-100.

Yoshioka, Y. et al. (1975), "Isolation, purification and structure of components from acidic polysaccharides of Pleurotus ostreatus (Fr.) Quél", *Carbohydrate Research*, Vol. 43, No. 2, pp. 305-320.

Zadrazil, F. (1974), "Pleurotus-spores as allergens", *Naturwissenschaften*, Vol. 61, No. 10, pp. 456-457.

Zadrazil, F. (1973), "Pleurotus-Sporen-Allergie", *Champignon*, Vol. 13, pp. 9-10.

Zhang, Z.-M. et al. (2008), "A GC-MS study of the volatile organic composition of straw and oyster mushrooms during maturity and its relation to antioxidant activity", *Journal of Chromatographic Science*, Vol. 46, No. 8, pp. 690-696.

Zhang, M. et al. (2007), "Antitumor polysaccharides from mushrooms: A review on their isolation process, structural characteristics and antitumor activity", *Trends in Food Science & Technology*, Vol. 18, Issue 1, January, pp. 4-19.

Zhu, F. et al. (2011), "Assessment of heavy metals in some wild edible mushrooms collected from Yunnan Province, China", *Environmental Monitoring and Assessment*, Vol. 179, Issues 1-4, pp. 191-199, http://dx.doi.org/10.1007/s10661-010-1728-5.

Zurera-Cosano, G. et al. (1988), "Mercury content in different species of mushrooms grown in Spain", *Journal of Food Protection*, Vol. 51, pp. 205-207.

Zurera-Cosano, G. et al. (1987), "Lead and cadmium content of some edible mushrooms", *Journal of Food Quality*, Vol. 10, No. 5, pp. 311-317, http://dx.doi.org/10.1111/j.1745-4557.1988.tb00916.x.

Zuzek, M.C. et al. (2006), "Toxic and lethal effects of ostreolysin, a cytolytic protein from edible oyster mushroom (Pleurotus ostreatus), in rodents", *Toxicon*, Vol. 48, No. 3, pp. 264-271.

# List of OECD consensus documents on the safety of novel foods and feeds, 2002-14

| CONSENSUS DOCUMENT | LEAD COUNTRY(IES) | YEAR ISSUED | VOLUME |
|---|---|---|---|
| **CROPS** | | | |
| **Alfalfa** (*Medicago sativa*) **and other temperate forage legumes** | Canada and the United Kingdom | 2005 | Vol. 1 |
| **Barley** (*Hordeum vulgare*) | Finland, Germany and the United States | 2004 | Vol. 1 |
| **Cassava** (*Manihot esculenta*) | South Africa | 2009 | Vol. 2 |
| **Cotton** (*Gossypium hirsutum* and *G. barbadense*) | United States | 2009 | Vol. 2 |
| **Cultivated mushroom** (*Agaricus bisporus*) | Sweden | 2007 | Vol. 1 |
| **Grain sorghum** (*Sorghum bicolor*) | United States and South Africa | 2009 | Vol. 2 |
| **Low erucic acid rapeseed (Canola)** | Canada | 2011 | Vol. 2 |
| **Maize** (*Zea mays*) | Netherlands and the United States | 2002 | Vol. 1 |
| **Oyster mushroom** (*Pleurotus ostreatus*) | Sweden | 2013 | Vol. 2 |
| **Papaya** (*Carica papaya*) | Thailand and the United States | 2010 | Vol. 2 |
| **Potato** (*Solanum tuberosum* ssp. *tuberosum*) | Germany | 2002 | Vol. 1 |
| **Rice*** (*Oryza sativa*) | Japan* | 2004* | Vol. 1 |
| **Sugar beet** (*Beta vulgaris*) | Germany | 2002 | Vol. 1 |
| **Sugarcane** (*Saccharum* ssp. hybrids) | Australia | 2011 | Vol. 2 |
| **Soybean** (*Glycine max*) | United States | 2012 | Vol. 2 |
| **Sunflower** (*Helianthus annuus*) | Canada, France, Germany and the U.S. | 2007 | Vol. 1 |
| **Sweet potato** (*Ipomea batatas*) | South Africa and Japan | 2010 | Vol. 2 |
| **Tomato** (*Lycopersicon esculentum*) | Greece | 2008 | Vol. 1 |
| **Wheat** (*Triticum aestivum*) | Australia | 2003 | Vol. 1 |
| **FACILITATING HARMONISATION** | | | |
| **Animal feedstuffs** derived from genetically modified plants | Canada and the United Kingdom | 2003 | Vol. 1 |
| **Unique Identifier for transgenic plants** (revised version) (guidance document) | Working Group on Harmonisation of Regulatory Oversight in Biotechnology | 2006 | Vol. 1 |
| **Molecular characterisation** of plants derived from modern biotechnology | Canada, *joint publication of the Biosafety Working Group and the Food/Feed Safety Task Force* | 2010 | Vol. 2 |

* Rice document under revision, new issue expected in 2015.

## Published in the Series on the Safety of Novel Foods and Feeds, by number

| | |
|---|---|
| 1 | Consensus Document on Key Nutrients and Key Toxicants in Low Erucic Acid Rapeseed (Canola) (2001) – ***REPLACED with revised Consensus Doc. No. 24 (2011)*** |
| 2 | Consensus Document on Compositional Considerations for New Varieties of Soybean: Key Food and Feed Nutrients and Anti-Nutrients (2001) – ***REPLACED with revised Consensus Doc. No. 25 (2012)*** |
| 3 | Consensus Document on Compositional Considerations for New Varieties of Sugar Beet: Key Food and Feed Nutrients and Anti-Nutrients (2002) |
| 4 | Consensus Document on Compositional Considerations for New Varieties of Potatoes: Key Food and Feed Nutrients, Anti-Nutrients and Toxicants (2002) |
| 5 | Report of the OECD Workshop on the Nutritional Assessment of Novel Foods and Feeds, Ottawa, Canada, February 2001 (2002) |
| 6 | Consensus Document on Compositional Considerations for New Varieties of Maize (*Zea mays*): Key Food and Feed Nutrients, Anti-Nutrients and Secondary Plant Metabolites (2002) |
| 7 | Consensus Document on Compositional Considerations for New Varieties of Bread Wheat (*Triticum aestivum*): Key Food and Feed Nutrients, Anti-Nutrients and Toxicants (2003) |
| 8 | Report on the Questionnaire on Biomarkers, Research on the Safety of Novel Foods and Feasibility of Post-Market Monitoring (2003) |
| 9 | Considerations for the Safety Assessment of Animal Feedstuffs Derived from Genetically Modified Plants (2003) |
| 10 | Consensus Document on Compositional Considerations for New Varieties of Rice (*Oryza sativa*): Key Food and Feed Nutrients and Anti-Nutrients (2004) – ***Under revision*** |
| 11 | Consensus Document on Compositional Considerations for New Varieties of Cotton (*Gossypium hirsutum* and *Gossypium barbadense*): Key Food and Feed Nutrients and Anti-Nutrients (2004) |
| 12 | Consensus Document on Compositional Considerations for New Varieties of Barley (*Hordeum vulgare* L.): Key Food and Feed Nutrients and Anti-Nutrients (2004) |
| 13 | Consensus Document on Compositional Considerations for New Varieties of Alfalfa (*Medicago sativa*) and Other Temperate Forage Legumes: Key Feed Nutrients, Anti-Nutrients and Secondary Plant Metabolites (2005) |
| 14 | An Introduction to the Food/Feed Safety Consensus Documents of the Task Force for the Safety of Novel Foods and Feeds (2006) |
| 15 | Consensus Document on Compositional Considerations for New Varieties of the Cultivated Mushroom *Agaricus Bisporus*: Key Food and Feed Nutrients, Anti-Nutrients and Toxicants (2007) |
| 16 | Consensus Document on Compositional Considerations for New Varieties of Sunflower: Key Food and Feed Nutrients, Anti-Nutrients and Toxicants (2007) |
| 17 | Consensus Document on Compositional Considerations for New Varieties of Tomato: Key Food and Feed Nutrients, Anti-Nutrients, Toxicants and Allergens (2008) |
| 18 | Consensus Document on Compositional Considerations for New Varieties of Cassava (*Manihot esculenta* Crantz): Key Food and Feed Nutrients, Anti-Nutrients, Toxicants and Allergens (2009) |
| 19 | Consensus Document on Compositional Considerations for New Varieties of Grain Sorghum [*Sorghum bicolor* (L.) Moench]: Key Food and Feed Nutrients and Anti-Nutrients (2010) |
| 20 | Consensus Document on Compositional Considerations for New Varieties of Sweet Potato [*Ipomoea batatas* (L.) Lam.]: Key Food and Feed Nutrients, Anti-Nutrients, Toxicants and Allergens (2010) |
| 21 | Consensus Document on Compositional Considerations for New Varieties of Papaya (*Carica papaya* L.): Key Food and Feed Nutrients, Anti-Nutrients, Toxicants and Allergens (2010) |
| 22 | Consensus Document on Molecular Characterisation of Plants Derived from Modern Biotechnology (2010) |
| 23 | Consensus Document on Compositional Considerations for New Varieties of Sugarcane (*Saccharum* spp. hybrids.): Key Food and Feed Nutrients, Anti-Nutrients and Toxicants (2011) |
| 24 | Revised Consensus Document on Compositional Considerations for New Varieties of Low Erucic Acid Rapeseed (Canola): Key Food and Feed Nutrients Anti-Nutrients and Toxicants (2011) |
| 25 | Revised Consensus Document on Compositional Considerations for New Varieties of Soybean [*Glycine max* (L.) Merr.]: Key Food and Feed Nutrients, Anti-Nutrients, Toxicants and Allergens (2012) |
| 26 | Consensus Document on Compositional Considerations for New Varieties of Oyster Mushroom [*Pleurotus ostreatus*]: Key Food and Feed Nutrients, Anti-Nutrients and Toxicants (2013) |

***Note:*** **The individual documents composing the Safety of Novel Foods and Feeds Series, latest version, are available online at the OECD BIOTRACK website: www.oecd.org/biotrack.**
**The Series of Biosafety Consensus Documents (environmental safety), issued by the OECD Working Group on the Harmonisation of Regulatory Oversight in Biotechnology,as well as the OECD *Biotech Product Database*, are also available at the same address.**

# ORGANISATION FOR ECONOMIC CO-OPERATION AND DEVELOPMENT

The OECD is a unique forum where governments work together to address the economic, social and environmental challenges of globalisation. The OECD is also at the forefront of efforts to understand and to help governments respond to new developments and concerns, such as corporate governance, the information economy and the challenges of an ageing population. The Organisation provides a setting where governments can compare policy experiences, seek answers to common problems, identify good practice and work to co-ordinate domestic and international policies.

The OECD member countries are: Australia, Austria, Belgium, Canada, Chile, the Czech Republic, Denmark, Estonia, Finland, France, Germany, Greece, Hungary, Iceland, Ireland, Israel, Italy, Japan, Korea, Luxembourg, Mexico, the Netherlands, New Zealand, Norway, Poland, Portugal, the Slovak Republic, Slovenia, Spain, Sweden, Switzerland, Turkey, the United Kingdom and the United States. The European Commission takes part in the work of the OECD.

OECD Publishing disseminates widely the results of the Organisation's statistics gathering and research on economic, social and environmental issues, as well as the conventions, guidelines and standards agreed by its members.

OECD PUBLISHING, 2, rue André-Pascal, 75775 PARIS CEDEX 16
(97 2012 12 1 P) ISBN 978-92-64-18032-1 – 2015-02

www.ingramcontent.com/pod-product-compliance
Lightning Source LLC
LaVergne TN
LVHW081402110826
845149LV00010B/1636
* 9 7 8 9 2 6 4 1 8 0 3 2 1 *